W0111700

Phosphorus-Based Polymers
From Synthesis to Applications

RSC Polymer Chemistry Series

Series Editors:

Professor Ben Zhong Tang (Editor-in-Chief), *The Hong Kong University of Science and Technology, Hong Kong, China*
Professor Alaa S. Abd-El-Aziz, *University of Prince Edward Island, Canada*
Professor Stephen L. Craig, *Duke University, USA*
Professor Jianhua Dong, *National Natural Science Foundation of China, China*
Professor Toshio Masuda, *Fukui University of Technology, Japan*
Professor Christoph Weder, *University of Fribourg, Switzerland*

Titles in the Series:

How to obtain future titles on publication:
A standing order plan is available for this series. A standing order will bring delivery of each new volume immediately on publication.

For further information please contact:
Book Sales Department, Royal Society of Chemistry, Thomas Graham House, Science Park, Milton Road, Cambridge, CB4 0WF, UK
Telephone: +44 (0)1223 420066, Fax: +44 (0)1223 420247
Email: booksales@rsc.org
Visit our website at www.rsc.org/books

Phosphorus-Based Polymers
From Synthesis to Applications

Edited by

Sophie Monge
University of Montpellier 2, Institut Charles Gerhardt de Montpellier, France
Email: Sophie.Monge-Darcos@univ-montp2.fr

and

Ghislain David
Institut Charles Gerhardt de Montpellier, France
Email: Ghislain.David@enscm.fr

THE QUEEN'S AWARDS
FOR ENTERPRISE:
INTERNATIONAL TRADE
2013

RSC Polymer Chemistry Series No. 11

ISBN: 978-1-84973-646-6
eISBN: 978-1-78262-452-3
ISSN: 2044-0709

A catalogue record for this book is available from the British Library

© The Royal Society of Chemistry 2014

All rights reserved

Apart from fair dealing for the purposes of research for non-commercial purposes or for private study, criticism or review, as permitted under the Copyright, Designs and Patents Act 1988 and the Copyright and Related Rights Regulations 2003, this publication may not be reproduced, stored or transmitted, in any form or by any means, without the prior permission in writing of The Royal Society of Chemistry, or in the case of reproduction in accordance with the terms of licences issued by the Copyright Licensing Agency in the UK, or in accordance with the terms of the licences issued by the appropriate Reproduction Rights Organization outside the UK. Enquiries concerning reproduction outside the terms stated here should be sent to The Royal Society of Chemistry at the address printed on this page.

The RSC is not responsible for individual opinions expressed in this work.

Published by The Royal Society of Chemistry,
Thomas Graham House, Science Park, Milton Road,
Cambridge CB4 0WF, UK

Registered Charity Number 207890

For further information see our web site at www.rsc.org

Printed in the United Kingdom by CPI Group (UK) Ltd, Croydon, CR0 4YY, UK

Preface

Phosphorus-Based Polymers: From Synthesis to Applications aims at providing a broad overview of recent developments in the synthesis and applications of phosphorus-containing polymers. Over the last few years, more and more research papers have been published on this field. Polymerization of different kinds of phosphorus-based monomers using various methods has been carried out: (meth)acrylates, (meth)acrylamides, vinylphosphonic acid, styrenic, and allylic monomers. The resulting phosphorus-based materials have found applications in different domains: biomedical, complexation with metals, fire retardant additives, fuel cell membranes, *etc.*

Each chapter of this book provides precise and complete information on the state of current knowledge in the field of phosphorus materials. Contributions were written by experts that we deeply acknowledge for sharing their expertise and knowledge, making this book understandable for a broad readership. Chapters 1 to 7 are devoted to the polymerization of various phosphorus-based monomers, whereas Chapters 8 to 13 describe applications of the resulting materials.

The book begins with Chapter 1 by Robin and co-workers, which describes the synthesis and polymerization of phosphorus-containing (meth)acrylate monomers. This class of polymers is probably the most studied in the field of phosphorus-containing polymers, since the corresponding polymers may contain phosphonated ester, phosphonic acid, or phosphoric ester functions. This chapter starts with a thorough description of the wide range of (meth)acrylate monomers carrying different types of phosphorus groups, *i.e.* phosphonate, aminobisphosphonate, *etc.* Then, both conventional and controlled radical polymerization (mainly reversible addition–fragmentation chain transfer) are discussed in the following sections, as well as the photopolymerization of these monomers for use as dental applications.

RSC Polymer Chemistry Series No. 11
Phosphorus-Based Polymers: From Synthesis to Applications
Edited by Sophie Monge and Ghislain David
© The Royal Society of Chemistry 2014
Published by the Royal Society of Chemistry, www.rsc.org

Chapter 2 by Monge *et al.* details the synthesis and polymerization of phosphorus-containing (meth)acrylamide monomers. Compared to their (meth)acrylate homologues, this class of monomer is more hydrolytically stable and thus more interesting for a large variety of applications. Nevertheless, these monomers are less studied and most of the results mainly report their photopolymerization in order to develop stable self-etching dental primers. Future research on phosphorus-based (meth)acrylamide monomers is also discussed in this chapter and specifically the synthesis of block copolymers by controlled radical polymerization is investigated.

Chapter 3 by David and Negrell-Guirao reviews phosphorus-containing allyl and vinyl monomers. The synthetic strategies to obtain such monomers are described, covering Arbuzov reactions, transetherification, *etc*. The radical copolymerization of these monomers in the presence of electron-donating monomers is thoroughly discussed, as well as chain transfer reactions leading to chain scission. The use of the resulting copolymers in the field of flame retardants is also covered.

Chapter 4 by Macarie and Ilia aims at discussing the synthesis and polymerization of vinylphosphonic acid (VPA), as well as the properties of the resulting poly(vinylphosphonic acid). Interestingly, the main synthesis path of VPA is based on 2-chloroethylphosphonic acid, a plant growth regulator obtained on an industrial scale. According to its poor reactivity in radical polymerization, its polymerization is very challenging. Furthermore, the authors afford a deep insight into the structure of poly(vinylphosphonic acid), which is influenced by the polymerization reactions of VPA. The proton conductivity as well as the thermal and mechanical properties of poly(VPA) are detailed.

Chapter 5 by Ishihara and Fukazawa focuses on polymers obtained from 2-methacryloyloxyethyl phosphorylcholine (MPC) monomer. Indeed, the molecular design of MPC polymers with significant functions for biomedical and medical applications is summarized in detail. It is especially shown that some MPC polymers can provide artificial cell membrane-like structures at the surface as excellent interfaces between artificial systems and biological systems. In the clinical medicine field, MPC polymers have been used for surface modification of medical devices, including long-term implantable artificial organs to improve biocompatibility. Thus some MPC polymers have been provided commercially for these applications.

Montembault and Fontaine highlight the synthesis and development of polyphosphoesters in Chapter 6. The synthetic strategies that can be used for polyphosphoesters synthesis, including poly(alkylene H-phosphonate)s, poly(alkylene phosphate)s, poly(alkylene phosphoramidate)s, and poly-(alkylene phosphonate)s, are reviewed. This chapter also summarizes recent developments employing new and efficient chemistries, such as organo-catalyzed ring-opening polymerization and "click" chemistry, to prepare complex polyphosphoester-based macromolecular architectures (block, graft, and hyperbranched copolymers). These phosphorus-containing materials are very useful, especially for biomedical applications such as tissue engineering or drug delivery.

Chapter 7 by Allcock and Pugh describes the family of polyphosphazene polymers that have a structure based on a backbone of alternating phosphorus and nitrogen atoms, and with two organic side groups attached to each phosphorus. The authors of this chapter show that polyphosphazenes are mostly synthesized by a macromolecular substitution method in which a reactive intermediate, polydichlorophosphazene, is used as a substrate for the replacement of the chlorine atoms by organic nucleophiles. The potential uses for these polymers are in the field of aerospace, fire resistant polymers, fuel cells, and also for biomedical applications.

The following chapters are devoted to applications of phosphorus-based materials. Thus Chapter 8 by Mozsner and Catel deals with the use of polymerizable phosphonic acids (PAs) and dihydrogen phosphates (DHPs) for dental applications. Several PAs and DHPs were synthesized to notably improve the shear bond strength to dentin and enamel, the stability of the adhesive formulation, and the chemical adhesion to tooth tissues. Some of these monomers are nowadays included in commercial dental adhesives.

Chapter 9 by Wentrup-Byrne *et al.* reports on the biomedical applications of phosphorus-containing polymers such as poly[2-(acryloyloxy)ethyl phosphate], poly[2-(methacryloyloxy)ethyl phosphate], and polyphosphoesters, especially tissue repair and regeneration, medical device development, and tissue engineering. A broad overview is achieved, as applications considered include cardiovascular, ophthalmological, drug and gene delivery, as well as orthopedics and bone-interface repair, their surface chemistry, and the subsequent biomineralization processes.

Anticorrosion properties are highlighted in the Chapter 10 by David *et al.* Recent trends concern the development of organic coatings containing phosphonic acid moieties, as adhesion of the latter towards metallic surfaces has lead to the obtaining of polymeric coatings with strong anticorrosive properties.

Chapter 11 by Popa *et al.* is devoted to the use of chelating resins obtained by the chemical modification of their surface with different phosphorus pendant groups for the removal of heavy metals from various industrial waste effluents. It is shown that modification of the polymeric matrix through phosphorylation of its surface with different phosphorus pendant groups leads to an increase in the adsorption efficiency of the polymer in the removal of metal ions from various aqueous solutions.

The influence of phosphorus incorporated through a reactive approach on the flame retardancy of polymers is reviewed in Chapter 12 by Sonnier *et al.* It is notably demonstrated that phosphorus has an influence on the degradation pathway of the polymer and hence its thermal stability and charring.

Finally, Chapter 13 by Jannasch and Bingöl deals with proton conducting phosphonated polymers and membranes for fuel cells. Polymers functionalized with phosphonic acids enable both intrinsic and water-assisted proton conductivity and are thus attractive for use as electrolyte materials in electrochemical devices.

We expect that both specialist and non-specialist readers will find *Phosphorus-Based Polymers: From Synthesis to Applications* very valuable to update or discover the newest trends in phosphorus-containing materials.

Sophie Monge
Ghislain David

Contents

RSC Polymer Chemistry Series No. 11
Phosphorus-Based Polymers: From Synthesis to Applications
Edited by Sophie Monge and Ghislain David
© The Royal Society of Chemistry 2014
Published by the Royal Society of Chemistry, www.rsc.org

Chapter 6 Polyphosphoesters 97
Véronique Montembault and Laurent Fontaine

Chapter 7 Phosphazene High Polymers 125
Harry R. Allcock

CHAPTER 1

Polymerization of Phosphorus-Containing (Meth)acrylate Monomers

SOPHIE MONGE,* BENJAMIN CANNICCIONI,
GHISLAIN DAVID AND JEAN-JACQUES ROBIN

Institut Charles Gerhardt de Montpellier, UMR5253 CNRS-UM2-ENSCM-UM1, Equipe Ingénierie et Architectures Macromoléculaires, Université Montpellier II, cc1702, Place Eugène Bataillon, 34095 Montpellier, France
*Email: sophie.monge-darcos@univ-montp2.fr

1.1 Introduction

In recent years, phosphorus-based polymers have been widely studied[1-6] as they exhibit very unusual and interesting properties.[7] Whereas the ester forms are the most available compounds, monoacids and diacids can be easily obtained under mild conditions with the use of bromotrimethylsilane, opening the way to a wide range of polymers showing different properties. The latter can be explained in part by the ionization potential of phosphonic acids, which is intermediate between that of sulfonic and carboxylic acids due to their intermediate pK_a.

Phosphorus-containing materials can be employed for a wide range of technological applications.[8] For instance, they are extensively used in industry, notably to bind metals.[9-12] Indeed, phosphorus-based materials show interesting complexing properties[13,14] and are used as dispersants, corrosion inhibiting agents, or for preventing deposit formation.[15] They are also involved

RSC Polymer Chemistry Series No. 11
Phosphorus-Based Polymers: From Synthesis to Applications
Edited by Sophie Monge and Ghislain David
© The Royal Society of Chemistry 2014
Published by the Royal Society of Chemistry, www.rsc.org

in flame retardancy,[16,17] where phosphorus is known to be particularly useful. An important industrial application deals with their use in the biomedical field,[18] as they are biodegradable, blood compatible, show reduced protein adsorption, and lead to strong interactions with dentin, enamel, or bones. As a consequence, various syntheses of monomers and polymers are carried out following different procedures: (i) introduction of the phosphorinated moieties onto polymers by (co)polymerization of monomers bearing the phosphorus atom or (ii) grafting of phosphorus-based groups onto the polymer.

In this contribution, we will focus on the polymerization of phosphorus-based (meth)acrylates. Different kinds of monomers will be considered, as a function of the targeted applications. The latter will be more thoroughly discussed in the second part of this book (Chapters 8–13). The resulting phosphorus-based poly(meth)acrylates are mainly used for anticorrosion, flame retardancy, tissue engineering, and dental applications. Concerning anticorrosion, polymers have been involved in corrosion protective coatings which require maintenance of adhesion under environmental exposure. Adhesion between galvanized steel plates and the polymer depends on the chemical structures of both substrate and coating and reduction of adhesion causes water penetration at the coating/metal interface, leading to a significant reduction of adhesion. Incorporation of phosphonic groups, known as adhesion promoters, into polymer structures allows improvement of the adhesion properties of polymers on metallic surfaces.[19] Furthermore, research has also been carried out on the development of new halogen-free flame retardants. In this context, phosphorus is known to be efficient with or without other elements like nitrogen or sulfur. Concerning tissue engineering, polymeric materials, especially phosphonated polymers, have already proved to be of great interest. Tissue engineering typically involves the seeding of biodegradable polymeric scaffolds with differentiated or pluripotent cells *in vitro*, followed by implantation of the whole cell–scaffold system into the region of tissue loss or damage. It has been shown that protein interactions were favored when phosphorus-grafted polymeric surfaces were used. Finally, among all self-etching adhesive systems developed in dentistry for bonding of resin composite to enamel or dentin, primers containing phosphonated or phosphonic acid groups have been quite widely considered. Indeed, such derivatives are potentially interesting as the incorporation of a phosphonic function would result in an increase of the biocompatibility and in the adhesion due to chelation with calcium ions at the tooth surface[20] because of complex formation with calcium in hydroxyapatite.[21] This mineral is partially dissociated by phosphonic acid[22] to give brushite acting as a macromolecular crosslinker[23] and the bond strength depends on the alkyl chain length of the monomer.[24,25] For this last application, bis(meth)acrylate monomers are also considered. Finally, special attention will be paid to the controlled radical polymerization of dimethyl [(methacryloyloxy)methyl]phosphonate, very recently reported in the literature, demonstrating that it is possible to prepare well-controlled architectures involving phosphorus-based monomers.

1.2 Synthesis by Free or Controlled Radical Polymerization of Phosphorus-Based Poly(meth)acrylates as Adhesion Promoters

Free-radical graft copolymerization of phosphonated (meth)acrylates onto polymers is carried out with dimethyl [2-(meth)acryloxyethyl]phosphonate (DAP) or corresponding mono- or diacid derivatives (DAP monoacid or DAP diacid) (Scheme 1.1).[26] These monomers are obtained using a simple synthetic method starting from commercial compounds. DAP is obtained by an esterification reaction involving acryloyl chloride and dimethyl (2-hydroxyethyl)phosphonate in dichloromethane in the presence of triethylamine. The synthesis of phosphonic acid monomer derivatives (DAP monoacid and DAP diacid) is achieved in two steps: (i) silylation of a dimethyl phosphonate function by bromotrimethylsilane, followed by (ii) hydrolysis with an excess of methyl alcohol. Using an equivalent amount of brominated silane and DAP leads to a mixture of three functional phosphonated groups (notably with DAP monoacid), whereas an excess of bromotrimethylsilane quantitatively converts phosphonated moieties into phosphonic diacid groups (DAP diacid). Then, ozone-oxidized poly(vinylidene difluoride) (PVDF) is grafted with phosphonated acrylate monomers and acrylic acid by two methods: solution or bulk polymerization. Polymerization in solution is achieved using dimethylformamide (DMF) as solvent at 90 °C for 24 h, or deionized water at 60 °C for 12 h. For bulk polymerization, the reaction is carried out at 130 °C in an air atmosphere for the prescribed time. Then, the graft copolymer is dissolved in DMF and the homopolymer is removed by precipitation in ethyl alcohol. Adhesion of the graft copolymers was subsequently studied. The presence of phosphonic acid groups leads to a significant improvement in the adhesive bond to metal compared to both dialkyl phosphonate and carboxylic acid groups. The same kinds of experiment were carried out with low-density polyethylene powders[27] using phosphonated methacrylates by grafting in the presence of free-radical initiators or thermally induced graft copolymerization onto ozone-pretreated low-density polyethylene.

$n = 2$; R = H ; R' = CH$_3$; R'' = CH$_3$
$n = 2$; R = H ; R' = CH$_3$; R'' = H
$n = 2$; R = H ; R' = H ; R'' = H

$n = 2$; R = CH$_3$; R' = CH$_3$; R'' = CH$_3$
$n = 2$; R = CH$_3$; R' = CH$_3$; R'' = H
$n = 2$; R = CH$_3$; R' = H ; R'' = H

$n = 1$; R = CH$_3$; R' = CH$_3$; R'' = CH$_3$
$n = 1$; R = CH$_3$; R' = CH$_3$; R'' = H

MAC3NP2

MAC3P2

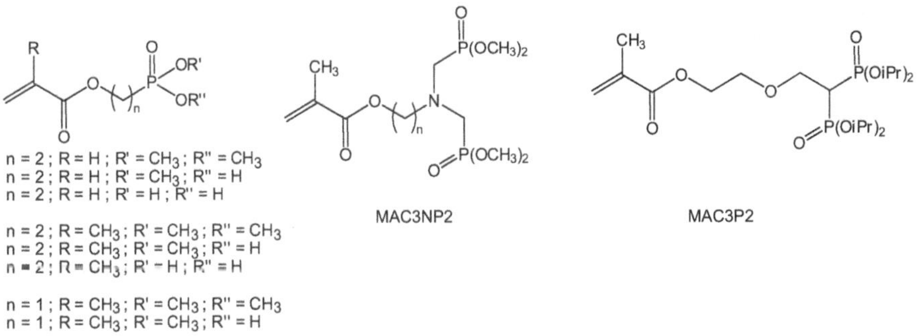

Scheme 1.1 Phosphorus-based (meth)acrylate monomers used for adhesion properties.

Poly(MMA)-*b*-poly(monophosphonic methacrylate) (MMA = methyl methacrylate) diblock copolymers are also prepared by atom transfer radical polymerization (ATRP) and used as additives in PVDF coatings to protect steel against corrosion.[28] ATRP of dimethyl [(methacryloyloxy)methyl]phosphonate (MAPC1), prepared by the Pudovic reaction,[29] was investigated in toluene, in the presence of methyl 2-bromoisobutyrate as the initiator, and using different metal and ligand systems. Polymerization proceeded with very low monomer conversion, which was attributed to the ability of phosphorus to complex the copper ions, removing copper ions from the original ligand and stopping the MAPC1 polymerization. As a result, another strategy was chosen for an efficient synthesis of poly(MMA)-*b*-poly(phosphonate acrylate) diblock copolymer (without phosphonated-based monomer), which was efficiently obtained by a four-step reaction, with first the synthesis of poly(MMA)-*b*-poly(*tert*-butyl acrylate) diblock copolymer by ATRP. Then the *tert*-butyl groups were removed and phosphonate functions were incorporated by esterification. This new diblock copolymer was used as an additive for an anticorrosive coating, but no improvement (using the salt spray test technique) was observed compared with the statistical copolymer with the same acid content.

Blends of phosphorus-containing copolymers with PVDF powders were also investigated to enhance adhesive and anticorrosive properties of fluoro polymer coatings.[30] Statistical copolymers were first prepared copolymerizing MMA and dimethyl [(2-methacryloyloxy)ethyl]phosphonate, [(2-methacryloyloxy)ethyl]phosphonic diacid, or methyl [(2-methacryloyloxy)ethyl]phosphonic hemiacid. Concerning the synthesis of the monomers involved, dimethyl (2-hydroxyethyl)phosphonate reacted with methacryloyl chloride in the presence of triethylamine in dichloromethane. The resulting dimethyl [(2-methacryloyloxy)ethyl]phosphonate was isolated by distillation under vacuum $(2 \times 10^{-3}$ mbar) at 85 °C. The acid derivatives were synthesized by methanolysis using bromotrimethylsilane and methanol. Copolymerizations were then carried out at 70 °C either in THF or DMF in the presence of AIBN. The resulting copolymers were introduced into PVDF as adhesion promoters and anticorrosion inhibitors. Good dry and wet adhesion properties onto galvanized steel plates were obtained with blends containing a high content of phosphonic acid groups. A salt spray exposure test also showed that the phosphonic acid groups prevented the spread of corrosion at the PVDF coating/galvanized steel interface. MAPC1, with only one methylene spacer, was copolymerized with MMA. Radical copolymerizations of both MAPC1 and its acidic form, MAPC1(OH), with MMA were carried out in acetonitrile at 80 °C, initiated by AIBN. When MAPC1 was used, the phosphonated ester moieties were then hydrolyzed. When blended with PVDF and coated onto a galvanized steel plate, the copolymer led to very good adhesion in both wet and dry states and provided good anticorrosion properties.

The reversible addition–fragmentation chain transfer (RAFT) polymerization of vinylidene chloride, methyl acrylate, and a phosphonated monomer named MAUPHOS $[CH_2=C(CH_3)C(O)O(CH_2)_2NHC(O)O(CH_2)_2P(O)(OCH_3)_2]$

has been achieved. The phosphonated methacrylate was synthesized by reacting isocyanoethyl methacrylate and the appropriate phosphonated alcohol.[31] Unlike usual methods (esterification and transesterification), this synthesis was quantitative and easy to realize. RAFT polymerization was carried out at 70 °C in benzene under pressure using 1-(ethoxycarbonyl)ethyl dithiobenzoate as a chain transfer agent in an autoclave because of the low boiling point of the vinylidene chloride, and yielded a gradient terpolymer which was then hydrolyzed. The latter was used as polymeric additive in coating formulations based on a poly(vinylidene chloride-*co*-methyl acrylate) copolymer matrix.[4] Different formulations were spun cast on stainless steel surfaces and the coatings were analyzed by electron microscopy coupled with X-ray analyses. Phosphonic acid groups induced preferential organization of the coating as the additive segregated and migrated toward the metal interface. As a consequence, the matrix at the air interface acted as a barrier to gas while the additive ensured adhesion at the polymer/metal interface. Another terpolymer was also evaluated, combining a methacrylic phosphonated monomer copolymerized with MMA and butyl methacrylate. The phosphonated monomer was obtained from diethyl 1,2-epoxyprop-2-ylphosphonate and methacrylic acid in the presence of tetrabutylammonium chloride in toluene at 80 °C, with 76% yield. Then, radical terpolymerization took place with AIBN in acetonitrile at 80 °C. In the case of the resulting terpolymer, high resistance towards corrosion was observed as no corrosion occurred even after 1100 h of the salt spray test.[32] Innovative multifunctional phosphonated monomers were incorporated into photopolymerizable mixtures of (meth)acrylates and showed, after UV hardening, good adhesion onto steel and excellent resistance to corrosion under salt spray.[33] The phosphonated monomers were propyl N,N-tetramethylbis(phosphonate)-2-hydroxylbismethyleneamine methyl methacrylate (for $n = 3$) (MAC3NP2) and 2-[2,2-bis(diisopropoxyphosphoryl)ethoxy]methyl methacrylate (MAC3P2) (see Scheme 1.1). The amine-containing methacrylate MAC3NP2 was synthesized *via* a three-step procedure, where the corresponding aminodiphosphonate hydroxy compound was prepared by the Kabachnick–Field process. It was subsequently esterified and the obtained aminobisphosphoryl methacrylate monomer was finally hydrolyzed, resulting in the bisphosphoryl methacrylate monomer in 60% yield. The synthesis of MAC3P2 was achieved with a similar synthetic strategy to the one used for the MAC3NP2 monomer, leading to MAC3P2 in a yield of 85%.

1.3 Synthesis by Radical Polymerization of Phosphorus-Based Poly(meth)acrylates for Flame Retardancy

One of the main drawbacks of the usual polymeric materials is their flammability and the emission of very hazardous gases and smoke during combustion. Only a few polymers are able to self-extinguish themselves. Among all possibilities, phosphorus-based materials have proved to be

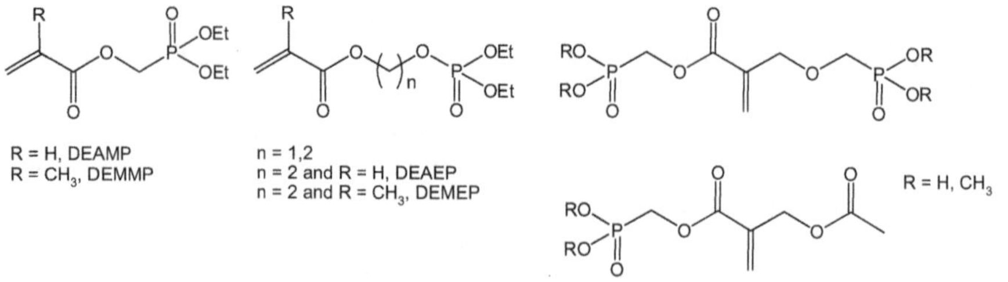

Scheme 1.2 Examples of phosphorus-based molecules used as comonomers for flame retardancy.

efficient in this respect. However, the design of new phosphonated halogen-free retardant additives is not an easy task. No universal compound exists and each additive is associated with one polymer type. This observation comes from the degradation mechanisms that differ from one polymer to another. Indeed, some polymers are able to self-generate chars when burning while others do not have this property. In the case of char formation, some phosphorus-based compounds can be easily integrated in this char and have a positive action in the formation of this porous and protective foamed material. Results reported in the literature have dealt with the co-polymerization of (meth)acrylic monomers bearing phosphonate or phosphate groups, namely diethyl [(methacryloyloxy)methyl]phosphonate (DEMMP), diethyl [(acryloyloxy)methyl]phosphonate (DEAMP), diethyl (methacryloyloxy)ethyl phosphate (DEMEP), and diethyl (acryloyloxy)ethyl phosphate (DEAEP) (Scheme 1.2).[34,35]

These studies clearly showed that the chemical structure of the phosphorus atom plays a major role in the flame retardancy, with phosphonate groups proving to be more active than the corresponding phosphate groups. Furthermore, phosphonated groups linked to the polymer backbone and obtained by copolymerization had a low effect on mechanical and physical properties in comparison with phosphonated compounds added to poly-styrene (PS), where a decrease of these properties was obtained. The additives acted in the vapor phase, whereas phosphonated copolymers were active both in the vapor and condensed phases.

Phosphorus-based (meth)acrylates were also combined with poly(MMA).[36] In this work, DEMMP was synthesized by the condensation of methacryloyl chloride with diethyl (hydroxymethyl)phosphonate in the presence of tri-ethylamine in anhydrous dichloromethane under argon at room temperature. Plaques of MMA and DEMMP copolymers, containing different molar ratios of DEMMP, were prepared by heating nitrogen-bubbled mixtures of MMA, DEMMP, and AIBN in an oven. Copolymers of MMA and DEMMP were also prepared in solution at 60 °C using AIBN as initiator. Copolymers made in solution were then pressed into plaques. It was shown that DEMMP acted in the condensed phase but also in the vapor phase, and char formation as well as the autoignition temperature were greatly increased for 10 mol% content.

Some mechanisms operating during the formation of char were proposed in a complementary study.[36] The well-known mechanism of depolymerization of PMMA was disturbed in the presence of DEMMP. Free additives containing PMMA afforded lower flame retardancy properties than those of additives incorporated in PMMA chains.[37] Furthermore, the authors showed that phosphonate monomers were more efficacious than the corresponding phosphate ones.[38] More recently, the same authors studied in detail the mechanisms occurring during thermal degradation of PMMA and showed the effect of the chemical structure of the phosphonated comonomer (acrylate or methacrylate) on the PMMA backbone decomposition.[35]

Thermosetting resins like UV curable resins have also been studied and the effect of phosphonated monomers was evaluated. Monomers bearing two phosphonic acid groups (Scheme 1.2, right) showed a good effect on the flame retardancy.[39–43] Phosphorus-containing acrylate monomers were synthesized from the reaction of ethyl (chloromethyl)acrylate and *tert*-butyl (bromomethyl)acrylate with triethyl phosphite. The selective hydrolysis of the ethyl ester monomer with bromotrimethylsilane gave a phosphonic acid monomer. The attempted bulk polymerizations of the monomers at 57–60 °C with AIBN were unsuccessful, but bulk copolymerization with methyl methacrylate at 60 °C with AIBN was performed. Additionally, α-(chloromethyl)acryloyl chloride was reacted with diethyl (hydroxymethyl)phosphonate to obtain another monomer. This monomer was hydrolyzed, homopolymerized, and copolymerized with MMA. For instance, the photopolymerization of this monomer gave 90% conversion in 60 s. More generally, the relative reactivities of the synthesized monomers in photopolymerizations were determined and compared with those of the other phosphorus-containing acrylate monomers. Changing the monomer structure allowed control of the polymerization reactivity so that new phosphorus-containing polymers with desirable properties could be obtained.

1.4 Synthesis by Free and Controlled Radical Polymerization of Phosphorus-Based Poly(meth)acrylates for Tissue Engineering

Phosphonated polymers are of great interest in the field of tissue engineering and scaffolding. Researchers have tried to promote interactions of biomaterials with bone cells and the protein interaction was favored when phosphonated-grafted polymeric surfaces were used. Methacrylate-based crosslinked hydrogels were synthesized and showed very interesting properties in water transport mechanisms for medical applications.[24,44–49]

Among (meth)acrylate-based monomers, (monoacryloxy)ethyl phosphate (MAEP, R = H) and 2-(methacryloyloxy)ethyl phosphate (MOEP, R = Me) (Scheme 1.3) were polymerized through RAFT-mediated polymerization to achieve low molecular weight polymers.[49] The MAEP and MOEP

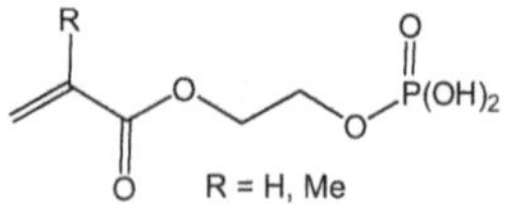

Scheme 1.3 Ethylene glycol (meth)acrylates used in tissue engineering.

homopolymerizations were carried out in the presence of either 1-phenyl-ethyl phenyldithioacetate (1-PEPDA) or cumyl dithiobenzoate (CDB) with AIBN as initiator in methanol at 60 °C.

PMAEP and PMOEP polymers showed a calcium phosphate (CaP) layer, and a secondary CaP mineral growth with a typical hydroxyapatite (HAP) globular morphology was found on the PMOEP gel. Crosslinked copolymers of 2-hydroxyethyl methacrylate (HEMA) and MOEP were also studied.[45] The incorporation of phosphate pendant groups into a crosslinked PHEMA hydrogel significantly altered the mechanism of water transport into the polymer matrix; 3–20 mol% MOEP polymers appeared to be the most promising materials for biomedical applications. Block copolymers consisting of PMAEP or PMOEP and poly[2-(acetoacetoxy)ethyl methacrylate] (PAAEMA) were successfully prepared. The homopolymerization of AAEMA with CDB gave a molecular weight close to theory with a low dispersity (1.13). This polymer was further chain extended with MAEP and MOEP. Molecular weights were double that of theory, and the dispersities were close to 1.38. A copolymer series using increasing concentrations of MOEP with (diethylamino)ethyl methacrylate (DEAEMA) and 1-vinylpyrrolid-2-one (VP) were also synthesized[50] to study the influence of phosphate content and distribution on the capacity to constitute calcium-rich layers. The homopolymers PMOEP, PDEAEMA, and PVP and two series of copolymers, MOEP–DEAEMA and MOEP–VP, obtained by using increasing concentrations of MOEP, were prepared. For all polymerizations, ethylene glycol dimethacrylate (EGDMA) was added as crosslinking agent. Reactions were carried out in an inert atmosphere at 80 °C for 8 h. The amounts of calcium increased when higher concentrations of MOEP were used and the best results were obtained with MOEP–VP copolymers. The amine group might favor the attraction of phosphorus, creating another way for the nucleation of CaP crystals. Finally, polymeric grafting with phosphate-containing monomers such as MAEP was carried out onto poly(tetrafluoroethylene) (PTFE) membranes, which are already used in facial augmentation and as a craniofacial implant material.[46] PTFE membranes and poly(tetrafluoroethylene-*co*-hexafluoropropylene) (FEP) films were modified by graft copolymerization with MAEP in an aqueous solution at ambient temperature using γ-irradiation. The surface modification led to an increase in hydrophilicity on the surface. Moreover, a MAEP-modified membrane with an external surface coverage of 44% or above was demonstrated to be promising as an improved biomaterial because these modified materials were able to induce CaP nucleation.[47]

1.5 Synthesis by Photopolymerization of Phosphorus-Based Poly(meth)acrylates for Dental Applications

Dental materials based on (meth)acrylic monomers are favored polymers for dentists owing to their high reactivity under UV radiation.[51] In general, (meth)acrylates lead to high photopolymerization rates, the latter being greatly modified in the presence of oxygen. The shrinkage occurring all along the polymerization process of monomers under thermal or UV activation leads to the formation of cracks in the materials and also at the interface between dentin (or enamel) and the polymeric composite. This phenomenon notably allows the passage of liquids and bacteria into the gaps, leading to discolorations or hydrolysis of the dental restoration. In some cases, bacteria included in the cracks generated at the joint rupture can initiate new caries. To avoid this damage, researchers have developed new polymeric compositions with no or low shrinkages during polymerization and with better adhesion onto enamel or dentin. The first technique was based on acid-etching of enamel using strong acids like phosphoric acid.[52] The treatment results in the formation of microporous sites, making the surface rough so that micromechanical-based adhesion is obtained.[53,54] Ever since the initial development of new techniques using polymers in dental restoration, phosphorus-containing polymers have been widely explored. One of the problems to be solved was the adhesion of polymeric matrixes onto hydroxyapatite. The most often used compounds were phosphoric acid esters, phosphonic esters, or phosphonic acids. Fu *et al.*[21] showed using FTIR analyses that phosphoric acid esters were able to decalcify hydroxyapatite (HA) and to form two different complexes with calcium ions.[55] This phenomenon occurred simultaneously with the formation of chemical bonding with HA. As a result, etching and chemisorbing led to better adhesion than that obtained with carboxylic acid groups.

Photopolymerization of phosphorus-based (meth)acrylic monomers was largely investigated for dental applications (Scheme 1.4).[20,21,56–69] The monomers bore one or two polymerizable groups and phosphoric acid ester, phosphonate, and phosphonic acid moieties were evaluated. When the phosphorus atom was directly linked to a hydrocarbon chain (*i.e.*, phosphonate ester), the monomers were more resistant to hydrolysis in comparison with phosphoric acid esters.

The hydrolysis of the esters of phosphonic acids resulted in increased reactivity of these monomers in some cases. This result was not clearly explained but was attributed to higher hydrogen-bonding in the case of acidic groups with respect to ester groups. Chain termination could decrease as a result of an increase in viscosity due to these bonds[60,70] and also to a pre-organization of molecules leading to double bonds close to each other, increasing the polymerization rates. This phenomenon has been studied by Avci *et al.*, who synthesized different phosphonic acid-based monomers.[21,41,66,71]

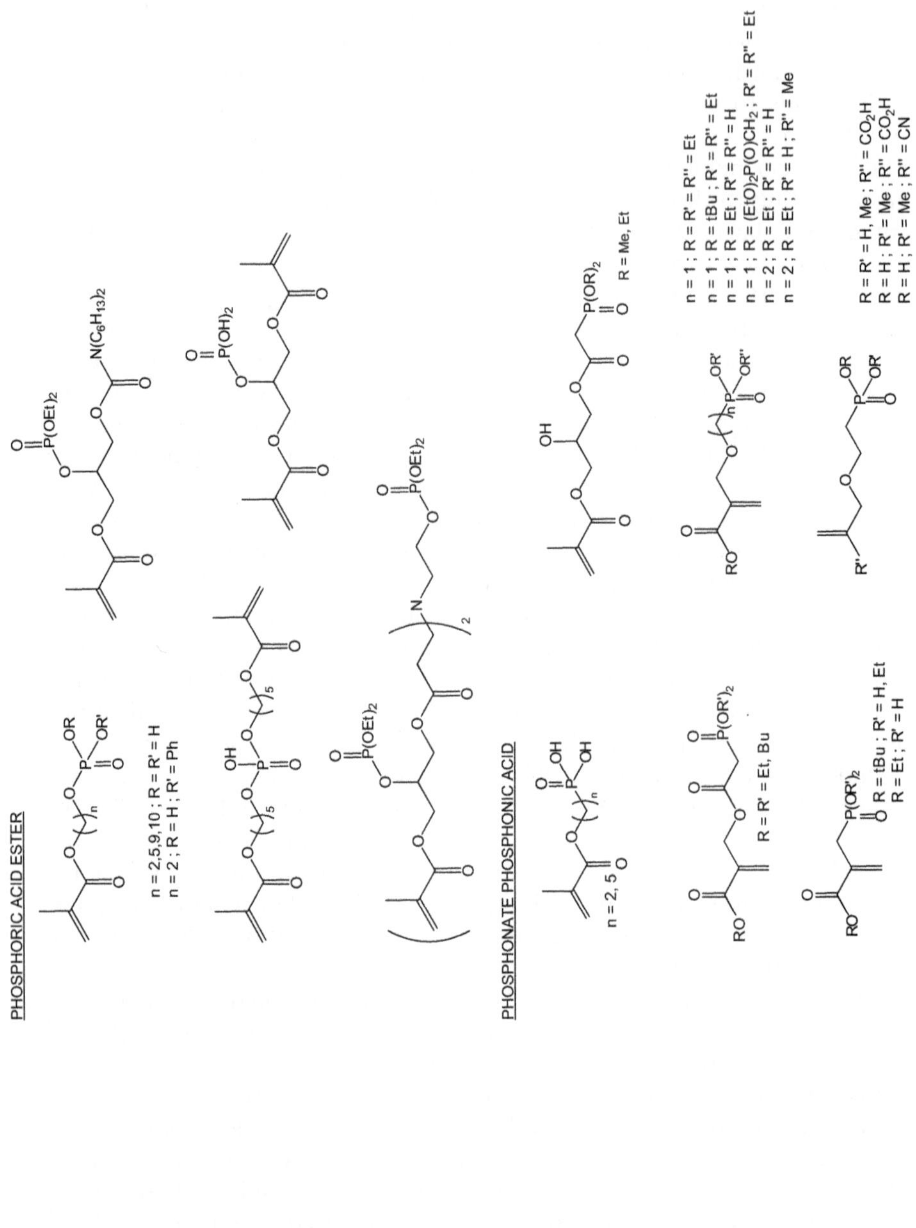

PHOSPHORIC ACID ESTER

n = 2,5,9,10 ; R = R' = H
n = 2 ; R = H ; R' = Ph

PHOSPHONATE PHOSPHONIC ACID

n = 2, 5

R = R' = Et, Bu

R = tBu ; R' = H, Et
R = Et ; R' = H

R = Me, Et

n = 1 ; R = R' = R'' = Et
n = 1 ; R = tBu ; R' = R'' = Et
n = 1 ; R = Et ; R' = R'' = H
n = 1 ; R = (EtO)₂P(O)CH₂ ; R' = R'' = Et
n = 2 ; R = Et ; R' = R'' = H
n = 2 ; R = Et ; R' = H ; R'' = Me

R = R' = H, Me ; R'' = CO₂H
R = H ; R' = Me ; R'' = CO₂H
R = H ; R' = Me ; R'' = CN

AROMATIC PHOSPHONATE PHOSPHONIC ACID

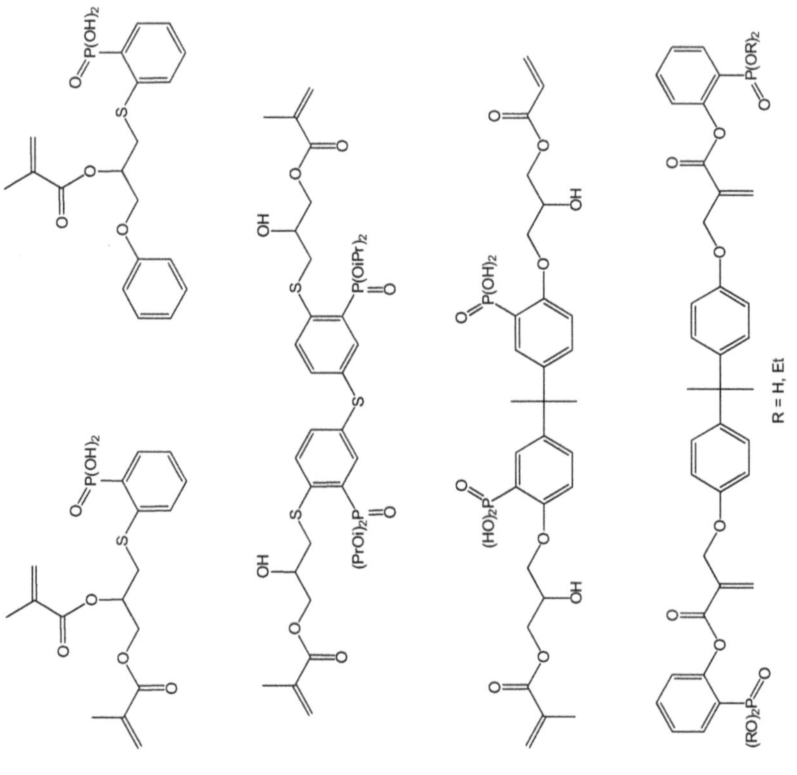

Scheme 1.4 Phosphorus-based monomers used as dental additives.

The authors studied the effect of the position of the PO(OH)$_2$ group in the molecule, and the influence of steric hindrance on the polymerization rate. Nevertheless, some contradictory results were published[72] with very close structures,[20] where the authors found lower conversion rates.

1.6 Living Radical Polymerization of Phosphonated Methacrylate

Whereas an important number of phosphorus-based monomers have been evaluated by free radical polymerization or photopolymerization, fewer examples have dealt with the "living" radical polymerization (LRP) of such monomers, whereas these polymerization techniques allow the synthesis of well-defined polymers with tunable structures, compositions, and properties. Among all possible chemical environments of the phosphorus atom, some monomers containing phosphate, phosphinic, and phosphonated functions have been evaluated. Polymerization of MAEP and MOEP by the RAFT process was successfully achieved, enabling the synthesis of block copolymers by association with a poly[(2-acetoacetoxy)ethyl methacrylate] segment.[49] ATRP of deprotonated MOEP[73] and dimethyl (1-ethoxy-carbonyl)vinyl phosphate[74] was also carried out, leading to polymer brush formation on a gold surface and to block copolymers, respectively. RAFT polymerization of phenylphosphinic acid monomer led to diblock copolymers which self-assembled in water[75] or to phenylphosphinic acid-functionalized polystyrene microspheres by an emulsion process.[76] The controlled radical polymerization by RAFT of MAPC1 and its phosphonic acid analogue (Scheme 1.5) was recently reported in the literature.[77] This monomer is interesting as its synthesis requires less hazardous and cheaper reactants than the preparation of most other phosphonated monomers.

Homopolymerization has been thoroughly investigated under different experimental conditions. The reaction was first carried out in DMF at different temperatures; 70 °C was chosen as it allowed a good control over the molecular weight, with a relatively high rate of polymerization and a short induction period. Two chain transfer agents were successfully used: 2-cyanoisopropyl dithiobenzoate and 4-cyanopentanoic acid dithiobenzoate {4-cyano-4-[(phenylcarbonothioyl)sulfanyl]pentanoic acid}. Then, it was also shown that it was possible to synthesize PMAPC1 of different molecular weights, ranging from 8000 to 24 000 g mol^{-1}. The effect of the solvent was also evaluated. Among the other polar solvents used, it was possible to control the polymerization in *N,N*-dimethylacetamide whereas 1,4-dioxane led to a termination reaction, and dimethyl sulfoxide and water to transfer reactions. This result was explained by the polarity of the different solvents. To be well controlled, the polarity has to be high enough to ensure a good rate of polymerization without being too high to avoid transfer. The "living" character of the produced poly{dimethyl [(methacryloyloxy)methyl]phosphonate} was checked and it was demonstrated that the polymer end-group was

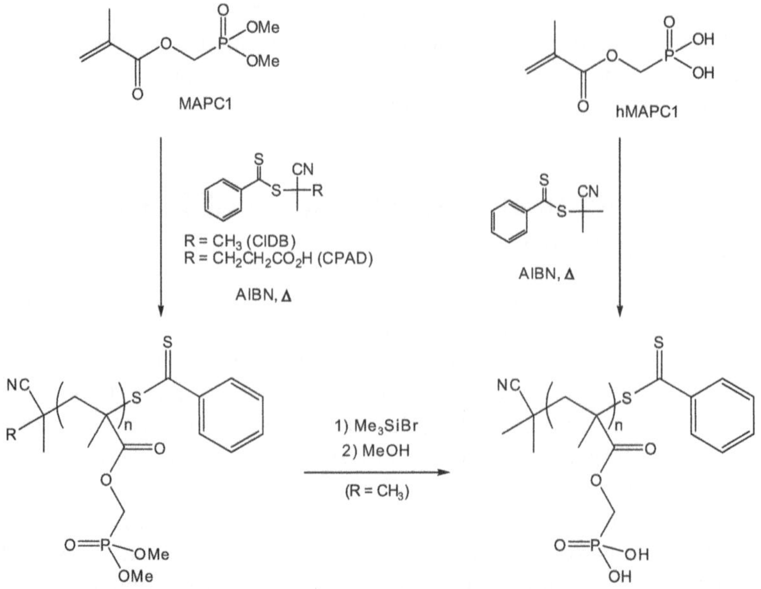

Scheme 1.5   Reaction pathway for (*left*) the RAFT polymerization of MAPC1 followed by the hydrolysis of phosphonate into phosphonic acid groups,
and (*right*) the RAFT polymerization of [(methacryloyloxy)methyl]phos-
phonic acid (hMAPC1).

a dithioester function, which allowed growth of the macromolecular
chains from the PMAPC1 macro chain transfer agent. It was shown that it was
possible to prepare complex architecture such as poly{dimethyl [(methacryloyloxy)methyl]phosphonate}-*b*-poly(MMA) (PMAPC1-*b*-PMMA)[78] or poly{di-
methyl   [(methacryloyloxy)methyl]phosphonate}-*b*-poly[methoxypoly(ethylene
oxide) methacrylate] (PMAPC1-*b*-PMPEOMA).[79] Finally, PMAPC1 was also
hydrolyzed to afford phosphonic acid functions that can be useful for many
applications and, in parallel, RAFT polymerization of (methacryloyloxy)methyl
phosphonic acid was also successfully performed. The control of the polymerization and of the chain-end functionality opens the way to numerous
chemical structures, and thus to the synthesis of innovative materials.

1.7   Conclusion

Phosphorus-based (meth)acrylates have been the subject of extensive research in recent years, mainly due to the properties brought by the phosphorus atom, which can be used for different purposes. Most of the prepared
monomers are phosphonated esters, whereas only limited numbers of
phosphoric esters have been reported in the literature. This can be explained
by easier hydrolysis of the latter in comparison with the phosphonate derivatives. Concerning the synthesis methodology, free radical polymerization
and photopolymerization have been the most investigated. Some examples

also dealt with the controlled radical polymerization of such kinds of monomers, by atom transfer radical polymerization or reversible addition–fragmentation transfer processes. These results are of great interest as "living" radical polymerization allows the synthesis of well-defined polymers with controlled architectures.

Finally, to conclude, it is also important to point out that phosphorus-based poly(meth)acrylates are successfully employed for many applications, including flame retardancy, anticorrosion, and in the biomedical field. As these applications are of great interest, we can assume that the development of other phosphorus-based (meth)acrylate monomers will continue in the future.

References

1. H. Ai, K. Xu, H. Liu, M. C. Chen and X. J. Zhang, *J. Appl. Polym. Sci.*, 2009, **113**, 541–546.
2. C. S. Zhao, L. Chen and Y. Z. Wang, *J. Polym. Sci., Part A: Polym. Chem.*, 2008, **46**, 5752–5759.
3. O. Senhaji, S. Monge, K. Chougrani and J. J. Robin, *Macromol. Chem. Phys.*, 2008, **209**, 1694–1704.
4. B. Rixens, R. Severac, B. Boutevin and P. Lacroix-Desmazes, *J. Polym. Sci., Part A: Polym. Chem.*, 2006, **44**, 13–24.
5. M. R. Pinto, B. M. Kristal and K. S. Schanze, *Langmuir*, 2003, **19**, 6523–6533.
6. B. Boutevin, Y. Hervaud, A. Boulahna and E. M. El Hadrami, *Polym. Int.*, 2002, **51**, 450–457.
7. M. V. Chaubal, A. Sen Gupta, S. T. Lopina and D. F. Bruley, *Crit. Rev. Ther. Drug Carrier Syst.*, 2003, **20**, 295–315.
8. S.-W. Huang and R.-X. Zhuo, *Phosphorus, Sulfur Silicon Relat. Elem.*, 2008, **183**, 340–348.
9. A. Clearfield, *Curr. Opin. Solid State Mater. Sci.*, 1996, **1**, 268–278.
10. A. Clearfield, *Curr. Opin. Solid State Mater. Sci.*, 2002, **6**, 495–506.
11. A. Clearfield, *J. Alloys Compd.*, 2006, **418**, 128–138.
12. M. Essahli, G. Colomines, S. Monge, J. J. Robin, A. Collet and B. Boutevin, *Polymer*, 2008, **49**, 4510–4518.
13. T. P. Knepper, *Trends Anal. Chem.*, 2003, **22**, 708–724.
14. E. Matczak-Jon and V. Videnova-Adrabinska, *Coord. Chem. Rev.*, 2005, **249**, 2458–2488.
15. L. Herrera-Taboada, M. Guzmann, K. Neubecker and A. Goethlich, *PCT Int. Appl.*, 2008, 28.
16. J. Canadell, B. J. Hunt, A. G. Cook, A. Mantecon and V. Cadiz, *Polym. Degrad. Stab.*, 2007, **92**, 1482–1490.
17. H. Singh and A. K. Jain, *J. Appl. Polym. Sci.*, 2009, **111**, 1115–1143.
18. S. Monge, B. Canniccioni, A. Graillot and J. J. Robin, *Biomacromolecules*, 2011, **12**, 1973–1982.
19. H. S. Yang, H. Adam and J. Kiplinger, *JCT CoatingsTech*, 2005, **2**, 44–52.

20. L. Y. Mou, G. Singh and J. W. Nicholson, *Chem. Commun.*, 2000, 345–346.
21. D. Avci and L. J. Mathias, *Polym. Bull.*, 2005, **54**, 11–19.
22. U. Salz, A. Mucke, J. Zimmermann, F. R. Tay and D. H. Pashley, *J. Adhes. Dent.*, 2006, **8**, 143–150.
23. M. A. Bayle, G. Gregoire and P. Sharrock, *J. Dent.*, 2007, **35**, 302–308.
24. N. Nishiyama, M. Aida, K. Fujita, K. Suzuki, F. R. Tay, D. H. Pashley and K. Nemoto, *Dent. Mater. J.*, 2007, **26**, 382–387.
25. K. L. Van Landuyt, Y. Yoshida, I. Hirata, J. Snauwaert, J. De Munck, M. Okazaki, K. Suzuki, P. Lambrechts and B. Van Meerbeek, *J. Dent. Res.*, 2008, **87**, 757–761.
26. C. Brondino, B. Boutevin, J.-P. Parisi and J. Schrynemackers, *J. Appl. Polym. Sci.*, 1999, **72**, 611–620.
27. M. Gaboyard, J.-J. Robin, Y. Hervaud and B. Boutevin, *J. Appl. Polym. Sci.*, 2002, **86**, 2011–2020.
28. G. David, C. Negrell, A. Manseri and B. Boutevin, *J. Appl. Polym. Sci.*, 2009, **114**, 2213–2220.
29. Z. E. Zhor, K. Chougrani, C. Negrell-Guirao, G. David, B. Boutevin and C. Loubat, *J. Polym. Sci., Part A: Polym. Chem.*, 2008, **46**, 4794–4803.
30. C. Bressy-Brondino, B. Boutevin, Y. Hervaud and M. Gaboyard, *J. Appl. Polym. Sci.*, 2002, **83**, 2277–2287.
31. B. Rixens, G. Boutevin, A. Boulahna, Y. Hervaud and B. Boutevin, *Phosphorus, Sulfur Silicon Relat. Elem.*, 2004, **179**, 2617–2626.
32. O. A. Lam, G. David, Y. Hervaud and B. Boutevin, *J. Polym. Sci., Part A: Polym. Chem.*, 2009, **47**, 5090–5100.
33. K. Chougrani, B. Boutevin, G. David, S. Seabrook and C. Loubat, *J. Polym. Sci., Part A: Polym. Chem.*, 2008, **46**, 7972–7984.
34. D. Price, K. J. Bullett, L. K. Cunliffe, T. R. Hull, G. J. Milnes, J. R. Ebdon, B. J. Hunt and P. Joseph, *Polym. Degrad. Stab.*, 2005, **88**, 74–79.
35. D. Price, L. K. Cunliffe, K. J. Bullet, T. R. Hull, G. J. Milnes, J. R. Ebdon, B. J. Hunt and P. Joseph, *Polym. Adv. Technol.*, 2008, **19**, 710–723.
36. J. R. Ebdon, B. J. Hunt, P. Joseph, C. S. Konkel, D. Price, K. Pyrah, T. R. Hull, G. J. Milnes, S. B. Hill, C. I. Lindsay, J. McCluskey and I. Robinson, *Polym. Degrad. Stab.*, 2000, **70**, 425–436.
37. D. Price, K. Pyrah, T. R. Hull, G. J. Milnes, J. R. Ebdon, B. J. Hunt, P. Joseph and C. S. Konkel, *Polym. Degrad. Stab.*, 2001, **74**, 441–447.
38. D. Price, K. Pyrah, T. R. Hull, G. J. Milnes, J. R. Ebdon, B. J. Hunt and P. Joseph, *Polym. Degrad. Stab.*, 2002, **77**, 227–233.
39. S. Zhu and W. Shi, *Polym. Degrad. Stab.*, 2003, **82**, 435–439.
40. B. Youssef, L. Lecamp, W. El Khatib, C. Bunel and B. Mortaigne, *Macromol. Chem. Phys.*, 2003, **204**, 1842–1850.
41. D. Avci and A. Z. Albayrak, *J. Polym. Sci., Part A: Polym. Chem*, 2003, **41**, 2207–2217.
42. A. Asif, H. Liang, S. Zhu and W. Shi, *Photochem. UV Curing*, 2006, 355–371.
43. S. Zhu and W. Shi, *Polym. Int.*, 2004, **53**, 266–271.
44. R. A. Gemeinhart, C. M. Bare, R. T. Haasch and E. J. Gemeinhart, *J. Biomed. Mater. Res., A*, 2006, **78**, 433–440.

45. K. A. George, E. Wentrup-Byrne, D. J. T. Hill and A. K. Whittaker, *Biomacromolecules*, 2004, **5**, 1194–1199.
46. L. Grondahl, F. Cardona, K. Chiem and E. Wentrup-Byrne, *J. Appl. Polym. Sci.*, 2002, **86**, 2550–2556.
47. L. Grondahl, F. Cardona, K. Chiem, E. Wentrup-Byrne and T. Bostrom, *J. Mater. Sci.: Mater. Med.*, 2003, **14**, 503–510.
48. L. Grondahl, S. Suzuki and E. Wentrup-Byrne, *Chem. Commun.*, 2008, 3314–3316.
49. S. Suzuki, R. M. Whittaker, L. Grondahl, J. M. Monteiro and E. Wentrup-Byrne, *Biomacromolecules*, 2006, **7**, 3178–3187.
50. I. C. Stancu, R. Filmon, C. Cincu, B. Marculescua, C. Zaharia, Y. Tourmen, M. F. Basle and D. Chappard, *Biomaterials*, 2004, **25**, 205–213.
51. K. Ikemura, Y. Kadoma and T. Endo, *Dent. Mater. J.*, 2011, **30**, 769–789.
52. M. G. Buonocore, *J. Dent. Res.*, 1955, **34**, 849–853.
53. J. Perdigao, E. J. Swift, G. E. Denehy, J. S. Wefel and K. J. Donly, *J. Dent. Res.*, 1994, **73**, 44–55.
54. B. Van Meerbeek, G. Vanherle, P. Lambrechts and M. Braem, *Curr. Opin. Dent.*, 1992, **2**, 117–127.
55. A. Z. Albayrak, Z. S. Bilgici and D. Avci, *Macromol. React. Eng.*, 2007, **1**, 537–546.
56. N. Moszner, U. Salz and J. Zimmermann, *Dent. Mater.*, 2005, **21**, 895–910.
57. F. A. Ogliari, E. D. da Silva, G. D. Lima, F. C. Madruga, S. Henn, M. Bueno, M. A. Ceschi, C. L. Petzhold and E. Piva, *J. Dent.*, 2008, **36**, 171–177.
58. M. D. Francis and W. W. Briner, *Calcif. Tissue Res.*, 1973, **11**, 1–9.
59. M. Anbar and E. P. Farley, *J. Dent. Res.*, 1974, **53**, 879–888.
60. B. Yeniad, A. Z. Albayrak, N. C. Olcum and D. Avci, *J. Polym. Sci., Part A: Polym. Chem.*, 2008, **46**, 2290–2299.
61. N. Sibold, P. J. Madec, S. Masson and T. N. Pham, *Polymer*, 2002, **43**, 7257–7267.
62. G. Sahin, D. Avci, O. Karahan and N. Moszner, *J. Appl. Polym. Sci.*, 2009, **114**, 97–106.
63. N. Moszner, F. Zeuner, S. Pfeiffer, I. Schurte, V. Rheinberger and M. Drache, *Macromol. Mater. Eng.*, 2001, **286**, 225–231.
64. E. P. Farley, R. L. Jones and M. Anbar, *J. Dent. Res.*, 1977, **56**, 943–952.
65. N. Moszner, F. Zeuner, U. K. Fischer and V. Rheinberger, *Macromol. Chem. Phys.*, 1999, **200**, 1062–1067.
66. S. Salman, A. Z. Albayrak, D. Avci and V. Aviyente, *J. Polym. Sci., Part A: Polym. Chem*, 2005, **43**, 2574–2583.
67. Y. Catel, V. Besse, A. Zulauf, D. Marchat, E. Pfund, T. N. Pham, D. Bernache-Assolant, M. Degrange, T. Lequeux, P. J. Madec and L. Le Pluart, *Eur. Polym. J.*, 2012, **48**, 318–330.
68. Y. Catel, L. Le Pluart, P. J. Madec and T. N. Pham, *J. Appl. Polym. Sci.*, 2010, **117**, 2676–2687.

69. S. Edizer, G. Sahin and D. Avci, *J. Polym. Sci., Part A: Polym. Chem*, 2009, **47**, 5737–5746.
70. G. Sahin, A. Z. Albayrak, Z. Sarayli and D. Avci, *J. Polym. Sci., Part A: Polym. Chem.*, 2006, **44**, 6775–6781.
71. D. Avci and L. J. Mathias, *J. Polym. Sci., Part A: Polym. Chem.*, 2002, **40**, 3221–3231.
72. G. Adusei, S. Deb, J. W. Nicholson, L. Y. Mou and G. Singh, *J. Appl. Polym. Sci.*, 2003, **88**, 565–569.
73. F. Zhou and W. T. S. Huck, *Chem. Commun.*, 2005, 5999–6001.
74. J. Y. Huang and K. Matyjaszewski, *Macromolecules*, 2005, **38**, 3577–3583.
75. D. B. Hua, J. Tang, J. L. Jiang and X. L. Zhu, *Polymer*, 2009, **50**, 5701–5707.
76. D. B. Hua, J. Tang, J. L. Jiang, X. L. Zhu and R. K. Bai, *Macromolecules*, 2009, **42**, 8697–8701.
77. B. Canniccioni, S. Monge, G. David and J. J. Robin, *Polym. Chem.*, 2013, **4**, 3676–3685.
78. B. Canniccioni, S. Monge, G. David and J. J. Robin, in preparation, 2014.
79. B. Canniccioni, C. Gerardin, G. David and S. Monge, in preparation, 2014.

CHAPTER 2

Polymerization of Phosphorus-Containing (meth)acrylamide Monomers

SOPHIE MONGE,* ALAIN GRAILLOT AND
JEAN-JACQUES ROBIN

Institut Charles Gerhardt de Montpellier, UMR5253 CNRS-UM2-ENSCM-UM1,
Equipe Ingénierie et Architectures Macromoléculaires, Université
Montpellier II, cc1702, Place Eugène Bataillon, 34095 Montpellier, France
*Email: sophie.monge-darcos@univ-montp2.fr

2.1 Introduction

Among all phosphorus-type monomers considered in the literature, phosphonated (meth)acrylamides have attracted much interest in recent years, mainly due to the stability of the amide function toward hydrolysis in acidic aqueous solutions compared to the usual (meth)acrylate monomers. Phosphorus-containing materials can be employed for a large range of technological applications. They were developed due to their aptitude to bind metals[1] as they showed interesting sorption properties, useful for preventing deposit formation. They are also employed as flame retardants,[2] where phosphorus is known to be efficient. Polymeric materials bearing phosphonic sites were also developed in alternative energy production, and have been involved in proton-conducting fuel-cell membranes.[3] It was shown that phosphonic acid moieties had higher chemical and thermal stability than sulfonic acid materials, usually used for such purposes. A final but

RSC Polymer Chemistry Series No. 11
Phosphorus-Based Polymers: From Synthesis to Applications
Edited by Sophie Monge and Ghislain David
© The Royal Society of Chemistry 2014
Published by the Royal Society of Chemistry, www.rsc.org

important development of phosphorus-based materials was their employ-ment in the biomedical field.[4]

Whereas phosphonated (meth)acrylates have been widely investigated, only a few examples have dealt with the synthesis and polymerization of (meth)acrylamide derivatives. To date, the latter have been mainly employed to obtain self-etching enamel/dentin adhesives, which are commonly used to achieve a strong bond between a restorative composite and the tooth sub-strate, *i.e.* enamel and dentin.[5] Indeed, the incorporation of phosphonic acid groups into monomer structures was proved to increase the adhesion to the tooth because of the formation of complexes with calcium in hydro-xyapatite.[6,7] In that context, the interest of the amide function is explained by its better stability. Most of the currently available self-etching primers/adhesives are (meth)acrylates, but as strong acidic conditions are required (pH value ranging from 1.5 to 2.5), esters are hydrolytically degraded[8] which is an important drawback.

Phosphonated (meth)acrylamide monomers contained in self-etching adhesives can be divided into two different groups: (i) self-etching adhesive monomers or (ii) crosslinking monomers. To be efficient, they need to fill some general requirements[9] among which, for instance, are a high rate of polymerization and a low oral toxicity and cytotoxicity. Additionally, as already mentioned, monomers and the resulting polymers must exhibit sufficient stability, not only against water and acidic media but also toward premature polymerization and degradation by oxygen, heat, light, and water during storage.

In general, these monomers are polymerized by photopolymerization or free radical polymerization. Very recently, phosphorus-based poly(acrylamide) have been synthesized by reversible addition-fragmentation transfer (RAFT) polymerization, leading to well-defined homopolymers, with control over the molecular weight and low dispersity.[10] In this case, the phosphonated block was attached to a poly(*N-n*-propylacrylamide) moiety, leading to thermo-sensitive copolymers. To our knowledge, this contribution represents the only example of a controlled radical polymerization involving phosphonated acrylamide monomers. The authors assumed that the synthesis of innova-tive phosphonated polyacrylamide-based polymers and copolymers with well-defined architectures will provide chemically very stable materials suitable for a wide range of applications.

This chapter deals with the synthesis of phosphonated-based (meth)acrylamide monomers and polymers reported in the literature. Syntheses will be described as a function of the polymerization process: photo-polymerization, free radical polymerization, and controlled radical poly-merization. When available, details will be given on the targeted application.

2.2   Photopolymerization

UV radiation is a simple and convenient energy source that does not require expensive devices and requires only a low temperature for an energy efficient process.[11,12] Thanks to its high output, this special form of polymer processing is applied at an industrial scale for coatings, paints or printing inks, adhesives, composite materials and for dental restorative formulations, and many other uses. Liquid resins can be converted into solid resins in a few tenths of a second, making this process very attractive for the scientific community for almost three decades. The photopolymerization can be achieved by polyaddition of double bonds under either radical or cationic initiation.

Photopolymerization of phosphonated acrylamide monomers was achieved using aliphatic or aromatic spacers (Scheme 2.1).[13,14] Concerning aliphatic moieties, (*N*-methylacrylamido)alkylphosphonic acid monomers were prepared in four steps[13] from different dibromoalkanes. First, the bromophosphonates were synthesized according to a Michaelis-Arbuzov reaction. Then, reaction with a large excess of methylamine allowed the synthesis of aminophosphonates. Finally, the acryloyl group was introduced and the phosphonated ester groups were hydrolyzed, leading to phosphonic acid acrylamide monomers.

In the case of aromatic spacers, the synthesis was carried out in five steps with an overall yield of 56%. 4-(Bromomethyl)benzonitrile was first reduced using diisobutylaluminium hydride and followed by hydrolysis, to obtain the corresponding aldehyde. The latter was then reacted with triethyl phosphite. Subsequent reaction with methylamine and reduction with sodium borohydride of the resulting phosphonate allowed synthesis of the phosphonated secondary amine. Finally, the 4-(*N*-methylacrylamido-methyl)benzylphosphonic acid monomer was obtained by an acryloylation reaction, and hydrolysis of the phosphonated ester group with bromotrimethylsilane and methanol finally led to the desired phosphonic acid moiety.

All prepared acrylamide monomers were fully characterized by high-resolution mass spectrometry (HRMS), and ^{1}H, ^{13}C, and ^{31}P NMR in deuterated water or chloroform. The hydrolytic stability of the phosphonic acid monomers has also been investigated by NMR using 20% weight solutions of monomers in an ethanol-d_6/D$_2$O mixture after storage at 37 °C. No signals relating to degradation were detected in the NMR spectra during the 13 weeks of investigation. These results showed that the use of an (*N*-methyl)acrylamide group instead of a methacrylate group significantly

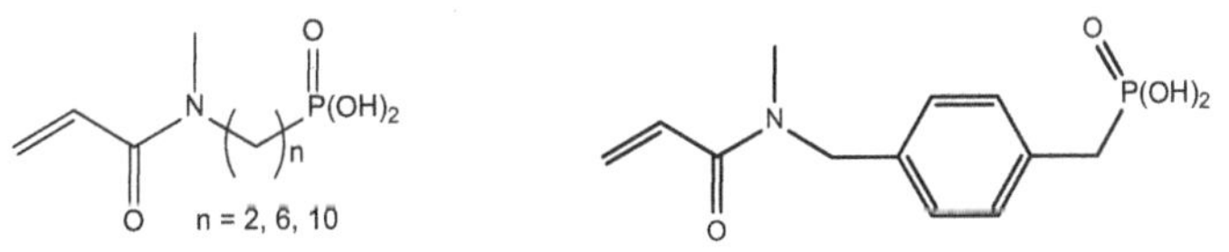

(*N*-Methylacrylamido)ethylphosphonic acid (*n* = 2)   4-(*N*-Methylacrylamidomethyl)benzylphosphonic acid

Scheme 2.1 Phosphonic acid-containing acrylamide monomers used to achieve photopolymerization or free radical polymerization.

improved the stability of the phosphonic acid monomers in acidic aqueous conditions. This parameter appeared as an essential innovation for the development of stable self-etching dental primers based on phosphonic acid (meth)acrylamide monomers.

The photopolymerization of (*N*-methylacrylamido)alkylphosphonic acids and 4-(*N*-methylacrylamidomethyl)benzylphosphonic acid monomers was carried out. The reactivities of the different monomers were compared by copolymerizing each of them with *N*,*N*′-diethyl-1,3-bis(acrylamido)propane (DEBAAP) crosslinking monomer, known to be hydrolytically stable and already used in commercial adhesives. It was shown that the aromatic monomer was the most reactive, but final double bond conversions were about the same. Additionally, dentin shear bond strength measurements have shown that primers based on (*N*-methylacrylamido)alkylphosphonic acids lead to a strong bond between the tooth substance and a dental composite. It was also found that adhesion was improved with the spacer length of the acidic monomer. On the other hand, the 4-(*N*-methylacryl-amidomethyl)benzylphosphonic acid monomer did not show adhesion. Another study dealt with photopolymerization kinetics of (*N*-methylacryl-amido)hexylphosphonic acid and ester as new dentin adhesives.[14] The acrylamide phosphonic acid was found to accelerate the polymerization rate, but similar monomers bearing a phosphonate ester group had a much smaller effect.

Methacrylamide monomers were also considered (Scheme 2.2). 2-Metha-crylamidoethylphosphonic acid was prepared from the expensive (2-ami-noethyl)phosphonic acid in one step.[15] Other monomers (Scheme 2.2, monomers **1** and **2**) were synthesized from bisphenol A diglycidyl ether. The ring-opening of epoxides was achieved with aqueous ammonia at high temperature and pressure and the subsequent reaction with diethyl (2-bromoethyl)phosphonate led either to mono- or disubstituted product. Targeted monomers were finally obtained after (i) reaction with methacryloyl chloride and (ii) hydrolysis of the phosphonated ester groups.

All methacrylamide monomers were characterized by electrospray mass spectrometry (ESMS), Fourier transform infrared (FTIR) spectroscopy, and NMR spectroscopy. The hydrolytic stability was also evaluated. These monomers proved to be stable, as ^{1}H and ^{31}P NMR chemical shifts remained unchanged after storage at 60 °C for two months. Photopolymerization of 2-methacrylamidoethylphosphonic acid and monomer (**1**) was performed and allowed the fabrication of three experimental neat resins by the mixing of these monomers with 2-hydroxyethyl methacrylate (HEMA), notably in the presence of camphorquinone. It was shown that 2-methacrylamidoethyl-phosphonic acid had a lower degree of conversion than monomer (**1**) or 2-methacryloyloxyethyl phosphoric acid (MEP), in particular at shorter cur-ing times. This confirmed that methacrylamide monomers are less reactive than methacrylates. Moreover, it was explained that 2-methacrylami-doethylphosphonic acid was probably more reactive because it is more hydrophilic and hydroscopic than monomer (**1**). In conclusion, such

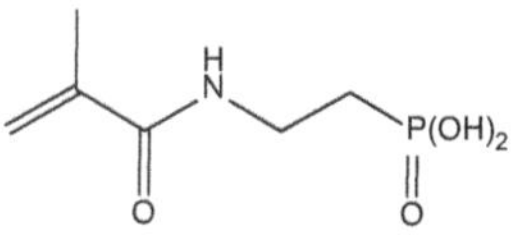

2-Methacrylamidoethylphosphonic acid

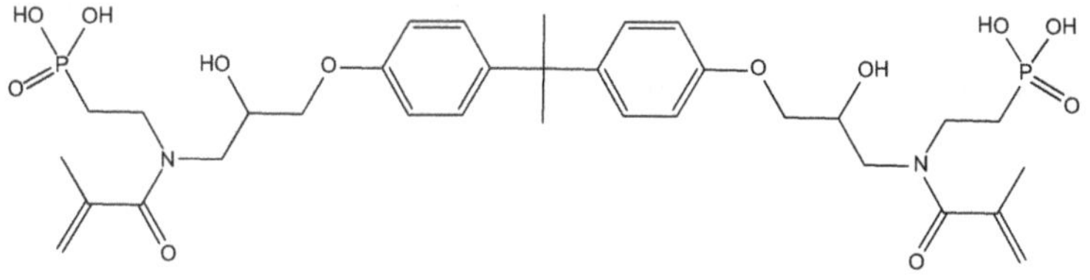

N,N'-[4,4'-(Propane-2,2-diyl)bis(phenoxy-2-hydroxypropyl)]bis(2-methacrylamidoethylphosphonic acid)

Monomer (**1**)

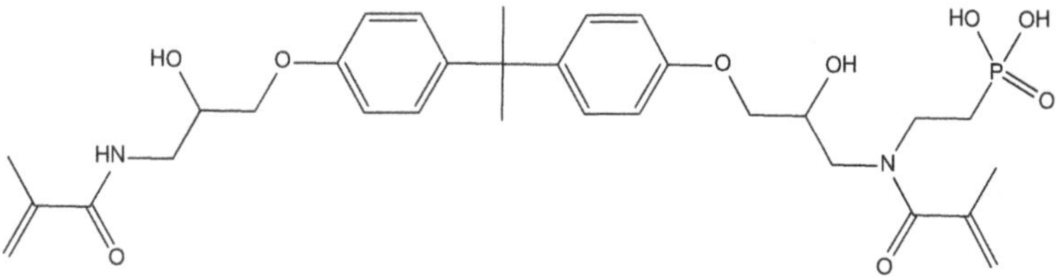

2-{*N*-[Hydroxy-3-(4-{2-[4-(2-hydroxy-3-methacrylamidopropoxy)phenyl]propan-2-

yl}phenoxy)propyl]methacrylamido}ethylphosphonic acid

Monomer (**2**)

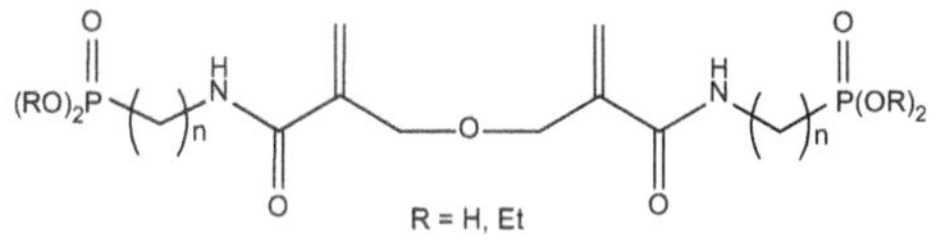

R = H, Et

Monomers (**3**) (R = Et) and (**4**) (R = H)

Scheme 2.2 Phosphonic acid-containing methacrylamide monomers used to achieve photopolymerization and free radical polymerization in the case of monomers (**3**) and (**4**).

monomers may be used in dental bonding agents, dental cements, and other dental materials.

Other phosphonated bis(methacrylamide)s were also studied (Scheme 2.2, monomers **3** and **4**).[16] Synthesis of the monomers was achieved in four (ester form) or five (acid form) steps. *tert*-Butyl acrylate reacted with para-formaldehyde in the presence of 1,4-diazabicyclooctane as catalyst; then conversion into the carboxylic acid was performed by cleavage of the *tert*-butyl groups using trifluoroacetic acid. The dicarboxylic acid reacted with oxalyl chloride and the resulting acid chloride was added to either 2-ami-noethylphosphonate or 1-aminomethylphosphonate. Phosphonated ester groups were then hydrolyzed with bromotrimethylsilane and, finally, four different monomers (phosphonated ester and acid with $n = 1$ or 2) were obtained. Characterization was achieved using ^{1}H, ^{13}C, and ^{31}P NMR, FTIR, and elemental analysis, confirming that the targeted structures were successfully synthesized. Monomers (3) and (4) were homopolymerized using photo-DSC with 2,2′-dimethoxy-2-phenylacetophenone as photoinitiator and their maximum rates of polymerization were higher than the ones of commercial 2,2-bis[4-(2-hydroxy-3-methacryloyloxypropoxy)phenyl]propane (Bis-GMA) or HEMA monomers, indicating their potential as reactive diluents or crosslinkers in dental materials. Copolymerization with monomer (3) (with $n = 1$) resulted in improvements in the photopolymerization kinetics of Bis-GMA and HEMA. On the other hand, monomers (4) (acid form) were much less reactive than their phosphonate-containing analogs. Addition of HEMA could result, as a function of the experimental conditions, in a decrease of both conversion and rate of polymerization. Nevertheless, because of their etching and binding ability, hydrolytic stability, and copolymerizability with dental monomers such as HEMA, monomers (4) are potentially interesting for use as adhesive derivatives in dental adhesives.

2.3   Free Radical Polymerization

Free radical polymerization (FRP) is a very widespread method of polymerization, which has been used for a long time in industry for many reasons: FRP is easy to carry out, can be performed in bulk, in solution in various solvents, and also in dispersed media (suspension, emulsion, *etc.*). It can be achieved over a large range of reaction temperatures (-100 to $+200\,^\circ$C). Additionally, this polymerization technique can be carried out with a lot of monomers, even functionalized ones. For example, monomers can bear "standard" functional groups or heteroatoms such as silicon[17,18] or phosphorus.[19,20]

Different phosphorus-based mono- and bisacrylamide monomers were polymerized by free radical polymerization. (N-Methylacrylamido)alkylphosphonic acid and 4-(N-methylacrylamidomethyl)benzylphosphonic acid monomers (Scheme 2.1) were evaluated. Polymerization was achieved at 65 °C in a water/ethanol mixture (1/2.5 v/v) using 2,2′-azobis(2-methylpropionamidine) dihydrochloride (AMPAHC) as radical initiator. The polymers were precipitated in THF, ethanol, or water, and were produced in good yield (65–91%). Concerning poly[(N-methylacrylamido)alkylphosphonic acid], size exclusion chromatography (SEC) with a multi-angle laser light scattering (MALLS) detector allowed the determination of the molecular weights of poly[(N-methylacrylamido)ethyphosphonic acid] and poly[(N-methylacrylamido)hexylphosphonic acid], which were equal to 200 000 [dispersity $(Đ) = 2.7$] and $32\,000\,\mathrm{g\,mol}^{-1}$ ($Đ = 1.9$), respectively. Thermogravimetric analyses showed that such kinds of polymers had good thermal stability, as weight losses only began above 160 °C. Additionally, differential scanning calorimetry (DSC) experiments permitted determination of the glass transition temperature (T_g). It was shown that the T_g logically decreased with the increase of the spacer length (from 98 °C for $n=2$ to 69 °C for $n=10$). Phosphorus-containing copolymers were also obtained by free radical co-polymerization of *N-n*-propylacrylamide with diethyl 2-(acrylamido)ethyl-phosphonate.[21] Copolymer syntheses were conducted in dioxane at 70 °C using 2,2'-azobisisobutyronitrile (AIBN) as radical initiator and variation of the feed content allowed formation of copolymers containing different pro-portions of phosphonated moieties. The similarity of both monomers, and thus of their reactivity ratios, permitted achievement of random statistical copolymers. The proportion of phosphonated groups largely influenced the lower critical solution temperature (LCST) of the corresponding copolymers and also enabled reaching the sorption properties for metallic cations.

In conclusion, all the results were consistent with efficient free radical polymerization of such monomers, despite the presence of phosphonic substituents. Free radical polymerization of other phosphorus-containing monomethacrylamide monomers (Scheme 2.3) was considered in the literature.[9,22] Nevertheless, no significant results were reported, notably on adhesion properties, as work achieved on this kind of monomer was mainly described in patents.[23,24]

For phosphonated bis(methacrylamide)s, thermal homopolymerization of monomers (3) (Scheme 2.2) in bulk and in solution using AIBN at 80 °C re-sulted in crosslinked polymers. On the other hand, copolymerization with 2-(2-carboxyallyloxymethyl)acrylic acid led to soluble polymers, obtained at low monomer and initiator concentrations. Molecular weights varied bet-ween 13 000 and 49 000 g mol^{-1}. Thermal copolymerization of phosphonic acid derivatives (monomers 4) was carried out (i) with acrylamide in water at 65 °C using V-50 as initiator or (ii) with HEMA in ethanol at 50 °C. Cross-linked polymers were obtained in both cases.

A last example of a bisacrylamide which has to be considered deals with the synthesis and radical polymerization of a phosphate monomer, namely the 1,3-bis(methacrylamido)propan − 2-yl dihydrogen phosphate (Scheme 2.4).[22,25]

Synthesis of the monomer was achieved in two steps. In the first step, 1,3-diaminopropan-2-ol was reacted with methacrylic anhydride to obtain 1,3-bis(methacrylamido)-2-hydroxypropane. In the second step, the latter was phosphorylated with phosphorus oxychloride in the presence of

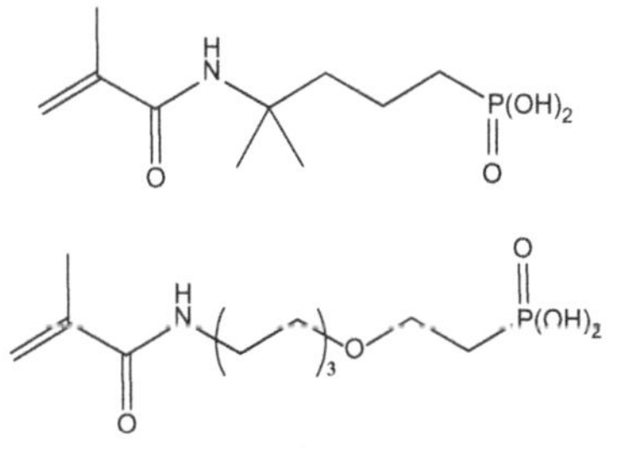

Scheme 2.3   Two phosphonic acid-containing methacrylamides described in patents.

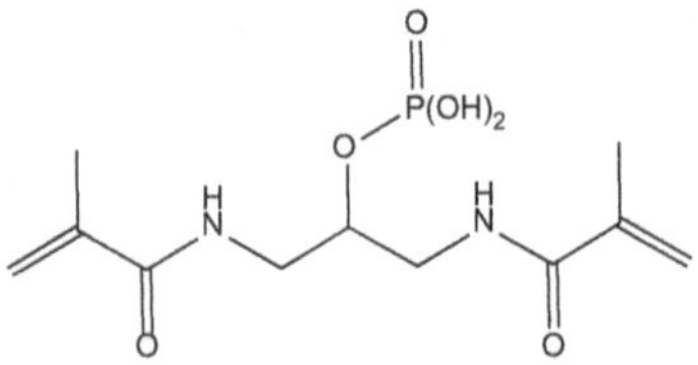

Scheme 2.4 Phosphate-containing methacrylamide monomer used to achieve free radical polymerization.

triethylamine. The resulting 1,3-bis(methacrylamido)propan-2-yl dihydrogen phosphate was characterized by ^{1}H, ^{13}C, and ^{31}P NMR, FTIR, mass spectrometry, and elemental analysis. This monomer showed improved hydrolytic stability compared to the corresponding methacrylate-based dihydrogen phosphate. It was homopolymerized in ethanol at 55–75 °C, using AIBN as initiator. The experimental results revealed the very high freeradical polymerizability of the bis(acrylamide), which could be an advantage in the case of an application as a photo-curing adhesive. As expected, gels were obtained, indicating that the polymers were crosslinked. This was confirmed by thermogravimetric analysis, as relevant degradation started to occur at 238 °C. Finally, it is also interesting to note that adhesive properties of the 1,3-bis(methacrylamido)propan-2-yl dihydrogen phosphate were investigated by measuring the shear bond strength of a corresponding selfetching enamel/dentin model adhesive containing the synthesized monomer. It was shown that the latter was able to generate a strong bond between the enamel/dentin surface and a restorative composite under self-etching conditions. In summary, 1,3-bis(methacrylamido)propan-2-yl dihydrogen phosphate showed exceptional performance, including good solubility in various solvents, hydrolytic stability, high polymerization reactivity, crosslinking properties, noncytotoxic effects, and self-etching behavior.

To conclude, phosphorus-based (meth)acrylamide monomers were involved in free radical polymerization and the resulting polymers proved to have interesting applications, in particular as new dentin adhesives. Nevertheless, the main drawback of conventional radical polymerization is the lack of control over the molecular weight, and the impossibility to achieve complex polymeric structures. As a result, controlled radical polymerization of phosphonated-containing (meth)acrylamides was also investigated.

2.4   Controlled Radical Polymerization

The synthesis of well-defined macromolecules with controlled compositions, architectures, and functionalities has emerged as an important aspect of polymer science. The development of controlled/living radical polymerization (CRP) permitted achievement of well-defined polymers using a radical process easy to carry out. In the last 20 years, techniques such as atom transfer radical polymerization (ATRP),[26] nitroxide mediated polymerization

(NMP),[27] and reversible addition-fragmentation chain-transfer polymerization (RAFT)[28] have led to the synthesis of important new polymers, with control over the molecular weight, low dispersity indexes, and complex architectures, notably due to the possible functionalization of the terminal end groups.[29] All the controlled radical polymerization processes are based on a fast equilibrium between active and dormant species. The concentration of active species has to remain low throughout the polymerization, minimizing termination and transfer reactions. The kinetics and thus control of the polymerization is easily determined by taking samples throughout the polymerization or by using appropriate probes.[30] To date, examples reported in the literature dealing with the controlled radical polymerization of phosphorus-based acrylamides are rare. Actually, the only example described in detail in the literature dealt with the RAFT polymerization of the diethyl 2-(acrylamido)ethylphosphonate (DAAmEP) monomer (Scheme 2.5).[10]

The synthesis of DAAmEP was conducted in three steps. The first step was a Michaelis-Arbuzov reaction between *N*-(2-bromoethyl)phthalimide and an excess of triethyl phosphite under reflux at 160 °C for 12 h. In the second step, the reaction between the resulting phthalimide and hydrazine led to the formation of (2-aminoethyl)phosphonic acid diethyl ester. The final step consisted of the reaction between the acryloyl chloride and the diethyl ester, which was achieved in the presence of poly(4-vinylpyridine), 2% crosslinked, to trap the hydrochloric acid produced during the acryloylation. The final DAAmEP monomer was obtained after purification by chromatography on silica gel. All intermediate products were characterized by ^{1}H and ^{31}P NMR. The ^{1}H NMR spectrum of the DAAmEP monomer allowed validation of the expected chemical structure and as a consequence the synthetic pathway. RAFT homopolymerization of DAAmEP was carried out with two different trithiocarbonate chain transfer agents (see Scheme 2.6).

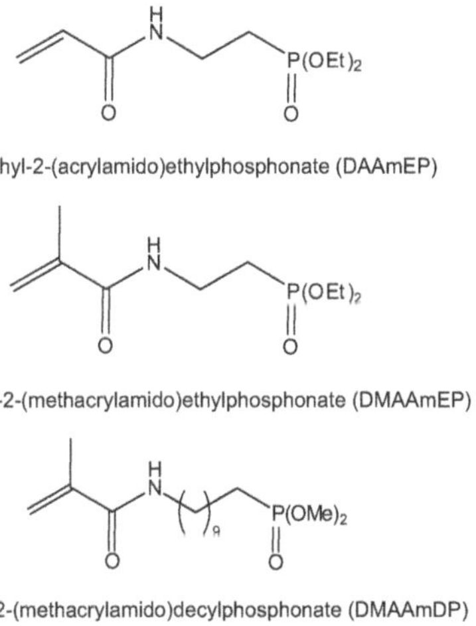

Diethyl-2-(acrylamido)ethylphosphonate (DAAmEP)

Diethyl-2-(methacrylamido)ethylphosphonate (DMAAmEP)

Dimethyl-2-(methacrylamido)decylphosphonate (DMAAmDP)

Scheme 2.5 Phosphorus-containing (meth)acrylamide monomers used to achieve RAFT polymerization.

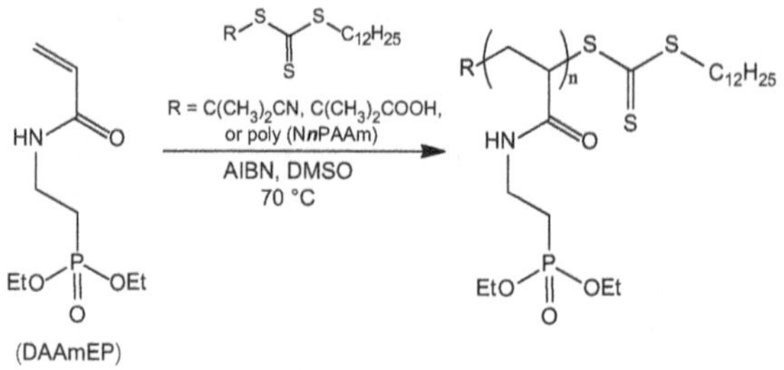

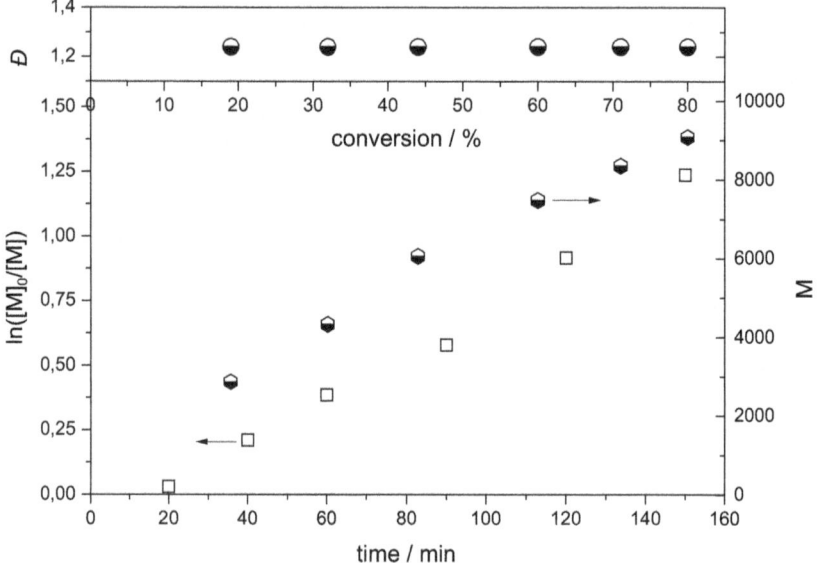

Scheme 2.6 Synthetic pathway for the RAFT polymerization of DAAmEP monomer using different chain transfer agents [2-cyano-2-propyl dodecyl trithiocarbonate, 2-(dodecylthiocarbonothioylthio)-2-methylpropionic acid, or P(N*n*PAAm)].

Figure 2.1 Kinetic plots and evolution of the molecular weight and dispersity with conversion for RAFT polymerization of DAAmEP monomer using 2-(dodecylthiocarbonothioylthio)-2-methylpropionic acid as chain transfer agent. The theoretical molecular weight was equal to 16 000 g mol^{-1}.

The reaction was achieved at 70 °C in DMSO in the presence of AIBN and the [CTA]/[AIBN] ratio was equal to three. Different molecular weights were targeted: 4000, 8000, and 16 000 g mol^{-1}, at 80% conversion. As an example, Figure 2.1 shows (i) the evolution of ln([M]$_0$/[M]) as a function of time and (ii) the evolution of the molecular weight and the dispersity ($Đ$) as a function of the conversion for a theoretical molecular weight equal to 16 000 g mol^{-1}, when 2-(dodecylthiocarbonothioylthio)-2-methylpropionic acid was used as the chain transfer agent. The conversion was measured by ^{1}H NMR in deuterated water, comparing the signals for the reactive double bond

(5.55–5.61 ppm) to the signal for 1,3,5-trioxane (5.1 ppm). Molecular weights and dispersity were determined by SEC in dimethylacetamide (DMA) at 50 °C.

RAFT polymerization of DAAmEP monomer gave first-order kinetic plots. An induction period of 25 min was measured for all targeted molecular weights. Linear evolution of the number-average molecular weight (M_n) *versus* conversion was also obtained, showing that no transfer reaction occurred during the polymerization. Finally, the dispersity remained below 1.5 throughout the reaction, meaning that all chains grew simultaneously. After purification by precipitation in cold hexane, the dispersities were less than 1.35. The experimental results allowed the conclusion that the RAFT homopolymerization of DAAmEP was controlled.

Apart from 2-(dodecylthiocarbonothioylthio)-2-methylpropionic acid, poly(N-n-propylacrylamide) with a targeted molecular weight of 10 000 g mol^{-1} was used as a macro-chain transfer agent to produce well-defined diblock copolymers. Poly(N-n-propylacrylamide) (PNnPAAm) was first prepared. After its precipitation in hot water, the resulting polymer was characterized by SEC in DMA to determine the number-average molecular weight and the dispersity ($M_{n,exp}$ = 15 100; $Đ$ = 1.25). The ^{1}H NMR spectrum in deuterated chloroform confirmed the synthesis of the targeted structure. Then, different P(NnPAAm-b-DAAmEP) copolymers were prepared, the theoretical molecular weight of the P(DAAmEP) block being equal to 1250, 2500, 5000, and 10 000 g mol^{-1} at 80% conversion. The reaction was carried out using experimental conditions already determined for the homopolymerization of DAAmEP. It was shown that polymerization rates for chain extension were similar in all cases and also equivalent to those obtained for homopolymerization of the DAAmEP monomer. The kinetic results showed the linearity of both curves representing (i) the evolution of $\ln([M]_0/[M])$ as a function of time and (ii) the evolution of the molecular weight *versus* conversion. Additionally, the dispersity remained low during the whole polymerization process. An induction period of 25 min was measured. The diblock copolymers were precipitated in cold hexane. The comparison between the ^{1}H NMR spectrum of P(NnPAAm) and P(NnPAAm-b-DAAmEP) demonstrated that the synthesis of the diblock copolymers was successful. The presence of the phosphorus atom was confirmed by ^{31}P NMR analysis. The copolymers were finally analyzed by SEC and 2D-DOSY NMR in order to confirm that the DAAmEP-based second block had grown on the P(NnPAAm) macro-chain transfer agent. SEC showed monomodal peaks (Figure 2.2) and a lower elution time was obtained for the P(NnPAAm-b-DAAmEP) diblock copolymers compared to the P(NnPAAm) macro-chain transfer agent, concomitant with an increase of the molecular weight. In all cases, dispersity of the diblock copolymers remained lower than 1.30.

From all these results it was concluded that the polymerization was controlled and that it was possible to prepare well-defined copolymers with a phosphonated-containing polyacrylamide moiety. It is also interesting to note that homo and diblock copolymers were successfully hydrolyzed, thus producing phosphonic diacid groups, using bromotrimethylsilane and methanol.

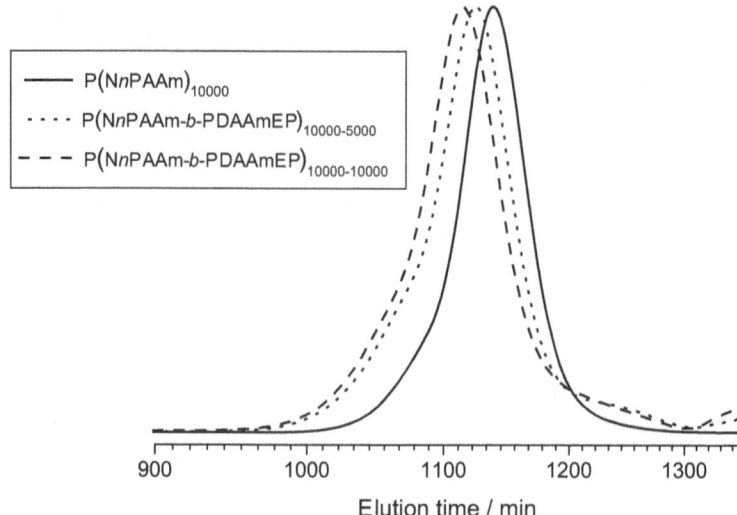

Figure 2.2 Size exclusion chromatograms of poly(*N-n*-propylacrylamide) [P(N*n*PAAm)] macro-chain transfer agent and poly[(*N-n*-propylacrylamide)-*b*-(diethyl 2-(acrylamido)ethylphosphonate)] [P(N*n*PAAm-*b*-DAAmEP)] diblock copolymer (DMA eluent + LiCl 0.1% weight at 0.8 mL min⁻¹, with PS standard at 50 °C).

Methacrylamide monomers were also considered, notably the diethyl 2-(methacrylamido)ethylphosphonate (DMAAmEP) and the dimethyl 2-(methacrylamido)decylphosphonate (see Scheme 2.5).[31] The goal was to study the difference in reactivity between the methacrylamide and its acrylamide analog. RAFT polymerization of DMAAmEP was first carried out with 2-(dodecylthiocarbonothioylthio)-2-methylpropionic acid as chain transfer agent, at 70 °C in DMSO in the presence of AIBN, and the [CTA]/[AIBN] ratio was equal to three. The theoretical molecular weight was 6000 g mol⁻¹. Unfortunately, under the experimental conditions similar to those used for the polymerization of the corresponding acrylamide, no polymerization occurred. So the temperature was raised to 80 °C and, at this temperature, DMAAmEP polymerized but termination reactions were observed at high conversion. As a consequence, the polymerizations were stopped at 60% conversion to keep control of the polymerization (Figure 2.3). Additionally, the rate of polymerization was lower in the case of the methacrylamide monomer even if the reaction temperature was higher.

Similar results were obtained with the DMAAmDP monomer (see Scheme 2.5). RAFT polymerization was carried out with 2-(dodecylthiocarbonothioylthio)-2-methylpropionic acid as chain transfer agent, at 80 °C in DMSO in the presence of AIBN, and the [CTA]/[AIBN] ratio was equal to three. The targeted molecular weights were 3000 and 6000 g mol⁻¹, at 40% conversion to avoid side reactions. Figure 2.4 shows (i) the evolution of ln([*M*]₀/[*M*]) as a function of time and (ii) the evolution of the molecular

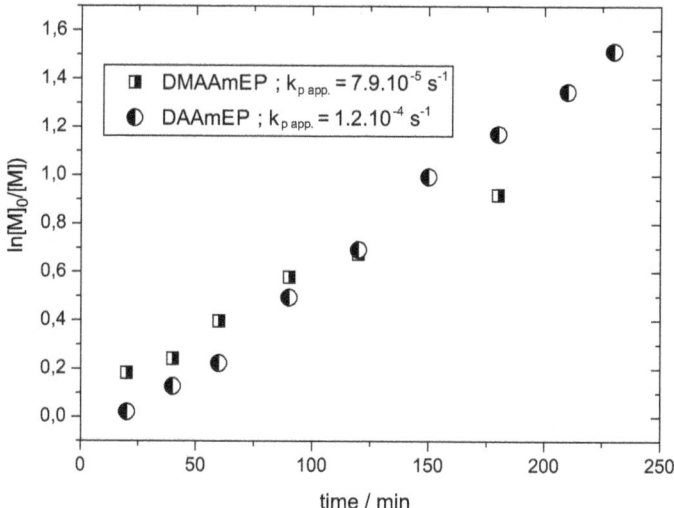

Figure 2.3 Kinetic plot for the RAFT polymerization of both diethyl 2-(methacrylamido)ethylphosphonate (DMAAmEP) and diethyl 2-(acrylamido)ethylphosphonate (DAAmEP), at 70 and 80 °C, respectively. The theoretical molecular weight was equal to 5000 g mol^{-1}, at 60 and 80% conversion for DMAAmEP and DAAmEP, respectively.

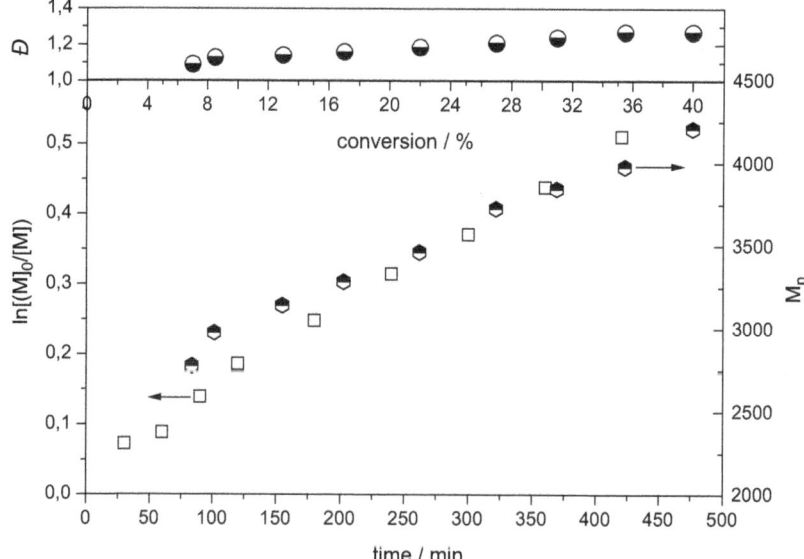

Figure 2.4 Kinetic plots and evolution of the molecular weight and dispersity with conversion for RAFT polymerization of DAAmDP monomer using 2-(dodecylthiocarbonothioylthio)-2-methylpropionic acid as chain transfer agent. The theoretical molecular weight is equal to 6000 g mol^{-1}.

weight and the dispersity as a function of the conversion for a theoretical molecular weight equal to 6000 g mol^{-1}.

RAFT polymerization of the DMAAmDP monomer gave first-order kinetic plots. There was no induction period. Linear evolution of the number-average molecular weight (M_n) *versus* conversion was also obtained, showing that no transfer reaction occurred during the polymerization. Finally, the dispersity remained below 1.27 during all the reaction. The experimental results permitted the conclusion that the RAFT homopolymerization of DMAAmDP was controlled for the low molecular weights targeted (similar results were obtained for a theoretical molecular weight of 3000 g mol^{-1}). On the other hand, it is obvious that it will not be possible to prepare high molecular weight homopolymers using phosphonated-containing meth-acrylamides. Indeed, even when low molecular weight polymers are targeted, the apparent rate constant was only equal to 1.9×10^{-5} s^{-1}, not surprisingly lower than the one obtained in the case of DMAAmEP methacrylamide monomer (7.9×10^{-5} s^{-1}).

2.5 Conclusion and Perspectives

Phosphorus-based (meth)acrylamides are very interesting monomers, mainly because of their hydrolytic stability, which is greatly improved in comparison with (meth)acrylate derivatives. Nevertheless, as shown in this contribution, polymerization of such monomers has not been widely studied and only a limited number of references exists to date in the literature. Most of them deal with polymerizations (by photopolymerization, free radical polymerization, controlled radical polymerization) of phosphonated or phosphonic acid compounds, only one example being focused on a phos-phate-based monomer. Concerning the polymerization process, the photo-polymerization reaction was the most discussed, notably because phosphonic acid derivatives show good adhesion properties, particularly useful for the development of stable self-etching dental primers. Free radical polymerization was only slightly considered, mostly in relation to photo-polymerization to evaluate polymerization rates. Additionally, the main drawback of conventional radical polymerization is the lack of control over the molecular weight, and the impossibility to achieve complex polymeric structures. For that reason, an interesting innovation to be developed in the next few years is the controlled radical polymerization (CRP) of phosphorus-based (meth)acrylamides. To date, only one publication reported in the lite-rature deals with the synthesis of a phosphonate-containing polyacrylamide by the CRP technique, using the RAFT process. In this contribution, diethyl 2-(acrylamido)ethylphosphonate was homopolymerized, with good control over the molecular weight which ranged from 4000 to 16 000 g mol^{-1}. The synthesis of diblock copolymers was also considered. In this case, it was possible to prepare well-defined structures. This is a great improvement in comparison with other polymers synthesized by photopolymerization or conventional radical polymerization. Indeed, controlling the polymerization of phosphorus-based (meth)acrylamides is very valuable, as phosphorus-containing materials can be used for a large variety of technological applications, *e.g.* to bind metals.[32–34] Indeed, organophosphonates exhibit interesting complexing properties[35–38] and are used as dispersants, corrosion inhibiting agents, or for preventing deposit formation.[39] Recently, the development of free halogen materials due to novel legislation has given new opportunities to phosphorinated polymers for use in flame retardancy,[40,41] where phosphorus is known to have a positive action. Polymeric materials containing phosphonic sites have also been developed in alternative energy production, such as in proton-conducting fuel-cell membranes.[3,42] To conclude, the synthesis of phosphorus-based (meth)acrylamide is expected to develop even more in the next few years, as the resulting polymers will combine valuable phosphorus-based moieties with the hydrolytic stability of amide groups.

References

1. M. Essahli, G. Colomines, S. Monge, J. J. Robin, A. Collet and B. Boutevin, *Polymer*, 2008, **49**, 4510–4518.
2. S. Chang, N. D. Sachinvala, P. Sawhney, D. V. Parikh, W. Jarrett and C. Grimm, *Polym. Adv. Technol.*, 2007, **18**, 611–619.
3. J. Parvole and P. Jannasch, *Macromolecules*, 2008, **41**, 3893–3903.
4. S. Monge, B. Canniccioni, A. Graillot and J. J. Robin, *Biomacromolecules*, 2011, **12**, 1973–1982.
5. M. J. Tyas and M. F. Burrow, *Aust. Dent. J.*, 2004, **49**, 112–121.
6. N. Sibold, P. J. Madec, S. Masson and T. N. Pham, *Polymer*, 2002, **43**, 7257–7267.
7. G. Sahin, A. Z. Albayrak, Z. Sarayli and D. Avci, *J. Polym. Sci., Part A: Polym. Chem.*, 2006, **44**, 6775–6781.
8. U. Salz, J. Zimmermann, F. Zeuner and N. Mozner, *J. Adhes. Dent.*, 2005, **7**, 107–116.
9. N. Moszner, U. Salz and J. Zimmermann, *Dent. Mater.*, 2005, **21**, 895–910.
10. A. Graillot, S. Monge, C. Faur, D. Bouyer and J. J. Robin, *Polym. Chem.*, 2013, **4**, 795–803.
11. E. Andrzejewska, *Prog. Polym. Sci.*, 2001, **26**, 605–665.
12. G. Oster and N. L. Yang, *Chem. Rev.*, 1968, **68**, 125–151.
13. Y. Catel, M. Degrange, L. Le Pluart, P. J. Madec, T. N. Pham and L. Picton, *J. Polym. Sci., Part A: Polym. Chem.*, 2008, **46**, 7074–7090.
14. V. Besse, L. Le Pluart, W. D. Cook, T. N. Pham and P. J. Madec, *J. Polym. Sci., Part A: Polym. Chem.*, 2013, **51**, 149–157.
15. X. M. Xu, R. B. Wang, L. Ling and J. O. Burgess, *J. Polym. Sci., Part A: Polym. Chem.*, 2007, **45**, 99–110.
16. B. Akgun, E. Savci and D. Avci, *J. Polym. Sci., Part A: Polym. Chem.*, 2012, **50**, 801–810.
17. F. Gayet, L. Viau, F. Leroux, S. Monge, J. J. Robin and A. Vioux, *J. Mater. Chem.*, 2010, **20**, 9456–9462.

18. F. Gayet, L. Viau, F. Leroux, F. Mabille, S. Monge, J. J. Robin and A. Vioux, *Chem. Mater.*, 2009, **21**, 5575–5577.
19. V. Zoulalian, S. Zurcher, S. Tosatti, M. Textor, S. Monge and J. J. Robin, *Langmuir*, 2010, **26**, 74–82.
20. V. Zoulalian, S. Monge, S. Zurcher, M. Textor, J. J. Robin and S. Tosatti, *J. Phys. Chem. B*, 2006, **110**, 25603–25605.
21. A. Graillot, S. Monge, C. Faur, D. Bouyer, C. Duquesnoy and J. J. Robin, *RSC Adv.*, 2014, DOI: 10.1039/C4RA00140K.
22. N. Moszner and U. Salz, *Macromol. Mater. Eng.*, 2007, **292**, 245–271.
23. W. Mühlbauer, S. Neffgen and C. Erdmann, *Eur. Pat.*, 1 169 996 A1, 2000, *Chem. Abstr.*, 2002, **136**, 91027.
24. J. Klee, U. Waltz and U. Lehmann, *World Pat.*, 03/035013 A1, 2001, *Chem. Abstr.*, 2003, **138**, 358522.
25. N. Moszner, J. Pavlinec, I. Lamparth, F. Zeuner and J. Angermann, *Macromol. Rapid Commun.*, 2006, **27**, 1115–1120.
26. K. Matyjaszewski and J. H. Xia, *Chem. Rev.*, 2001, **101**, 2921–2990.
27. C. J. Hawker, A. W. Bosman and E. Harth, *Chem. Rev.*, 2001, **101**, 3661–3688.
28. C. Barner-Kowollik (Ed.), *Hanbook of RAFT Polymerization*, Wiley-VCH, Weinheim, 2008.
29. S. Monge, O. Giani, E. Ruiz, M. Cavalier and J. J. Robin, *Macromol. Rapid Commun.*, 2007, **28**, 2272–2276.
30. V. Darcos, S. Monge and D. M. Haddleton, *J. Polym. Sci., Part A: Polym. Chem.*, 2004, **42**, 4933–4940.
31. C. Bouilhac, C. Travelet, A. Graillot, S. Monget, R. Borsali and J. J. Robin, *Polym. Chem.*, 2014, DOI: 10.1039/C3PY01512B.
32. A. Clearfield, *Curr. Opin. Solid State Mater. Sci.*, 1996, **1**, 268–278.
33. A. Clearfield, *Curr. Opin. Solid State Mater. Sci.*, 2002, **6**, 495–506.
34. A. Clearfield, *J. Alloys Compd.*, 2006, **418**, 128–138.
35. T. P. Knepper, *Trends Anal. Chem.*, 2003, **22**, 708–724.
36. E. Matczak-Jon and V. Videnova-Adrabinska, *Coord. Chem. Rev.*, 2005, **249**, 2458–2488.
37. A. Graillot, D. Bouyer, S. Monge, J. J. Robin and C. Faur, *J. Hazard. Mater.*, 2013, **244**, 507–515.
38. A. Graillot, S. Djenadi, C. Faur, D. Bouyer, S. Monge and J. J. Robin, *Water Sci. Technol.*, 2013, **67**, 1181–1187.
39. L. Herrera-Taboada, M. Guzmann, K. Neubecker and A. Goethlich, *PCT Int. Appl.*, 2008, 28.
40. J. Canadell, B. J. Hunt, A. G. Cook, A. Mantecon and V. Cadiz, *Polym. Degrad. Stab.*, 2007, **92**, 1482–1490.
41. H. Singh and A. K. Jain, *J. Appl. Polym. Sci.*, 2009, **111**, 1115–1143.
42. T. Bock, R. Muelhaupt and H. Moehwald, *Macromol. Rapid Commun.*, 2006, **27**, 2065–2071.

CHAPTER 3

Phosphorus-Containing Vinyl or Allyl Monomers

GHISLAIN DAVID* AND CLAIRE NEGRELL-GUIRAO

Institut Charles Gerhardt (ICGM)/Equipe Ingénierie et Architectures Macromoléculaires (IAM), ENSCM, 8 rue de l'école normale, 34296 Montpellier, France
*Email: ghislain.david@enscm.fr

3.1 Introduction

The specific features of functional synthetic polymers are provided by the presence of chemical functional groups dissimilar to those of the main chains. The functionality may be introduced as side groups, chain-ends, in-chain, block, or graft structures. Materials with special structures or architectures such as hyperbranched polymers or dendrimers are also of great interest. Functional polymers can be obtained following chemical modification of the already defined polymers or by direct polymerization of the functionalized monomers. Phosphorus-containing polymers functionalized either at the main chain (*e.g.* polyphosphazene, polyphosphoesters) or at the side chain have found applications in many fields. Recently, we have published a review paper[1] gathering the syntheses and radical (co)polymerizations of phosphonate-containing vinyl monomers, ranging from allyl-type to (meth)acrylic-type monomers. This article shows that the phosphonate group weakly changes the double bond reactivity, despite its electron-donating character, with the exception of *p*-benzylphosphonate monomers, where the phosphonate group enhances the reactivity of the styryl monomer in free-radical polymerization.[2] In this chapter, we especially focus on both

RSC Polymer Chemistry Series No. 11
Phosphorus-Based Polymers: From Synthesis to Applications
Edited by Sophie Monge and Ghislain David
© The Royal Society of Chemistry 2014
Published by the Royal Society of Chemistry, www.rsc.org

phosphorus vinyl- and allyl-type monomers. We will show that for most of them their radical homopolymerization results in the synthesis of oligomers, whereas their copolymerization with electron-accepting monomers affords high molecular weight copolymers with interesting properties.

3.2 Phosphorus-Containing Vinyl Monomers

3.2.1 Dialkyl Phosphonate-Containing Vinyl Monomers

Several synthetic pathways allow the preparation of dialkyl vinylphosphonate monomers; Scheme 3.1 shows some of these synthetic pathways. Among them, the Arbuzov reaction (Scheme 3.1a) is commonly used with vinyl chloride, in the presence of either platinum or nickel catalysts.[3] Interestingly, vinyl chloride (VC) can also react with dialkyl phosphate (Scheme 3.1c) to generate dialkyl vinylphosphonate monomers through a radical reaction. Nevertheless, dialkyl phosphites, acting as chain transfer agents (CTAs), must be used in excess to avoid polymerization of VC. Radical addition of dialkyl phosphites onto acetylene was also attempted but some addition of two dialkyl phosphate units was observed, leading to bisphosphonate compounds to some extent.

Another strategy concerns the esterification of the vinylphosphonic dichloride monomer (Scheme 3.1b) in the presence of hydroxyl compounds, catalyzed by primary amines. This reaction usually proceeds quantitatively, but requires the prior synthesis of the vinylphosphonic dichloride monomer. Alternative methods have also been developed to obtain the dialkyl vinylphosphonate monomers, but are based on multiple-step reactions.

With the exception of vinylphosphonic acid showing unusual behavior in radical polymerization[4–9] (polymerization was shown to proceed through a cyclopolymerization mechanism, resulting in either five- or six-membered rings), relatively few investigations into homopolymerization of dialkyl vinylphosphonate monomers have been reported. The few reports show the failure of vinylphosphonate monomers to homopolymerize, which usually results in polymers of low molecular weight. This behavior was ascribed to chain transfer onto either the polymers or monomers. Indeed, recently,

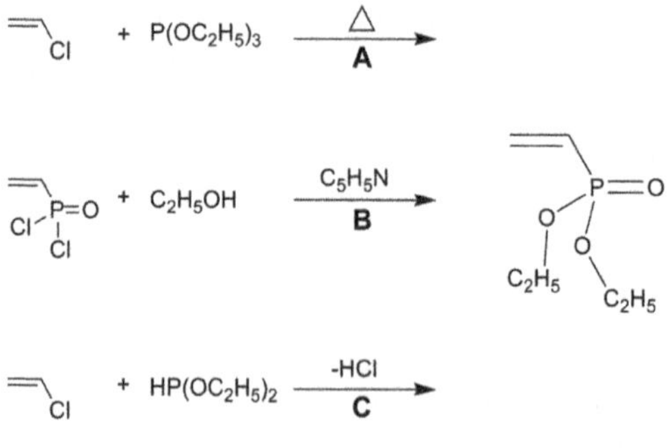

Scheme 3.1   Synthetic pathways for dialkyl vinylphosphonate monomers.

Bingöl et al.[10] have demonstrated, especially by means of electrospray ionization mass spectrometry (ESI-MS), that polymerization of vinylphosphonates is mainly dominated by transfer reactions; the predominant transfer occurs by intramolecular hydrogen transfer of phosphonate ester groups, which in consequence creates a P–O–C bond in the main chain (Scheme 3.2). Moreover P–O–C bonds are more thermally labile compared to phosphonate, and thus lead to chain scission reactions.

Polymerization of dialkyl vinylphosphonate monomers was nevertheless efficiently carried out in the presence of lanthanide derivatives and especially cyclopentadienyl lanthanide complexes, used both as initiators and catalysts.[11–14] Very recently, Shen et al.[15] performed the synthesis of poly (diethyl vinylphosphonate) using a lanthanide tris(borohydride) below 50 °C. The authors showed that the polymerization could be controlled and proceeds under pseudo-first-order kinetics, giving rise to high molecular weight polymers, *i.e.* ranging from 20 to 40 kDa with molecular weight dispersity below 1.7.

Finally, a thiol–ene reaction was used with vinylphosphonate monomers using CTAs[6,16] carrying hydrophobic chains, in order to improve the hydrophobic properties of metallic surfaces (Scheme 3.3). The phosphonic acid groups are used to promote adhesion of hydrophobic chains onto the metallic surfaces.

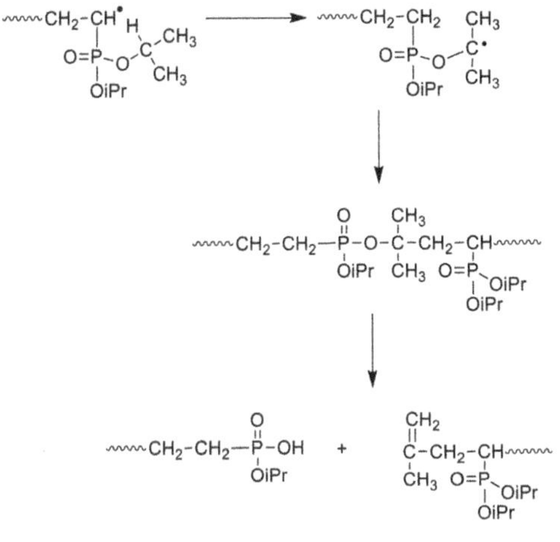

Scheme 3.2 Chain scission reaction of poly(diisopropyl vinylphosphonate).

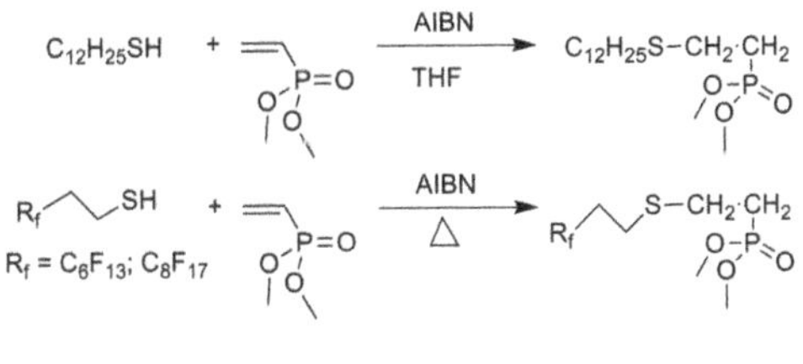

Scheme 3.3 Thiol–ene reaction with dimethyl vinylphosphonate.

Similarly, David *et al.*[4] carried out the thiol–ene radical coupling of vinylphosphonic acid in the presence of sulfanylacetic acid in water at 70 °C (Scheme 3.4). They showed that a low amount of the thiol compound was required, according to its very high chain transfer constant, and this reaction returned good yields.

Copolymers of vinylphosphonate monomers have been thoroughly studied and have found application in the electrical, transportation, and construction industries, thanks to their flame-retardant properties. For instance, radical copolymerization of diethyl vinylphosphonate (DEVP) with styrene carried out at 100 °C resulted in copolymers of high molecular weight;[17] the incorporation of diethyl vinylphosphonate units lowered the T_g value of polystyrene due to the steric hindrance of the phosphonate group. Noteworthy is that when DEVP was copolymerized with styrene or acrylonitrile in an emulsion, copolymers showed MW values up to 100 000 g mol^{-1}. In radical copolymerization, vinylphosphonate monomers are supposed to lead to alternated structures with electron-donating monomers, according to the electron-accepting character of the phosphonate group. However, some authors have noticed that the polarizing effect of P=O was much lower than that of C=O, *e.g.* for acrylate monomers, which indicates that the radicals are poorly stabilized.

3.2.2 Dialkyl Phosphonate-Containing Vinyl Ether Monomers

Vinyl ether monomers are good candidates[1] in order to reach high molecular weight polymers either by cationic homopolymerization[18] or by radical copolymerization (when associated with an electron-accepting monomer).[19,20] Consequently, the synthesis of dialkyl vinyl ether phosphonate monomers is very challenging. The first attempts were made through the Arbuzov reaction (Scheme 3.5) by reaction of chloroethyl vinyl ether with triethyl phosphite.[21] Diethyl 2-(vinyloxy)ethylphosphonate was obtained in good yield; however, diethyl ethylphosphonate is always obtained in non-negligible amounts, despite drastic reaction conditions.

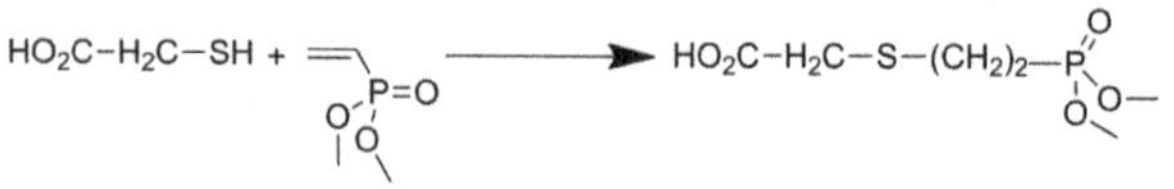

HO$_2$C–H$_2$C–SH + 　$\longrightarrow$ HO$_2$C–H$_2$C–S–(CH$_2$)$_2$–P

Scheme 3.4   Thiol–ene radical coupling using sulfanylacetic acid.

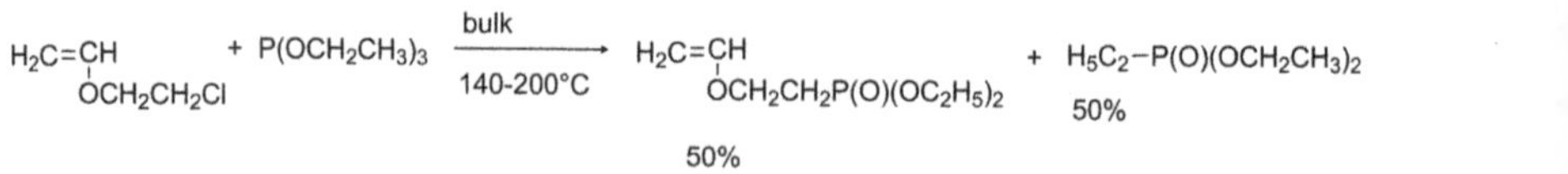

H$_2$C=CH　　+ P(OCH$_2$CH$_3$)$_3$ $\xrightarrow[\text{140-200°C}]{\text{bulk}}$ H$_2$C=CH　　+ H$_5$C$_2$–P(O)(OCH$_2$CH$_3$)$_2$
|OCH$_2$CH$_2$Cl |OCH$_2$CH$_2$P(O)(OC$_2$H$_5$)$_2$　　50%

50%

Scheme 3.5   Synthesis of diethyl 2-(vinyloxy)ethylphosphonate *via* an Arbuzov reaction.

More recently, Souzy *et al.*[22] synthesized a fluorinated vinyl ether-containing phosphonate moiety, *i.e.* dimethyl 4-[(α,β,β-trifluorovinyl)oxy]-phenylphosphonate, from a Michaelis–Arbuzov reaction in the presence of nickel(II) chloride as Lewis acid at 110 °C. Nevertheless, the reaction yield remains low (Scheme 3.6).

In order to improve the synthesis of dialkyl vinyl ether phosphonates, transetherification has been used between the hydroxyl compound and the vinyl ether. In this case, the phosphonate group is introduced by the hydroxyl compound. Transetherification efficiently occurs when catalyzed by mercury or palladium salts.[23-26] Watanabe showed that with an excess of alcohol and with about 3% of mercury(II) acetate, they were able to obtain a wide range of functional vinyl ethers,[26] according to the mechanism shown in Scheme 3.7.

Furthermore, acetals may be also formed, depending on both the efficiency and the stability of the catalyst. The rather complex mechanism of transetherification, catalyzed by palladium salts, was recently discussed by Muzart[27] and is given in Scheme 3.8.

This technique was thus applied by Iftene *et al.*[28] for the synthesis of dimethyl 2-(vinyloxy)ethylphosphonate monomers (Scheme 3.9). Prior to the transetherification of the vinyl ether, a series of phosphonate-containing hydroxyl

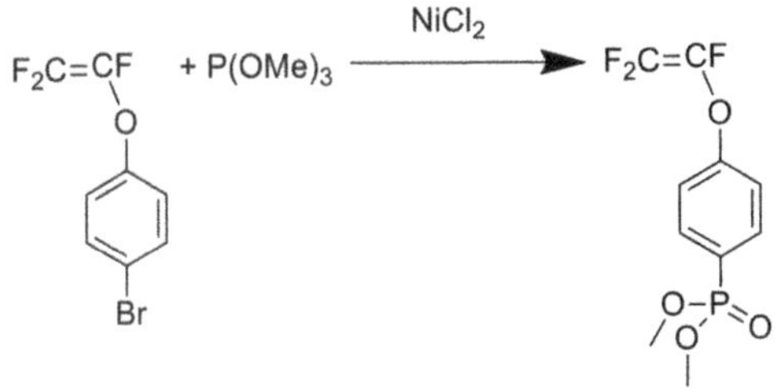

Scheme 3.6   Synthesis of dimethyl 4-[(α,β,β-trifluorovinyl)oxy]phenylphosphonate.

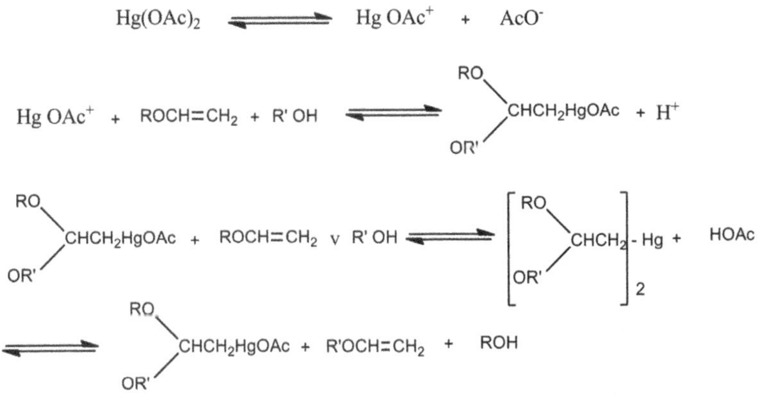

Scheme 3.7   Transetherification of vinyl ethers in the presence of a mercury catalyst.

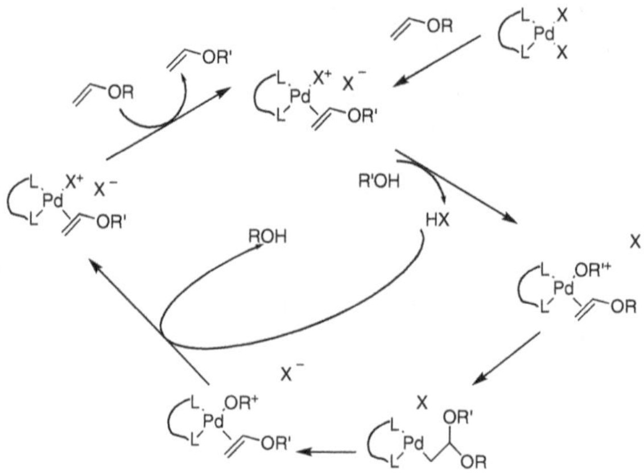

Scheme 3.8 Mechanism of transetherification catalyzed by palladium salts.

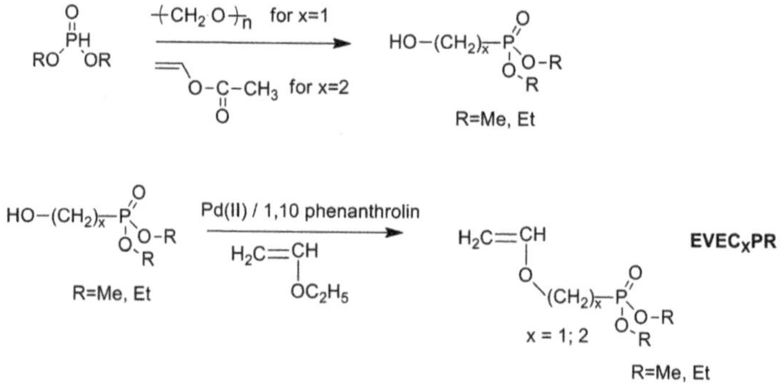

Scheme 3.9 Synthesis of dimethyl 2-(vinyloxy)ethylphosphonate by transetherification catalyzed by a palladium salt.

compounds was synthesized. Then, as the second step, 1,10-phenanthroline was used to avoid the formation of acetals during the transetherification, and phosphonate-bearing vinyl ethers were obtained in high yields.

Vinyl ether monomers are recognized as "good" electron-donating monomers and, thus, when associated with electron-accepting monomers such as maleimide, their radical copolymerization generally leads to alternating polymers.[29-31] To the authors' knowledge, there are only a few investigations about the radical (co)polymerizations of dialkyl vinyl ether phosphonate monomers. Iftene et al.[1,28] performed a kinetic investigation on the radical copolymerization of phosphonate-containing vinyl ethers (VEPs) with electron-accepting monomers by means of FTIR. Other functional vinyl ethers were also used as model compounds. They first showed that, when VEPs were copolymerized with maleic anhydride (MA), the phosphonate moieties tended to lower the rate of polymerization (Figure 3.1). The authors also pointed out that another factor has to be taken into account, relating to phosphonate moieties acting as radical inhibitors. Indeed, phosphorus

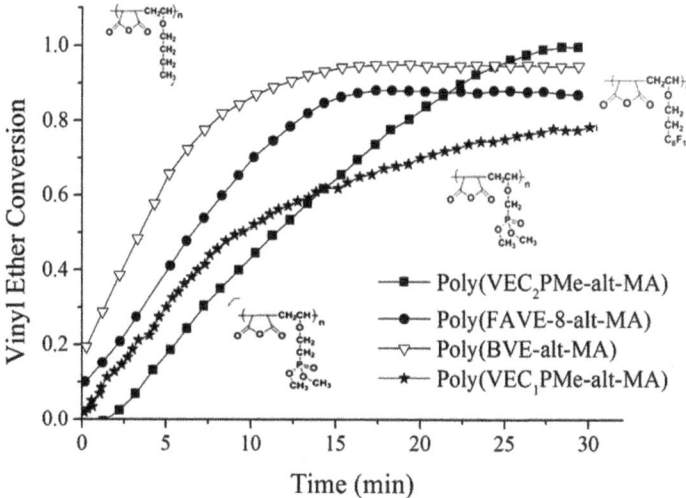

Figure 3.1 Vinyl ether conversion *vs.* time for radical copolymerization with maleic anhydride.

compounds are known to act as radical scavengers, especially in the gas phase when used as a flame retardant;[32] furthermore, retardation is often observed during the controlled radical polymerization of phosphonate-containing monomers due to radical initiator inhibition.[33] Finally, if the polymerization rates for both VEC$_1$PMe/MA and VEC$_2$PMe/MA copolymerizations are fairly similar, VEC$_2$PMe reaches 100% conversion after 30 min, whereas for VEC$_1$PMe the maximum conversion is 78%. This result enhances the negative effect of the spacer length: the shorter the spacer length, the higher the electron-withdrawing effect of the phosphonate group.

Finally, the same authors investigated the radical copolymerization of a dimethyl 2-(vinyloxy)ethylphosphonate with a range of electron-accepting monomers (Scheme 3.10).

From the size exclusion chromatography (SEC) traces (Figure 3.2), we can note that both M_w and $Đ$ values are dependent on the type of electron-accepting monomer. Indeed, electron-poor monomers, such as MA or maleimides, lead to copolymers of higher M_w and higher $Đ$ values. In contrast, monomers showing a lower electron-accepting effect, such as dibutyl itaconate or itaconic anhydride, lead to copolymers of low M_w and $Đ$ values. Furthermore, these monomers show a strong polarizing effect, unlike vinyl ether monomers. These opposite effects are detrimental for efficient radical copolymerization.

Kohli and Blanchard[21,34] also provided an interesting study regarding the radical copolymerization of *N*-phenylmaleimide with diisopropyl 2-(vinyloxy)ethylphosphonate, as indicated in Scheme 3.11.

According to the 1 : 1 stoichiometry between maleimide and vinyl ether, the polymers were AD (acceptor–donor) alternating copolymers. These new copolymers, being hemihydrolyzed, allowed the design and growth of robust

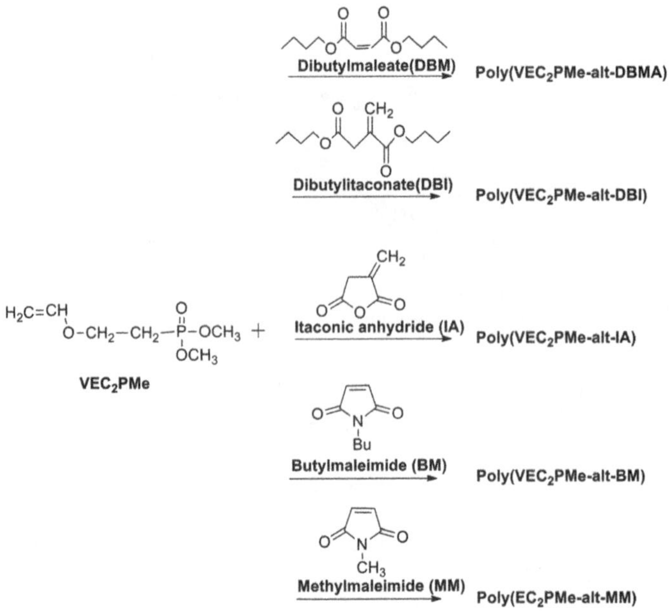

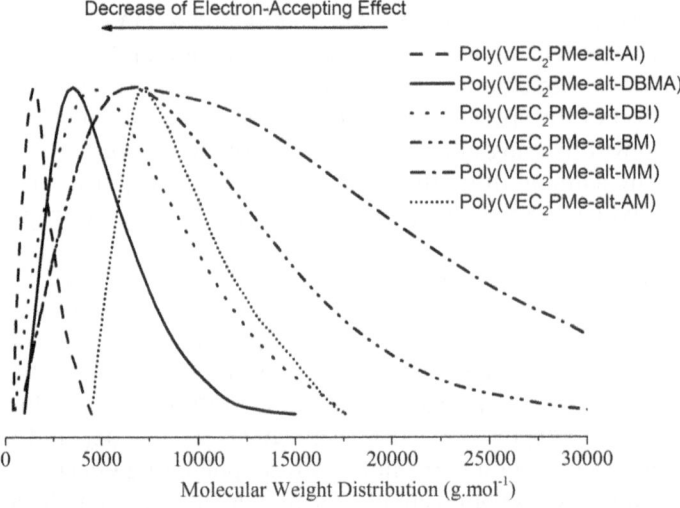

Scheme 3.10   Radical copolymerization of dimethyl 2-(vinyloxy)ethylphosphonate (VEC₂PMe) with a series of electron-accepting monomers.

Figure 3.2   Molecular weight distribution of copolymers obtained from radical copolymerization of dimethyl 2-(vinyloxy)ethylphosphonate (VEC₂PMe) with a series of electron-accepting monomers.

layered polymer assemblies.[34,35] In this study the authors performed the interlayer connections with the help of zirconium(II) chloride solution, thus allowing for growth of multiple polymer layers. The authors claim that these

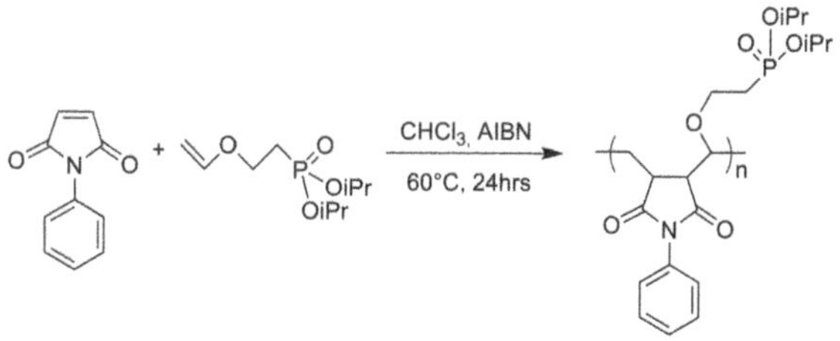

Scheme 3.11 Synthetic route for poly[*N*-phenylmaleimide-*co*-diisopropyl 2-(vinyl-oxy)ethylphosphonate].

novel materials will help in the design of chemically selective surfaces with controlled porosity.

3.3   Phosphorus-Containing Allyl Monomers

Only a few articles deal with the polymerization of phosphorus-containing allyl ether monomers. Most of them discuss radical polymerization; however, the strong stability of allyl ether radicals is probably a disadvantage, leading to longer polymerization times compared with those of the more reactive vinyl ethers.

3.3.1   Radical Homopolymerization

In an old patent,[36] Friedman prepared homopolymers from cyclic phosphites and phosphonates with allyl functions (Scheme 3.12). These compounds were homopolymerized or copolymerized with other monomers such as styrene or acrylate monomers in the presence of conventional initiators or from UV activation. The polymers produced quite tough, hard, colorless and clear non-burning plastics.

Negrell-Guirao *et al.*[37] performed radical homopolymerization of 5-(allyl-oxymethyl)-5-ethyl-2-oxo-1,3,2-dioxaphosphorinane carried out at 70 °C in the presence of 2,2′-azobisisobutyronitrile (AIBN) as initiator and acetonitrile as solvent (Scheme 3.13). The obtained polymers showed two SEC distributions, one composed of oligomers (with $M_n \approx 800$ g mol^{-1}) and one of high molecular weight polymers (with $M_n \approx 100\,000$ g mol^{-1}). The latter were hyperbranched polymers suffering from solubility limitations for future industrial applications. Hence, telomerization of one phosphonated monomer, using dimethyl phosphonate as a CTA, limited the hyperbranching through two concomitant mechanisms of polymerization, which are discussed. For $R_0 = 2$ (R_0 represents the initial [CTA]$_0$/[monomer]$_0$ molar ratio), the results showed a limited chain length, whereas monoadduct formation occurs for $R_0 = 10$.

Kahraman *et al.*[38] described the synthesis of aromatic phosphorus-containing allyl monomers such as allyldiphenylphosphine oxide (ADPPO) or {bis[4-(allyloxy)phenyl]}phenylphosphine oxide (DAPPO) (Scheme 3.14). The

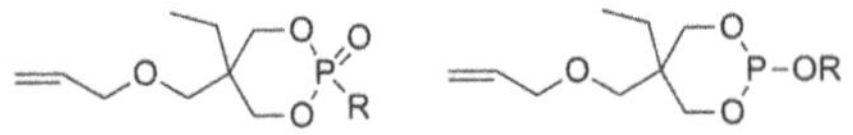

Scheme 3.12 Structures of allyl monomers containing cyclic phosphite or phosphonate.

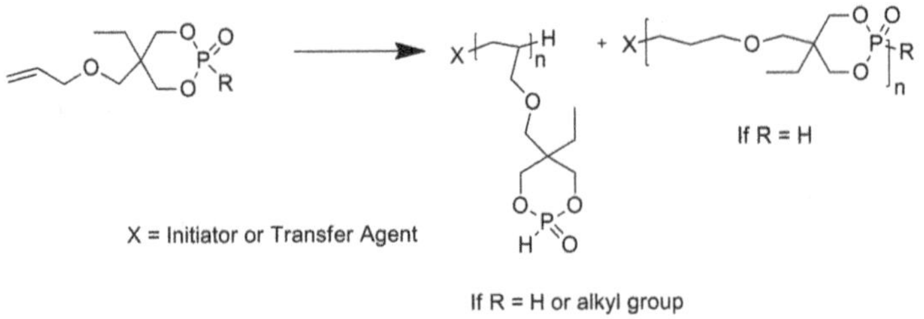
X = Initiator or Transfer Agent
If R = H
If R = H or alkyl group

Scheme 3.13 Polymerization of 2-R-5-(allyloxymethyl)-5-ethyl-2-oxo-1,3,2-dioxaphosphorinane by a radical solution procedure.

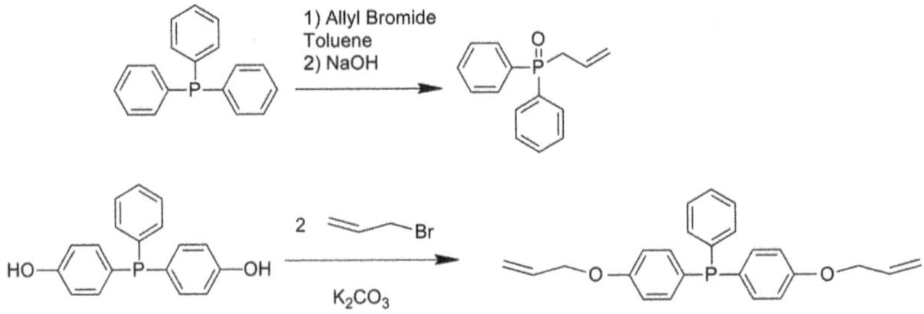
1) Allyl Bromide
Toluene
2) NaOH
2
Br
K_2CO_3

Scheme 3.14 Structures of allyldiphenyl phosphine oxide and {bis[4-(allyloxy)phenyl]}phenylphosphine monomers.

ADPPO compound was obtained by the reaction between triphenyl phosphine and allyl bromide according to a Michaelis–Arbuzov rearrangement, and the DAPPO compound by a Williamson reaction between an aromatic phosphine and 2 equivalents of allyl bromide. The phosphonated monomers were added separately in an epoxy acrylate resin cured by UV radiation in the presence of a photoinitiator.

Huang *et al.*[39] performed the synthesis of a nitrogen- and phosphorus-containing allyl monomer (Scheme 3.15), which was added to an acrylate adhesive resin to afford a polymer with flame-retardant properties.

Spooncer[40] proposed the synthesis of cyclic phosphorus-containing linear polymers prepared by polymerization of dialkenyl P-containing allyl compounds having at least one P–C linkage in the presence of azo-peroxide initiators. The obtained polymers were linear with a molecular weight ranging from 1000 to 30 000 g mol^{-1}, depending on the solvent.

Negrell-Guirao *et al.*[41] proposed the homopolymerization of nitrogen- and phosphorus-containing allyl monomers by a radical method in aqueous media. They demonstrated the possibility to produce phosphonated

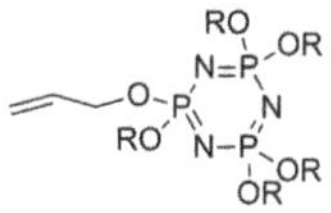

Scheme 3.15 Structure of a nitrogen- and phosphorus-containing allyl monomer.

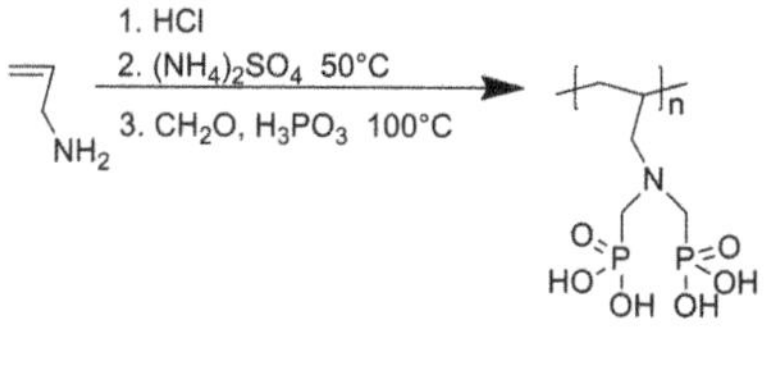

Scheme 3.16 Synthesis of phosphonated oligomers.

oligoallylamines, obtained by radical polymerization then by post-functionalization in one stage, shown in Scheme 3.16, and obtained original polymers with zwitterionic functions that can be utilized for many applications. Capillary electrophoresis has been found to be well suited to characterize the charged oligoallylamines, given their molar masses are about 1600 g mol^{-1} (M_n) to 3600 g mol^{-1} (M_w).

3.3.2 Radical Copolymerization

Under free-radical conditions, the homopolymerization of allyl monomers is unlikely, but if reaction does occur the monomer usually polymerizes at rather low rates. This effect is the result of degradative chain transfer; the weakness of the allyl C–H bond arises from the high resonance stability of the allyl radical, which is too stable to reinitiate polymerization and undergoes termination by a reaction with propagating radicals.[42] Negrell-Guirao *et al.*[43] proposed acceptor/donor copolymerization as a solution to the problem of reactivity. For the copolymerization of allyl ether monomers, most of the studies were achieved on couples of acceptor/donor monomers because allyl monomers have donor properties. A series of four phosphonate-bearing allyl monomers, namely diethyl allylphosphonate (AP), dimethyl allyloxymethylphosphonate (AOP), 5-(allyloxymethyl)-5-ethyl-2-oxo-1,3,2-dioxaphosphorinane, and 5-(allyloxymethyl)-2-benzyl-5-ethyl-2-oxo-1,3,2-dioxaphosphorinane, were synthesized and copolymerized. The authors described the synthesis of AOP, obtained by nucleophilic substitution of allyl bromide by a primary alcohol bearing a phosphonate group (Scheme 3.17). This reaction requires an equimolar amount of allyl bromide and primary alcohol. The functional alcohol, *i.e.* dimethyl hydroxymethylphosphonate, was previously synthesized from reaction of dimethyl phosphonate with paraformaldehyde and anhydrous potassium carbonate, as described elsewhere.[44] The AP monomer was obtained by reaction of allyl bromide (Scheme 3.17) with triethyl phosphite, *via* the Arbuzov rearrangement,[2] *i.e.* isomerization of

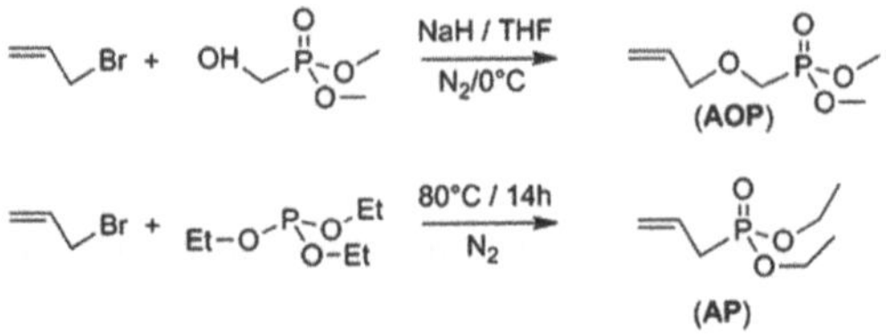

Scheme 3.17 Synthesis of dimethyl allyloxymethylphosphonate (AOP) and of diethyl allylphosphonate (AP).

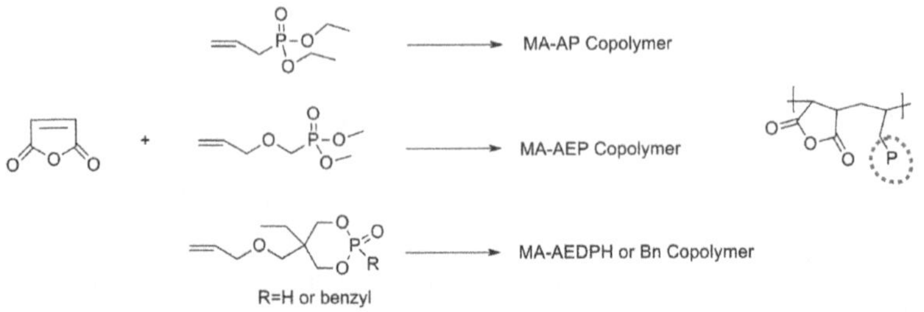

Scheme 3.18 Radical copolymerization of maleic anhydride with phosphonate-bearing allyl monomers.

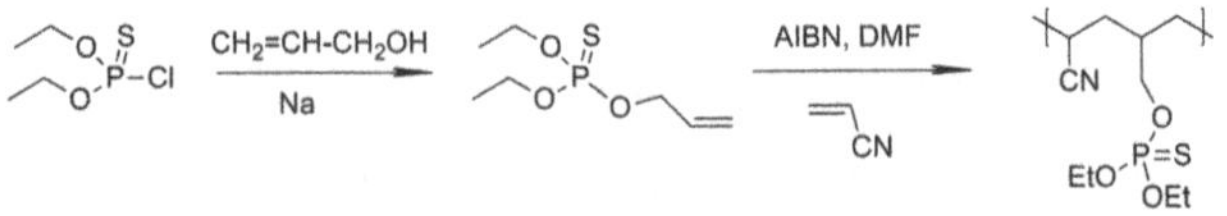

Scheme 3.19 Synthesis of a copolymer obtained by free radical copolymerization of *O*-allyl *O,O*-diethyl phosphorothioate and acrylonitrile.

$P(OR)_3$ to $(RO)_2P(O)R_0$ (where R_0 corresponds to an allyl group). The Arbuzov reaction of triethyl phosphite with commercial allyl bromide afforded AP in quantitative yield.

These four monomers were then copolymerized by free radical polymerization in the presence of maleic anhydride, as shown in Scheme 3.18, thus leading to alternating copolymers with phosphonate moieties with rather low molecular weights. Interestingly, the kinetics of the copolymerization were influenced by the phosphonate group of the allyl monomers: the bulkier the phosphonate moieties, the faster the polymerization rate and the higher the monomer conversion. Furthermore, the close proximity between the phosphonate moieties and the allyl double bond led to a decrease in the electron-donating character of the allyl monomer. The authors also showed the occurrence of allyl transfer, by comparison with model free radical copolymerization, thus explaining the low molecular weight values.

Ren *et al.*[45] described the synthesis of a new copolymer obtained by free radical copolymerization of *O*-allyl *O,O*-diethyl phosphorothioate (DATP) and acrylonitrile (Scheme 3.19). The original sulfur- and phosphorus-containing

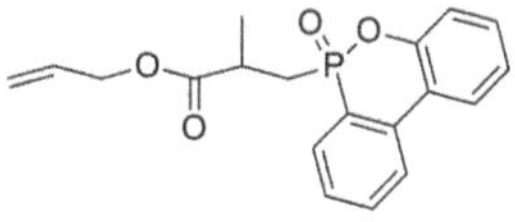

Scheme 3.20 Original methacrylate monomer with a phosphinate function.

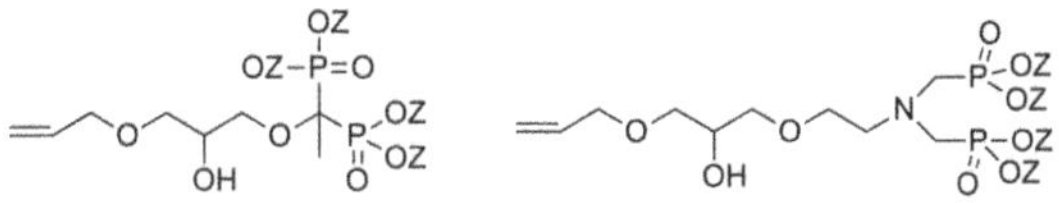

Scheme 3.21 Structures of ionic phosphonate allyl monomers (Z = H or cations).

allyl monomer DATP was synthesized at room temperature by a Williamson reaction of *O,O*-diethyl phosphorochloridothioate with allyl alcohol.

Mizutani and Hirashima[46] obtained by conventional radical polymerization a photosensitive resin using methacrylic acid, butyl acrylate, and a new phosphorus-containing allyl monomer (Scheme 3.20). To create the new monomer, they treated 9,10-dihydro-9-oxa-10-phosphaphenanthrene (DOPO) with allyl methacrylate at high temperature. By copolymerization with butyl acrylate, they obtained a photosensitive resin that showed good fire resistance.

The invention of Heath *et al.*[47] relates to novel phosphonate allyl monomers, made from reaction of an unsaturated oxirane with amine- or hydroxyl-functionalized phosphonic acids (Scheme 3.21). These monomers were copolymerized with other unsaturated species (acrylic acid, maleic acid, acrylamide, or monomer derivatives of sulfonic acid, *etc.*), yielding phosphonate polymers or oligomers. Phosphorus-containing monomers were incorporated at a ratio of 0.1–30% and polymerized in aqueous media. The final polymers had a molecular weight of 800–30 000 g mol^{-1}. With their phosphonic acids groups (free acid or salts forms), they are of particular use as oilfield scale inhibitors.

Phosphonate monomers of type 1 were made from the reaction of allyl glycidyl ether with hydroxy-functionalized phosphonic acid. To obtain azo-phosphonated products (type 2), multicomponent reactions (amine-, aldehyde/ketone-, or phosphorus-containing compounds) such as the Kabachnick–Fields,[48] Mannich,[49] or Moedritzer[50] reactions, were used. These reactions generated in a selective way the α-aminoalkylphosphonate products.

3.4 Conclusions

In this chapter we have gathered together the syntheses and radical (co)polymerizations of phosphonate-containing vinyl monomers, ranging from allyl-type to vinyl- and vinyl ether-type monomers. We have showed that the phosphonate group weakly changes the double bond reactivity, despite its electron-donating character. Considering both phosphonate allyl-type and vinyl-type monomers, we have showed that their radical homo-polymerization leads to oligomers in expected poor yields.[37,43] In contrast, their radical copolymerization with electron-accepting monomers has been more thoroughly investigated. For instance, radical copolymerization of diethyl vinylphosphonate with styrene, carried out at 100 °C, resulted in copolymers of high molecular weight.[17] The incorporation of diethyl vinyl-phosphonate units lowered the T_g value of polystyrene due to the steric hindrance of the phosphonate group. Finally, it was shown that vinyl ether monomers are good candidates in order to reach high molecular weight polymers by radical copolymerization[19,20] when associated with an electron-accepting monomer.

References

1. G. David, C. Negrell-Guirao, F. Iftene, B. Boutevin and K. Chougrani, *Polym. Chem.*, 2012, **3**, 265–274.
2. Z. Yu, W. X. Zhu and I. Cabasso, *J. Polym. Sci., Part A: Polym. Chem.*, 1990, **28**, 227–230.
3. P. Tavs and H. Weitkamp, *Tetrahedron*, 1970, **26**, 5529–5534.
4. G. David, B. Boutevin, S. Seabrook, M. Destarac, G. Woodward and G. Otter, *Macromol. Chem. Phys.*, 2007, **208**, 635–642.
5. G. David, C. Boyer, R. Tayouo, S. Seabrook, B. Ameduri, B. Boutevin, G. Woodward and M. Destarac, *Macromol. Chem. Phys.*, 2008, **209**, 75–83.
6. M. Essahli, M. El Asri, A. Boulahna, M. Zenkouar, M. Viguier, Y. Hervaud and B. Boutevin, *J. Fluorine Chem.*, 2006, **127**, 854–860.
7. R. D. Jackson and K. R. K. Matthews (Rhodia Consumer Specialties), *Br. Pat.*, 016319, 2003.
8. H. Steininger, M. Schuster, K. D. Kreuer, A. Kaltbeitzel, B. Bingoel, W. H. Meyer, S. Schauff, G. Brunklaus, J. Maier and H. W. Spiess, *Phys. Chem. Chem. Phys.*, 2007, **9**, 1764–1773.
9. B. Bingöl, W. Meyer, M. Wagner and G. Wegner, *Macromol. Rapid Commun.*, 2006, **27**, 1719–1724.
10. B. Bingöl, G. Hart-Smith, C. Barner-Kowollik and G. Wegner, *Macromolecules*, 2008, **41**, 1634–1639.
11. S. Salzinger and B. Rieger, *Macromol. Rapid Commun.*, 2012, **33**, 1327–1345.
12. S. Salzinger, U. B. Seemann, A. Plikhta and B. Rieger, *Macromolecules*, 2011, **44**, 5920–5927.
13. N. Zhang, S. Salzinger, F. Deubel, R. Jordan and B. Rieger, *J. Am. Chem. Soc.*, 2012, **134**, 7333–7336.
14. N. Zhang, S. Salzinger and B. Rieger, *Macromolecules*, 2012, **45**, 9751–9758.
15. J. Li, X. Ni, J. Ling and Z. Shen, *J. Polym. Sci., Part A: Polym. Chem.*, 2013, **51**, 2409–2415.

16. N. Pelaprat, G. Rigal, B. Boutevin, A. Manseri and M. Belbachir, *Eur. Polym. J.*, 1996, **32**, 1189–1197.

17. Q. Wu and R. A. Weiss, *J. Polym. Sci., Part B: Polym. Phys*, 2004, **42**, 3628–3641.

18. F. D'Agosto, M.-T. Charreyre, F. Delolme, G. Dessalces, H. Cramail, A. Deffieux and C. Pichot, *Macromolecules*, 2002, **35**, 7911–7918.

19. R. Auvergne, L. R. Saint, C. Joly-Duhamel, J. J. Robin and B. Boutevin, *J. Polym. Sci., Part A: Polym. Chem.*, 2007, **45**, 1324–1335.

20. R. Tayouo, G. David and B. Ameduri, *Eur. Polym. J.*, 2010, **46**, 1111–1118.

21. G. J. Blanchard and P. Kohli (Board of Trustees Operating Michigan State University), *U.S. Pat.*, 6 528 603, 2003.

22. R. Souzy, B. Ameduri, B. Boutevin and D. Virieux, *J. Fluorine Chem.*, 2004, **125**, 1317–1324.

23. H. Rampars and J. Takahara (Mitsubishi Chemical), *Jpn. Pat.*, 170 912, 2005.

24. J. L. Brice, J. E. Meerdink and S. S. Stahl, *Org. Lett.*, 2004, **6**, 4799.

25. M. Bosch and M. Schlaf, *J. Org. Chem.*, 2003, **68**, 5225–5227.

26. W. H. Watanabe and L. E. Conlon, *J. Am. Chem. Soc.*, 1957, **79**, 2828–2833.

27. J. Muzart, *Tetrahedron*, 2005, **61**, 5955–6008.

28. F. Iftene, G. David, B. Boutevin, R. Auvergne, A. Alaaeddine and R. Meghabar, *J. Polym. Sci., Part A: Polym. Chem.*, 2012, **50**, 2432–2443.

29. X. Zhang, Z.-C. Li, K.-B. Li, S. Lin, F.-S. Du and F.-M. Li, *Prog. Polym. Sci.*, 2006, **31**, 893–948.

30. K. G. Olson and G. B. Bulter, *Macromolecules*, 1984, **17**, 2480.

31. P. Kohli, A. B. Scranton and G. J. Blanchard, *Macromolecules*, 1998, **31**, 5681–5689.

32. H. Vahabi, C. Longuet, L. Ferry, G. David, J.-J. Robin and J.-M. Lopez-Cuesta, *Polym. Int.*, 2012, **61**, 129–134.

33. G. David, Z. El Asri, S. Reach, P. Castignolles, Y. Guillaneuf, P. Lacroix-Desmazes and B. Boutevin, *Macromol. Chem. Phys.*, 2009, **210**, 631.

34. P. Kohli and G. J. Blanchard, *Langmuir*, 1999, **15**, 1418–1422.

35. P. Kohli, M. C. Rini, J. S. Major and G. J. Blanchard, *J. Mater. Chem.*, 2001, **11**, 2996–3001.

36. L. Friedman (Union Carbide), *U.S. Pat.*, 3 194 795, 1965.

37. C. Negrell Guirao and B. Boutevin, *Macromolecules*, 2009, **42**, 2446–2454.

38. M. V. Kahraman, N. Kayaman-Apohan, N. Arsu and A. Guengoer, *Prog. Org. Coat.*, 2004, **51**, 213–219.

39. J. Huang, D. Yuan, A. Tang and X. Liao (Sichuan EM Technology), *China Pat.*, 102 719 206, 2012.

40. W. W. Spooncer (Shell Oil), *U.S. Pat.*, 3 239 492, 1966.

41. C. Negrell-Guirao, F. Carosio, B. Boutevin, H. Cottet and C. Loubat, *J. Polym. Sci., Part B: Polym. Phys.*, 2013, **51**, 1244–1251.

42. P. D. Bartlett and F. A. Tate, *J. Am. Chem. Soc.*, 1953, **75**, 91.

43. C. Negrell-Guirao, G. David, B. Boutevin and K. Chougrani, *J. Polym. Sci., Part A: Polym. Chem.*, 2011, **49**, 3905–3910.

44. Z. El Asri, K. Chougrani, C. Negrell-Guirao, G. David, B. Boutevin and C. Loubat, *J. Polym. Sci., Part A: Polym. Chem.*, 2008, **46**, 4794–4803.
45. Y. Ren, B. Cheng, L. Xu, A. Jiang and Y. Lu, *J. Appl. Polym. Sci.*, 2010, **115**, 1489–1494.
46. M. Mizutani and K. Hirashima (Nitto Denko), *Jpn. Pat.*, 153 193, 2011.
47. G. Moir, S. Heath, M. Archibald and J. Goulding (Clariant International), *Eur. Pat.*, 2 180 004, 2010.
48. M. I. Kabachnik and T. Y. Medved, *Dokl. Akad. Nauk SSSR*, 1952, **83**, 689–692, *Chem. Abstr.*, 1953, **CAN47**, 15755.
49. C. Mannich, B. Lesser and F. Silten, *Ber. Dtsch. Chem. Ges. B*, 1932, **65B**, 378–385, *Chem. Abstr.*, 1932, **CAN26**, 28310.
50. K. Moedritzer and R. R. Irani, *J. Org. Chem.*, 1966, **31**, 1603–1607.

CHAPTER 4

Synthesis and Polymerization of Vinylphosphonic Acid

LAVINIA MACARIE AND GHEORGHE ILIA*

Institute of Chemistry Timisoara of Romanian Academy, B-dul Mihai Viteazu 24, 300223 Timisoara, Romania
*Email: gheilia@yahoo.com; ilia@acad-icht.tm.edu.ro

4.1 Introduction

Our modern life is connected to the world of polymers. In recent decades, polymeric materials have appeared in all domains of human activity, having many applications. These materials can be designed with varied characteristics for different applications, depending on the imagination of their architects. Among these designed materials, polymers containing heteroatoms have attracted great interest and are increasingly becoming a focus of research activities. The number of available functionalities tailored by the introduction of heteroatoms is vast and, in combination with macromolecular organization and self-assembly principles, a wide range of new functional materials are accessible.

Phosphorus-containing polymers, especially those comprising vinylphosphonic acid (VPA) and vinylphosphonate moieties, have become attractive polymers on an industrial scale, due to their various applications based on the properties of the phosphonic group.[1] In early applications, poly(VPA) and its derivatives were noted as efficient scale inhibitors in cooling and boiler water systems, by inhibiting the formation of calcium sulfate, carbonate, and phosphate[2,3] and as flame retardants.[4,5]

RSC Polymer Chemistry Series No. 11
Phosphorus-Based Polymers: From Synthesis to Applications
Edited by Sophie Monge and Ghislain David
© The Royal Society of Chemistry 2014
Published by the Royal Society of Chemistry, www.rsc.org

Other important applications of poly(VPA) and its copolymers are as polymer electrolyte membranes for fuel cells,[6–10] in the medical field as components in dental cements,[11–14] bone reconstruction,[15–18] hydrogels for drug delivery,[19] and in ion exchange membranes.[20,21]

4.2   Synthesis of Vinylphosphonic Acid

The synthesis and characterization of VPA was previously described by Kabachnik and Medvedev.[22] They used vinylphosphonic dichloride as the starting reagent, which was prepared by catalytic dehydrochlorination of the already known 2-chloroethylphosphonic dichloride (Scheme 4.1) by passing its vapor through a quartz tube filled with barium chloride at high temperature.[23]

VPA was obtained as a viscous liquid by hydrolysis of vinylphosphonic dichloride in cool water, distillation of water and hydrochloric acid, and drying under vacuum (Scheme 4.2).

The acid is converted into the corresponding pyro acid when it is distilled under vacuum (Scheme 4.3).

By reacting chlorotrimethylsilane with dialkyl esters of VPA, followed by treatment of the silyl phosphonate with methanol, Rabinowitz[24] obtained VPA in good yield (Scheme 4.4).

A more elaborate method for obtaining VPA starts from phosphorus trichloride and ethylene oxide and follows five steps (Scheme 4.5).[25]

First obtained is tris(2-chloroethyl) phosphite, which undergoes an Arbuzov rearrangement to give bis(2-chloroethyl) 2-chloroethylphosphonate. This product reacts with thionyl chloride in the presence of a suitable

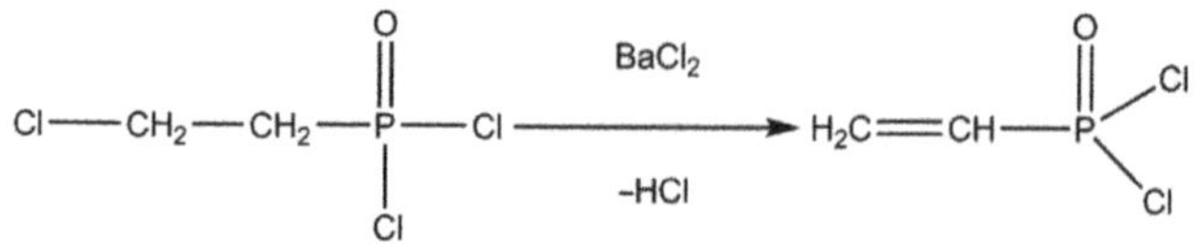

Scheme 4.1   Dehydrochlorination of 2-chloroethylphosphonic dichloride.

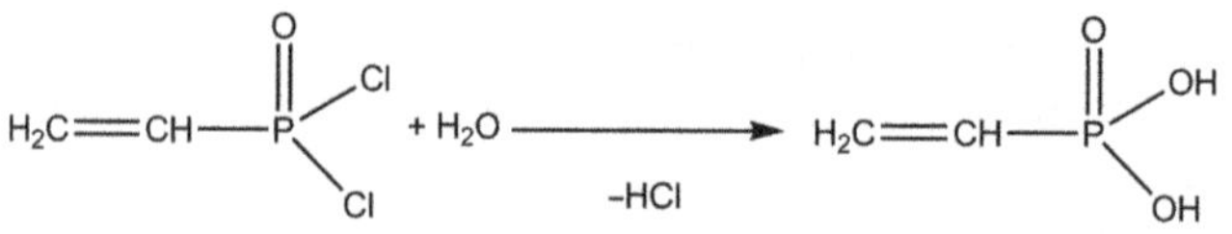

Scheme 4.2   Hydrolysis of vinylphosphonic dichloride.

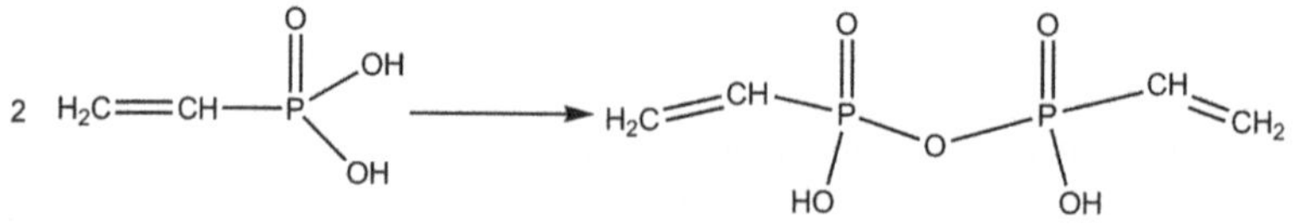

Scheme 4.3   Pyrolysis of vinylphosphonic acid.

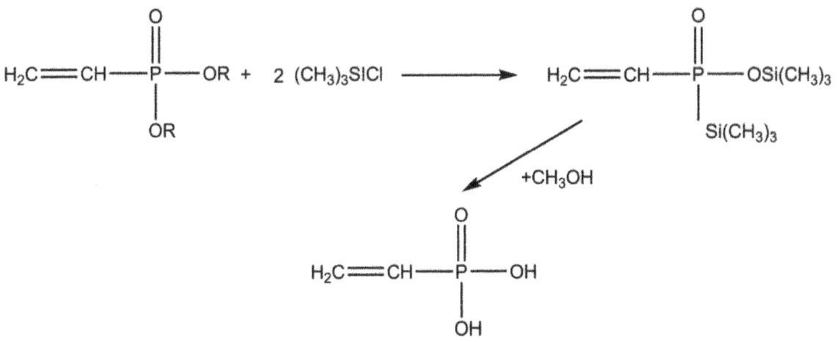

Scheme 4.4 Synthesis of vinylphosphonic acid by a mild hydrolytic route.

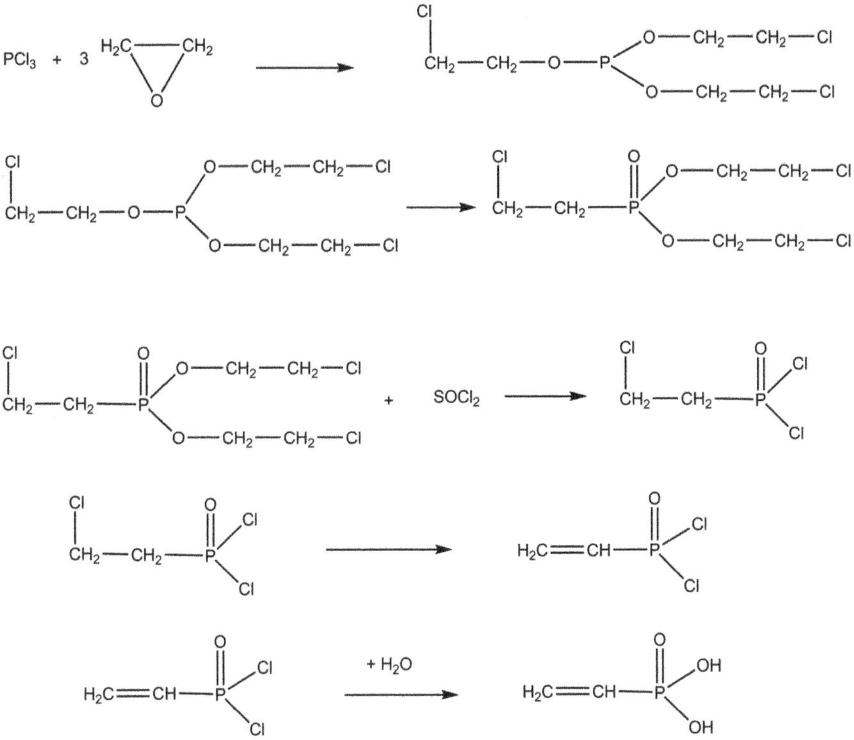

Scheme 4.5 Synthesis of vinylphosphonic acid in five steps.

catalyst (triphenylphosphine oxide) to give 2-chloroethylphosphonic dichloride. In the following step, the latter can be converted into VPA by eliminating HCl in the presence of $BaCl_2$ (as catalyst) and subsequent hydrolysis of the obtained vinylphosphonic dichloride. This procedure for the preparation of VPA involves a long reaction path and also requires corresponding equipment. The use of thionyl chloride as starting material in an industrial process requires expensive safety measures. An advantageous one-step method to obtain VPA from 2-chloroethylphosphonic acid was proposed

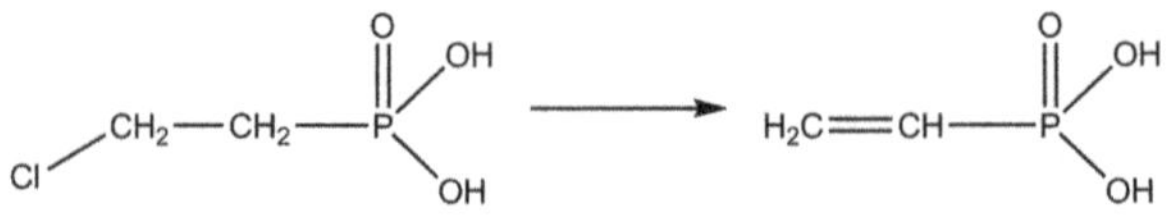

Scheme 4.6   Dehydrochlorination of 2-chloroethylphosphonic acid.

R=alkyl(C1-C4)

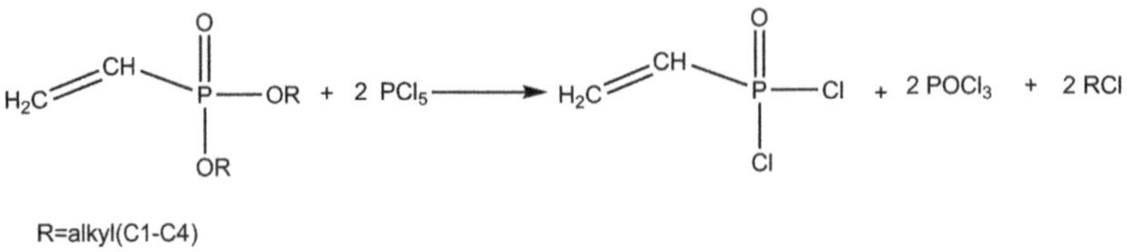

Scheme 4.7   Vinylphosphonic dichloride obtained from its diesters.

by Kleiner and Roscher.[26] 2-Chloroethylphosphonic acid is an available starting material since it is prepared on an industrial scale and widely used in agriculture as a plant growth regulator. This acid is heated at 270 °C in the presence of a catalyst such as a tertiary amine, quaternary ammonium salt, tertiary phosphine, heterocyclic nitrogen compound, quaternary phosphonium salt, or phosphine oxide. Hydrochloric acid and water are eliminated as gases (Scheme 4.6).

In order to avoid undesirable side reactions (formation of polymers), it may be advantageous to carry out the process in the presence of polymerization inhibitors, for example hydroquinone, hydroquinone monomethyl ether, or phenothiazine.

VPA can also be obtained by hydrolysis of vinylphosphonic dichloride in the presence of dichloromethane, at low temperature.[27] The resulting hydrogen chloride and dichloromethane are removed by vacuum distillation. It was found that the reaction of VPA esters with phosphorus pentachloride at higher temperature leads to the formation of vinylphosphonic dichloride (Scheme 4.7).

In this reaction, VPA esters can be used with short-chain aliphatic alcohols. The vinylphosphonic dichloride can be obtained from the reaction mixture in a pure state by distillation under reduced pressure. The use of metal halides as catalysts leads to vinylphosphonic dichloride in higher yield. It is advantageous to use iron(III) chloride as catalyst because the residue can be removed easily from the reaction vessel.

A simple and economical method to obtain VPA consists in heating 2-acetoxyethylphosphonic acid dialkyl esters (alkyl = methyl, ethyl) at a temperature around 200 °C, in the presence of an acidic (sulfuric acid, VPA) or basic [4-(dimethylamino)pyridine] catalyst.[28] Alkyl acetates are simultaneously distilled off. The remaining mixture is hydrolyzed at 160 °C and the alcohol is simultaneously distilled off. By this procedure is obtained a technical grade "crude VPA" or "ester containing crude VPA", *i.e.* a mixture containing more then 75% pure VPA and other compounds such as VPA derivatives, 2-hydroxyethylphosphonic acid or its derivatives, pyrophosphonic acid derivatives, and phosphoric acid, in different percentages. The "crude VPA" or "ester-containing crude VPA" obtained by this low polluting process, particularly inexpensive, is suitable for further application, for example for copolymerization with (meth)acrylic acid in aqueous solution, with applications as detergent and cleaning agents.[29] A similar procedure was elaborated in order to obtain VPA[30] starting from dimethyl 2-acetoxyethylphosphonate, which was heated with stirring at 220 °C in the presence of 4-(dimethylamino)pyridine. Methyl acetate and dimethyl ether were removed and crude monomethyl vinylphosphonate was obtained. This compound was mixed with trimethyl orthoacetate and the mixture was heated at 150 °C, while a mixture of methanol and methyl acetate distilled off. The residue was distilled under vacuum and dimethyl vinylphosphonate was obtained. The latter was mixed with VPA (9:1 w/w) and the mixture was heated at 160–175 °C. Water was metered in at the same time, while methanol distilled off and dimethyl ether was collected.

Substantially pure VPA in good yield is obtained if the hydrolysis of the bis(2-chloroethyl) ester of vinylphosphonic acid is carried out in the presence of a carbonyl compound (*e.g.* formaldehyde, paraformaldehyde, acetone, cyclohexanone) at 150 °C.[31]

An unconventional but very convenient method for the synthesis of VPA from the starting material 2-chloroethylphosphonic acid is by a microwave assisted technique.[32] The pyrolysis of technical grade 2-chloroethylphosphonic acid is performed in a reactor inserted in a microwave oven, with the pyrolysis temperature controlled at 220–230 °C. The reaction occurs in high yield within a few minutes instead of the 4–12 hours required by conventional methods.

4.3 Synthesis of Poly(vinylphosphonic acid)

Poly(vinylphosphonic acid) (PVPA) can be synthesized *via* radical polymerization of VPA in the presence of initiators and, optionally, several chain

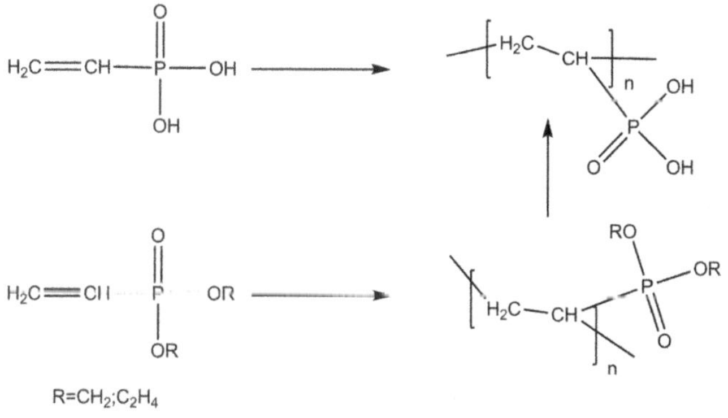

R=CH$_2$;C$_2$H$_4$

Scheme 4.8 The two routes for synthesis of poly(vinylphosphonic acid).

transfer agents (CTAs) with different functionality and reactivity. The other path to obtain PVPA consists of the polymerization of VPA derivatives followed by hydrolysis of the ester groups (Scheme 4.8).[33]

4.3.1 Synthesis of Poly(Vinylphosphonic Acid) from Vinylphosphonic Acid

One of the successful radical homopolymerizations of VPA was performed by Levin *et al.*[34] in DMF in the presence of AIBN as initiator, in a yield of 95%. PVPA was obtained in protic solvents both from pure VPA, "crude VPA", and "ester-containing crude VPA" in the presence of initiator.[35] Suitable protic solvents were water and aliphatic alcohols such as isopropanol, which keep the mixtures stirrable and workable. The free radical initiators that can be used are peroxides such as dibenzoyl peroxide, di-*tert*-butyl peroxide, *tert*-butyl peroxybenzoate, *tert*-butyl peroxy-2-ethylhexanoate, and ammonium or potassium persulfate. The amount of initiator necessary is 1.5–4% *versus* monomer and depends directly on the amount of diluent. The authors recommend that the calculated amount of initiator be added in equal portions during the reaction, after the reaction temperature has been reached, because the polymerization is highly exothermic at the start when the monomer concentration is high and relatively higher residual monomer content could be obtained. The reaction temperature was maintained between 80 °C and 110 °C and depends on the dissociation half-life of the initiator. A reaction time of between 5 and 12 h is approximately inversely proportional to the concentration of VPA monomer in solvent. The yield of PVPA varied from 32% to 60% when peroxide initiators were used and from 3% to 6% when ammonium or potassium persulfate were used. Pure PVPA can be obtained by precipitation. The homopolymerization of VPA in methanol did not occur.[36]

A detailed synthesis and characterization of PVPA has been presented by Bingol *et al.*[33] VPA was polymerized in distilled water in the presence of 2.5% w/w azobisisobutyramidine dihydrochloride (AIBA), under an argon atmosphere. The reaction mixture was heated to 80 °C for 3 h. The product was then dissolved in water and dialyzed in order to separate the polymer from oligomer and unreacted monomer, and the resulting polymer was freeze dried. The yield of polymer was approximately 70% and the molecular weight was found to be in the order of 60 000 g mol^{-1}, as characterized by static light scattering.

The authors proposed a mechanism for the radical polymerization of VPA based on the analysis of ^{1}H and ^{13}C NMR spectra of the polymer. They assumed that the polymerization occurs *via* cyclopolymerization of the vinylphosphonic anhydride, which may be present in equilibrium with the free acid in water at the polymerization temperature (Scheme 4.9). The formation of anhydrides of phosphonic acids in water at higher temperature is already known.[37] By intramolecular propagation can be formed the six-membered ring, generating the pattern of head-to-tail addition and formation of a

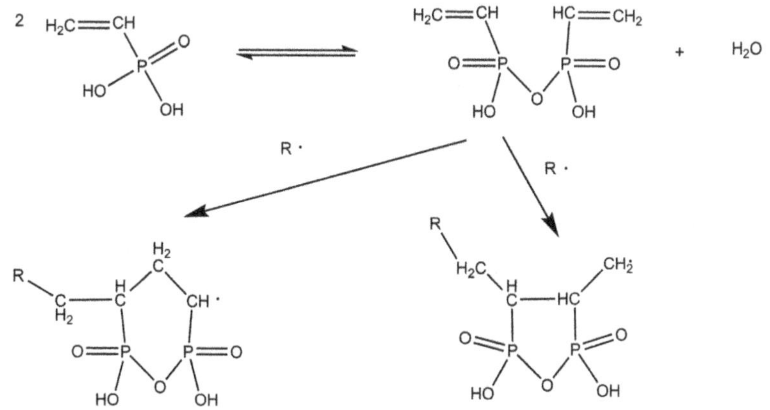

Scheme 4.9 Mechanism of cyclopolymerization of vinylphosphonic acid.

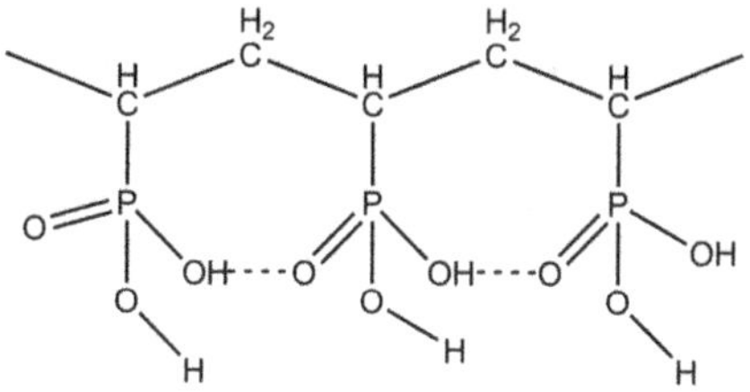

Scheme 4.10 Protons covalently bound to the matrix in poly(vinylphosphonic acid).

methine radical or a five-membered ring generating the pattern of head-to-head linkage with formation of a methylenic radical. Further, this radical reacts with the vinyl group in the β-position to form tail-to-tail links. The authors assumed that the structure of PVPA is atactic and contains a substantial fraction of head-to-head and tail-to-tail links over regular head-to-tail links of 23.5%.

The titration curves indicate that high molecular weight PVPA in its purified form behaves as a monoprotic acid in comparison with VPA monomer, which shows the behavior of a diprotic acid. Further, Lee *et al.* explained, by solid-state NMR under magic-angle spinning (^{1}H and ^{31}P MAS NMR), the local proton mobility and structure of PVPA.[38] The ^{1}H MAS NMR spectra revealed that the P–OH protons are mobile among the phosphonic moieties by hydrogen bonding, which allows proton transport *via* a Grotthuss-type mechanism (structural diffusion), where the protons are covalently bound to the matrix (Scheme 4.10).[39]

In the ^{31}P MAS NMR spectrum, two distinct resonances appear and are assigned to phosphonic acid groups and the phosphonic anhydride. ^{31}P double-quantum experiments show that there is no phase segregation

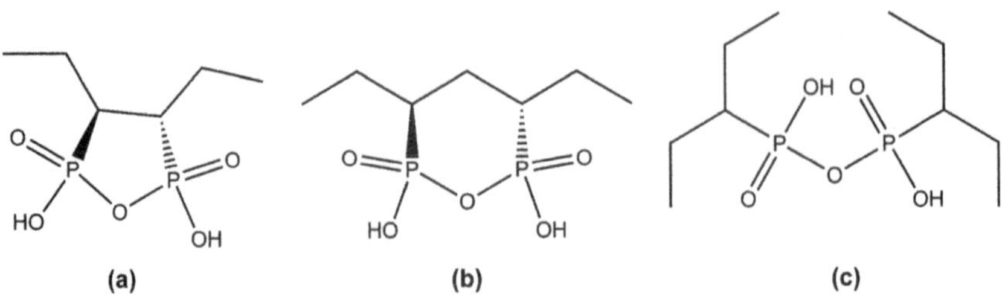

Scheme 4.11 Possible phosphonic anhydride structures.

between vinylphosphonic acid and vinylphosphonic anhydride; hence, the polymerization reaction occurs randomly (Scheme 4.11).

The ^{1}H and ^{31}P MAS NMR spectra were recorded for annealed PVPA at 150 °C to observe the effect of heat treatment on the NMR spectra. An increase in intensity of the ^{31}P signal assigned to phosphonic anhydride and a decrease in intensity of ^{1}H NMR signal assigned to the P–OH protons was observed. This indicated that the phosphonic anhydride is formed at high temperature. The presence of phosphonic anhydride in the structure of PVPA leads to a decrease of proton conductivity by reduction of the number of mobile P–OH protons (charge carriers) and obstruction of any rotational motions of the phosphonic acid group. It is important to minimize the condensation of phosphonic acid groups when the polymer is used as a proton conductor.[40]

The microstructure of PVPA prepared by free radical polymerization using a combination of one- and two-dimensional ^{1}H, ^{13}C, and ^{31}P NMR spectra recorded in D$_2$O solutions was investigated.[41] The authors found that PVPA obtained by radical polymerization contains only 17% of head-to-head and tail-to-tail units as regio-irregular structures, in comparison with the investigations of Bingol.[33]

The synthesis of PVPA from dried VPA in three solvents (ethyl acetate, dimethylformamide, and acetic anhydride) in the presence of 2,2′-azobisisobutyronitrile (AIBN) as initiator was performed[42] and the results were compared using ^{1}H, ^{13}C, and ^{31}P NMR spectra. The polymerization of a 40 wt% solution of VPA in ethyl acetate or in acetic anhydride containing 2 mol% AIBN was performed at 60 °C for up to 8 h and at 35 °C for up to 63 h. The resulting polymer was washed with acetic acid and filtered; the yield of PVPA was 90%. For the polymerization of VPA in DMF, a mixture was used of VPA and VPA anhydrides obtained by stirring VPA and acetic anhydride at 40 °C. This mixture contained VPA : VPA anhydride:VPA dianhydride of composition 1 : 3.3 : 0.8. A solution of 50 wt% in DMF of this mixture, including 2 mol% AIBN as initiator, was polymerized for 3 h at 60 °C. The ^{31}P NMR spectra gave information about a complex mixture of acetylated and non-acetylated (VPA anhydrides) new products formed during the reaction between VPA and acetic anhydride, by consecutive acetylation reactions, and the reaction of a mixed anhydride with a P–OH group under formation of P–O–P structures (Scheme 4.12).

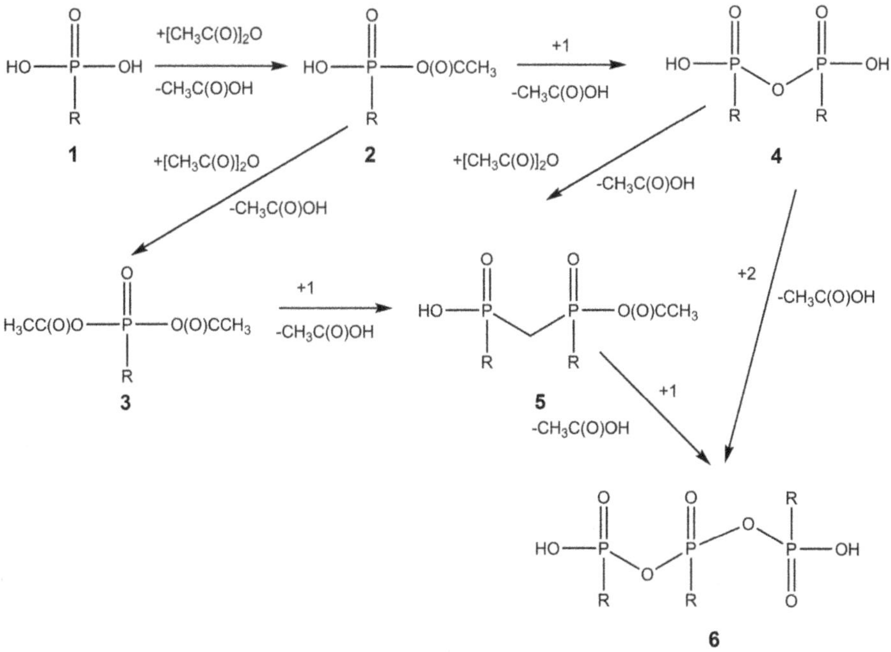

Scheme 4.12 Acetylation of vinylphosphonic acid.

Comparative experiments of the polymerization of VPA and VPA non-acetylated species (VPA anhydrides) in an organic solvent and the polymerization of VPA in DMF were performed in order to establish the relative reactivity of VPA and VPA derivatives. The VPA anhydrides showed a higher reactivity than VPA, explained by the higher reactivity of P–O–P five- and six-membered anhydrides as reactive radical species formed by a cyclopolymerization mechanism. In this case the yield of PVPA was 95%, with molecular weights in the range of 40 000–109 000 g mol^{-1}.

An interesting investigation regarding the polymerization of VPA, in the presence of several CTAs, called telogens, with different functionality and reactivity have been performed.[43] These experiments allowed the controlled polymerization of VPA oligomers, *i.e.* telomerization. Telomerization is a radical polymerization reaction initiated by common initiators, thermally, or by UV rays in which CTAs are used to control the molecular weight and chain-end functionality in order to introduce reactive terminal groups such as acids, alcohols, or halogens.[44]

Transfer agents, with the structure R–R′, such as sulfanylacetic acid (HSCH$_2$CO$_2$H), diphenyl disulfide (C$_6$H$_5$S–SC$_6$H$_5$), 1,1,1-trichloroethane (H$_3$CCCl$_3$), trichloroacetic acid (Cl$_3$CCO$_2$H), and bromotrichloromethane (BrCCl$_3$), have the ability to initiate VPA polymerization through the R group and stop the process by transferring R′ onto the polymer chain; hence the polymer chain has as extremities the R group at one side and the R′ group at

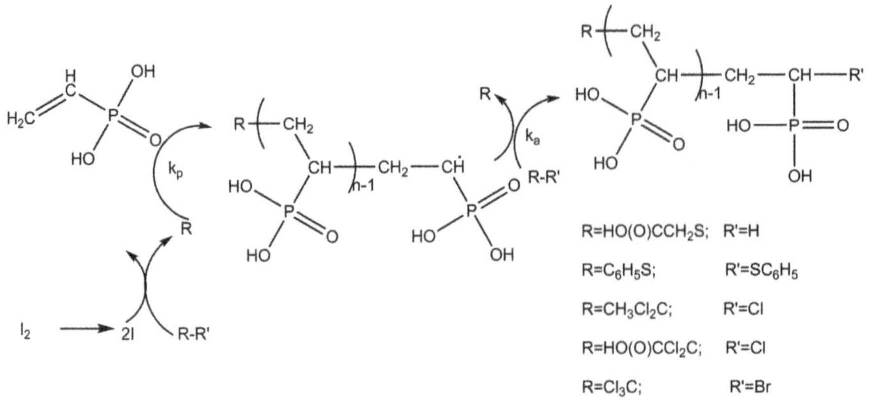

R=HO(O)CCH₂S; R'=H

R=C₆H₅S;   R'=SC₆H₅

R=CH₃Cl₂C;   R'=Cl

R=HO(O)CCl₂C; R'=Cl

R=Cl₃C;   R'=Br

Scheme 4.13   Chain transfer polymerization of vinylphosphonic acid.

the other. The polymerization is performed in water or DMF in the presence of AIBN as initiator. ^{1}H and ^{13}C NMR, mass spectrometry, and elemental analysis of the polymer revealed if the polymer was synthesized *via* transfer reactions (Scheme 4.13).

Sulfanylacetic acid is not a suitable CTA because a monoadduct of VPA/ sulfanylacetic acid bearing chain-end reactive phosphonic acid groups was obtained but not any oligomeric product. Dibenzyl disulfide did not act as a CTA for VPA, because the presence of aromatic groups at the chain-end of VPA oligomers was not detected. The efficiency of the halogenated telogens (1,1,1-trichloroethane, trichloroacetic acid, and bromotrichloromethane) depends on the nature of the R group.[45] Only the use of bromotri-chloromethane as a CTA in VPA polymerization led to obtaining telechelic VPA oligomers, with *n* in the range 6–60 and a yield up to 70%. Depending on chain-end functionality (phosphonic acid group, halogen), such products can be further used in condensation reactions or as macro-CTAs in a radical polymerization with other monomers, in order to obtain block copolymers for different applications.

The telomerization of VPA was investigated in the presence of per-fluorohexyl iodide (C₆F₁₃I), using AIBN as initiator and DMF as the solvent, at different temperatures in the range 60–100 °C.[46] The ^{1}H, ^{13}C, ^{31}P, and ^{19}F NMR spectra revealed that only one structure, C₆F₁₃(VPA)$_n$I, was obtained, indicating that the oligomers are synthesized from transfer reactions and that conventional radical initiation and termination did not occur (Scheme 4.14).

Polymerization of VPA was successfully performed by aqueous RAFT/ MADIX polymerization, controlled with *O*-ethyl xanthate as a transfer agent.[47] Reversible addition–fragmentation chain transfer (RAFT) is a poly-merization method which uses a chain transfer agent in the form of a thiocarbonylthio compound such as dithioesters, thiocarbamates, and xanthates having the role to mediate the polymerization *via* a reversible

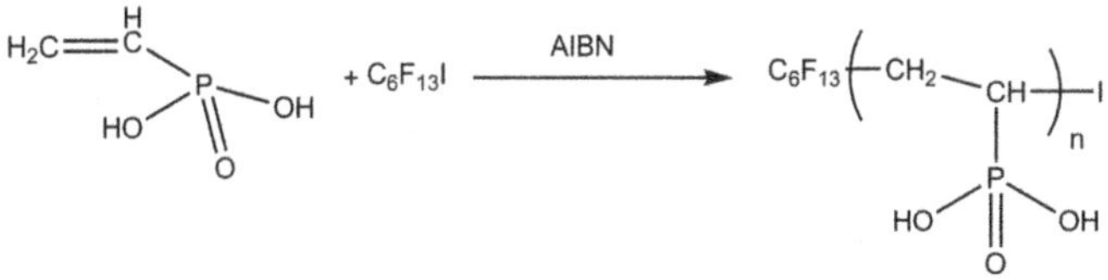

Scheme 4.14 Telomerization of vinylphosphonic acid.

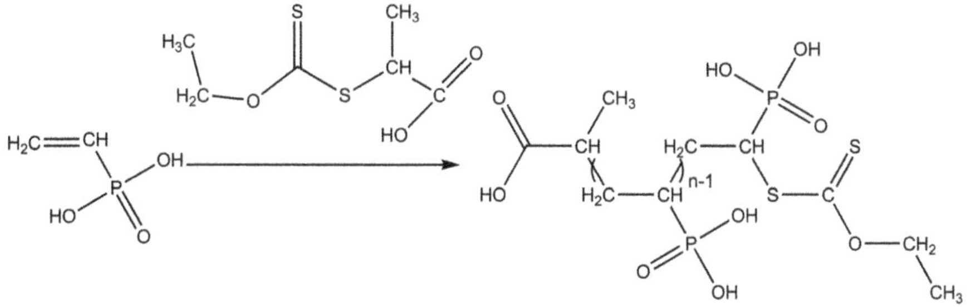

Scheme 4.15 RAFT/MADIX polymerization of vinylphosphonic acid.

chain transfer process and control over the generated molecular weight and polydispersity during a free radical polymerization.[48,49] The polymerization of VPA was performed in water at 65 °C using AIBA as initiator and in the presence of three different concentrations of xanthate at different reaction times (1, 4, 8, and 24 h) (Scheme 4.15), corresponding to molar mass (M_n) values of 1000, 3000, and 6000 g mol^{-1} for PVPA. The conversion of VPA reached 75–85% after 24 h. The presence of xanthate led to a narrower molar mass distribution; hence, it has the role of regulating average molar masses, in comparison with classical methods of polymerization.

The structural analysis by ^{31}P NMR of both PVPA and VPA/X 1 : 1 adducts confirmed that all the chains end with the dithiocarbonate group. Also, the xanthate terminal group could be reversibly transferred, leading to an increase of M_n during polymerization with an increase of reaction time. PVPA obtained from the monomer bearing unprotected phosphonic acid groups contains a significant proportion of regio-irregular structures, which is in accordance with previous studies of Bingol[33] and Komber.[41]

4.3.2 Synthesis of Poly(vinylphosphonic acid) from Vinylphosphonic Acid Derivatives

PVPA with high purity having good solubility characteristics and controllable molecular weight distribution was prepared by an improved method.[50] As initiator, dicumyl peroxide, dibenzoyl peroxide, di-*tert*-butyl peroxide, or AIBN, can all be used. In the first step, bis(2-chloroethyl) vinylphosphonate and an initiator at 0.5–2.0 wt% were heated with stirring under an inert atmosphere for 90 h (Scheme 4.16). The resultant viscous liquid was

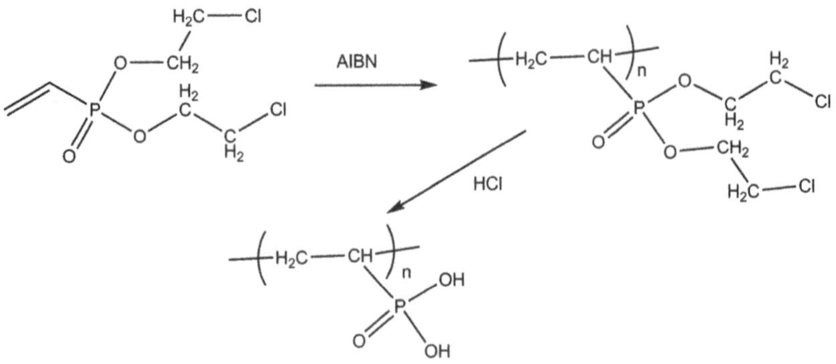

Scheme 4.16   Polymerization of bis(2-chloroethyl) vinylphosphonate and its hydrolysis.

dissolved in ethyl acetate or another solvent such as propyl acetate, butyl acetate, isobutyl acetate, or hexyl acetate. To this mixture was slowly added cyclohexane in order to separate out the poly[bis(2-chloroethyl) vinylphosphonate]s. The unreacted starting material remained in the solvent mixture. In the next step the polymer was mixed with an aqueous solution of hydrochloric acid and heated at reflux for 75 h. HCl was distilled off at reduced pressure and completely removed by washing the solution containing PVPA with deionized water. The aqueous PVPA solution was evaporated under vacuum and a white powder was obtained.

PVPA was also obtained using as starting material vinylphosphonic dichloride (VPDC).[51] VPDC was dissolved under a nitrogen atmosphere in 1,1,1-trichloroethane, with stirring, and polymerized at a reflux temperature of 70 °C, in the presence of AIBN as initiator at a level of 3 wt% of the monomer. If the polymerization is carried out at 70 °C for 18 h, the yield is 50–60%. However, if the temperature was set at 70 °C for 2 h, and then lowered to 40–45 °C for 16–18 h, the yield increased to 85–90%. A viscous, orange-brown solution was obtained, which is a solution of the crude polymer of VPDC. This polymer is hydrolyzed with a large volume of cool water and the resulting HCl and the organic solvent were removed by vacuum distillation. The analysis of final product typically yields 85–90% of the desired PVPA, a solid, together with 10–15% residual monomer.

PVPA can also be obtained by hydrolysis of poly(dimethyl vinylphosphonate) (PDMVP), which was synthesized both through free radical and anionic polymerization.[52] Structural analysis revealed that PDMVP obtained by anionic polymerization is isotactic-rich.

The complete hydrolysis of PDMVP was carried out in aqueous 14 M HCl at 100 °C for 12 h and after dialysis of the reaction mixture in water, PVPA was obtained from the dialyzed solution by freeze drying (Scheme 4.17).

PVPA is known as an atactic polymer. PVPA obtained by hydrolysis of PDMVP from radical polymerization has $M_n = 5000$ and a *meso* content $m = 52\%$, but PVPA obtained by hydrolysis of PDMVP from anionic

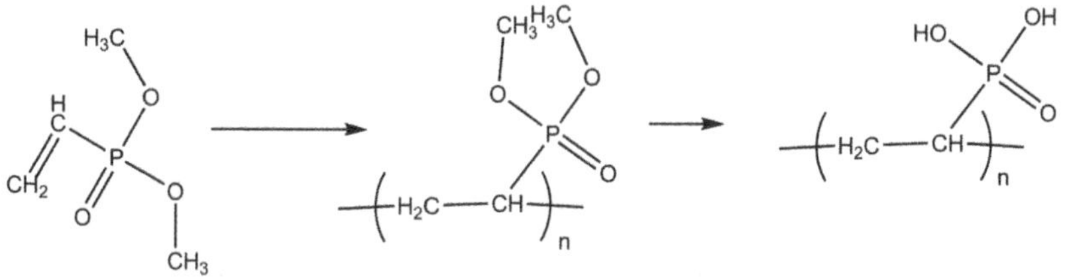

Scheme 4.17   Radical polymerization of dimethyl vinylphosphonate.

polymerization has $M_n = 5500$ and $m = 67\%$. If the aqueous solution of both PVPA products was cast on a glass plate, after 7 days the PVPA from PDMVP obtained by radical polymerization looked like a viscous solution while that from PDMVP obtained by anionic polymerization gave a transparent film due to the isotactic structure.

4.4   Properties of Poly(vinylphosphonic acid)

PVPA is a water-soluble white powder, insoluble in aprotic organic solvents. PVPA shows a similar structure to poly(acrylic acid) (PAA) and poly(vinyl-sulfonic acid) (PVSA). Whereas PAA is a weak electrolyte and PVSA is a strong electrolyte, PVPA is intermediate. It shows in reality only one step in neu-tralization, although should theoretically exhibit two distinctly different steps, like the VPA monomer.[33] Hence, PVPA behaves as a monobasic acid, because the second proton of the monomer unit cannot be ionized in aqueous media. It is assumed that the charge density of the fully ionized PVPA is very high and that the second ionization step does not occur.

The main property of PVPA is proton conductivity. PVPA is largely used as a component in polymer electrolyte membranes obtained either by co-polymerization or blending, in a low humidity environment at operating temperatures higher than 100 °C.

PVPA is a suitable and gainful alternative for phosphoric acid because, owing to the highest concentration of immobilized phosphonic acid groups, self-condensation at a temperature up to 150 °C is limited.[53] The proton transport mechanism consists in hydrogen bond-breaking and bond-form-ing between the neighboring phosphonic acid groups while the protons "hop" to the next molecule.[54,55] This mechanism is known as the "Grotthus mechanism" when working in anhydrous materials.

The conductivity of PVPA depends on temperature, water content (PVPA is a hygroscopic material), and self-condensation at higher temperature as a function of the relative humidity of air.[56] The relative humidity influences the water uptake: at higher humidity, PVPA becomes even a gel. Relative humidity and temperature determine the equilibrium between condensed and non-condensed phosphonic OH groups; also, water and self-conden-sation coexist in the polymer, both influencing the proton conductivity be-cause charge transport in PVPA involves water transferring protons between neighboring phosphonic acid sites. The conductivity of dried PVPA was measured from 20 °C to 150 °C, under nitrogen, with a short time for each step. Under these conditions the conductivity increased with the increase of temperature and the condensation degree was low and constant. However, above 100 °C, under water vapor pressure, in conditions corresponding to fuel cells, the proton conductivity steadily decreased with the increase of temperature and decrease of water content. At a temperature above 100 °C, self-condensation increased because the water evaporates and the number of phosphonic acid sites decrease. Therefore, to design new polyelectrolyte membranes it is necessary to add materials (spacers) in order to avoid self-condensation reactions of phosphonic acid groups at higher temperature.

In other experiments it was revealed that the conductivity of PVPA increases as the temperature increases from 20 to 80 °C, but decreases when the polymer was exposed to higher temperatures and longer times.[40] The presence of water improves the proton conductivity of PVPA at low temperatures. At higher temperature, when the water content is reduced by evaporation, the undesired formation of phosphonic anhydride species takes place and the conductivity decreases.[33]

The presence of water in PVPA seems to contribute to the conductivity of PVPA, at a temperature below the boiling point of water, by proton transport in additional proton solvents. Hence, the proton movement in PVPA can be explained by rapid transfer of protons *via* hydrogen bond-forming and bond-breaking (hopping mechanism) and by self-diffusion (vehicle mechanism).[57] Different conductivity values could be obtained due to a different water content. Also, recent studies have revealed that the behavior of PVPA as a polyelectrolyte is very similar to poly(acrylic acid) in aqueous salt solution under identical conditions.[58]

PVPA shows thermostability to at least 350 °C in an oxygen atmosphere. At this temperature the cleavage of the P–C bonds could take place. Under an inert atmosphere of nitrogen, this fragmentation occurs at 380 °C. At lower temperature the weight loss is assigned to water (humidity).[56,59]

4.5 Conclusions

Vinylphosphonic acid has been synthesized by several methods, starting from different reagents. The main path to obtain VPA utilizes 2-chloro-ethylphosphonic acid as starting material, which is heated at 270 °C in the presence of catalysts.

Other possibilities to obtain VPA are by hydrolysis of vinylphosphonic dichloride or from phosphorus trichloride, ethylene oxide, and thionyl chloride, following five steps. An economical method consists in the heating of 2-acetoxyethylphosphonic acid dialkyl esters (alkyl = methyl, ethyl) at a temperature around 200 °C, in the presence of an acidic or basic catalyst when a technical grade "crude VPA" or "ester containing crude VPA" is obtained.

PVPA can be synthesized *via* radical polymerization of VPA in the presence of initiators or from VPA derivatives (vinylphosphonic acid esters, vinyl-phosphonic acid chlorides). PVPA with a low molecular weight (oligomers) is obtained by radical polymerization in the presence of various chain transfer agents. The microstructure of PVPA was established by one- and two-dimensional ^{1}H, ^{13}C, and ^{31}P NMR spectra.

The most important property of PVPA is conductivity based on hydrogen bonding, which allows proton transport *via* a Grotthuss type mechanism (structural diffusion). The conductivity is the key property of PVPA, which is successfully used to obtain polymer electrolyte membranes by copolymerization or blending with appropriate materials. The applications of these materials have greatly expanded and now have many roles in the modern industrial economy.

References

1. L. Macarie and G. Ilia, *Prog. Polym. Sci.*, 2010, **35**, 1078.
2. L. W. Becker, *U.S. Pat.*, 4 446 046, 1984.
3. G. Woodward, G. P. Otter, K. P. Davis and R. E. Talbot, *Br. Pat.*, 085 616, 2001.
4. D. D. Jiang, Q. Yao, M. A. McKinney and C. A. Wilkie, *Polym. Degrad. Stab.*, 1999, **63**, 423.
5. T. Imamoglu and Y. Yagci, *Turk. J. Chem.*, 2001, **25**, 1.
6. M. Yamada and I. Honma, *Polymer*, 2005, **46**, 2986.
7. F. Sevil and A. Bozkurt, *J. Phys. Chem. Solids*, 2004, **65**, 1659.
8. J. Parvole and P. Jannasch, *Macromolecules*, 2008, **41**, 3893.
9. F. J. Jiang, H. J. Zhu, R. Graf, W. H. Meyer, H. W. Spiess and G. Wegner, *Macromolecules*, 2010, **43**, 3876.
10. A. Aslan and A. Bozkurt, *J. Power Sources*, 2009, **191**, 442.
11. G. O. Adusei, S. Deb and J. W. Nicholson, *Dent. Mater.*, 2005, **21**, 491.
12. K. Ikemura, F. R. Tay, N. Nishiyama, D. H. Pashley and T. Endo, *Dent. Mater. J.*, 2006, **25**, 566.
13. Y. E. Greish and P. W. Brown, *J. Am. Ceram. Soc.*, 2002, **85**, 1738.
14. J. Ellis and A. D. Wilson, *U.S. Pat.*, 5 179 135 A, 1993.
15. J. Tan, R. A. Gemeinhart, M. Ma and W. N. Saltzman, *Biomaterials*, 2005, **26**, 3663.
16. S. Downes, M. Zakikhani, A. K. Bassi and J. E. Gough, *Br. Pat.*, 0071 464, 2013.
17. A. K. Bassi, J. E. Gough and S. Downes, *J. Tissue Eng. Regener. Med.*, 2012, **6**, 833.
18. H. W. Choi, H. J. Lee, K. J. Kim, H.-M. Kim and S. C. Lee, *J. Colloid Interface Sci.*, 2006, **304**, 277.
19. B. Bingol, C. Strandberg, A. Szabo and G. Wegner, *Macromolecules*, 2008, **41**, 2785.
20. G. J. Schlichting and A. M. Herring, *Br. Pat.*, 0 227 006, 2011.

21. B. L. Rivas, E. Pereira, P. Gallegos, D. Homper and K. E. Geckeler, *J. Appl. Polym. Sci.*, 2004, **92**, 2917.
22. M. I. Kabachnik and T. Medved, *Russ. Chem. Bull.*, 1959, **8**, 2043.
23. M. I. Kabachnik and P. A. Rosiiskaya, *Izv. Acad. Nauk SSSR, Otdel. Khim. Nauk*, 1946, 295, *Chem. Abstr.*, 1948, **42**, 7242i.
24. R. Rabinowitz, *J. Org. Chem.*, 1963, **28**, 2975.
25. F. Rochlitz and H. Vilcsek, *Angew. Chem. Int. Ed. Engl.*, 1962, **1**, 652.
26. H.-J. Kleiner and G. Roscher, *U.S. Pat.*, 5 811 575, 1998.
27. K. Schimmelschmidt, W. Denk, *U.S. Pat.*, 3 098 865, 1963.
28. H.-J. Kleiner, *U.S. Pat.*, 4 386 036, 1983.
29. W. Dursch, J. Grosse, W. Gohla, F. Engelhardt and U. Riegel, *U.S. Pat.*, 4 749 758 A, 1988.
30. H.-J. Kleiner and W. Dursch, *U.S. Pat.*, 4 493 803, 1985.
31. R. D. Jackson and K. R. K. Matthews, *World Pat.*, 2003/016319, 2003.
32. S. B. Zotov, M. O. Tuzhikov, O. I. Tuzhikov and T. V. Khokhlova, *Russ. J. Appl. Chem.*, 2012, **85**, 639.
33. B. Bingol, W. Meyer, M. Wagner and G. Wegner, *Macromol. Rapid Commun.*, 2006, **27**, 1719.
34. Y. A. Levin, V. G. Romanov and B. Y. Ivanov, *Polym. Sci. U.S.S.R.*, 1975, **17**, 880, *Chem. Abstr.*, 1975, **83**, 98046f.
35. W. Dursch, F. E. Herwig and F. Engelhardt, *U.S. Pat.*, 1987, 4(696), 987.
36. H. Erdemi and A. Bozkurt, *Eur. Polym. J.*, 2004, **40**, 1925.
37. G. M. Kosolapoff, *J. Am. Chem. Soc.*, 1952, **74**, 3427.
38. Y. J. Lee, B. Bingol, T. Murakhtina, D. Sebastiani, W. H. Meyer, G. Wegner and H. W. Spiess, *J. Phys. Chem. B*, 2007, **111**, 9711.
39. W. Grot, *Chem. Ing. Tech.*, 1975, **47**, 617.
40. B. Bingol, *PhD Dissertation*, University of Mainz, 2007.
41. H. Komber, V. Steinert and B. Voit, *Macromolecules*, 2008, **41**, 2119.
42. M. Millaruelo, V. Steinert, H. Komber, R. Klopsch and B. Voit, *Macromol. Chem. Phys.*, 2008, **209**, 366.
43. G. David, B. Boutevin, S. Seabrook, M. Destarac, G. Woodward and G. Otter, *Macromol. Chem. Phys.*, 2007, **208**, 635.
44. B. Boutevin, *J. Polym. Sci., Part A: Polym. Chem.*, 2000, **38**, 3235.
45. M. Destarac, B. Pees and B. Boutevin, *Macromol. Chem. Phys.*, 2000, **201**, 1189.
46. G. David, C. Boyer, R. Tayouo, S. Seabrook, B. Ameduri, B. Boutevin, G. Woodward and M. Destarac, *Macromol. Chem. Phys.*, 2007, **209**, 75.
47. I. Blidi, R. Geagea, O. Coutelier, S. Mazieres, F. Violleau and M. Destarac, *Polym. Chem.*, 2013, **3**, 609.
48. K. Matyjaszewski, S. Gaynor and J.-S. Wang, *Macromolecules*, 1995, **28**, 2093.
49. J. Chiefari, Y. K. Chong, F. Ercole, J. Krstina, J. Jeffery, T. P. T. Le, R. T. A. Mayadunne, G. F. Meijs, C. L. Moad, G. Moad, E. Rizzardo and S. H. Thang, *Macromolecules*, 1998, **31**, 5559.
50. W. Rowe, S. Saraiya and A. Shah, *U.S. Pat.*, 4 822 861, 1989.
51. J. H. Braybrook, *U.S. Pat.*, 5 278 266, 1994.

52. T. Kawauchi, M. Ohara, M. Udo, M. Kawauchi and T. Takeichi, *J. Polym. Sci., Part A: Polym. Chem.*, 2010, **48**, 1677.
53. B. Bingol, G. Hart-Smith, C. Barner-Kowollik and G. Wegner, *Macromolecules*, 2008, **41**, 1634.
54. K.-D. Kreuer, *Chem. Mater.*, 1996, **8**, 610.
55. K.-D. Kreuer, S. J. Paddison, E. Spohr and M. Schuster, *Chem. Rev.*, 2004, **104**, 4637.
56. A. Kaltbeitzel, S. Schauff, H. Steininger, B. Bingol, G. Brunklaus, W. H. Meyer and H. W. Spiess, *Solid State Ionics*, 2007, **178**, 469.
57. K. A. Mauritz and R. B. Moore, *Chem. Rev.*, 2004, **104**, 4535.
58. C. Strandberg, C. Rosenauer and G. Wegner, *Macromol. Rapid Commun.*, 2010, **31**, 374.
59. D. D. Jiang, Q. Yao, M. A. McKinney and C. A. Wilkie, *Polym. Degrad. Stab.*, 1999, **63**, 423.

CHAPTER 5

2-Methacryloyloxyethyl Phosphorylcholine Polymers

KAZUHIKO ISHIHARA*[a,b] AND KYOKO FUKAZAWA[a]

[a] Department of Materials Engineering, School of Engineering,
The University of Tokyo, 7-3-1, Hongo, Bunkyo-ku, Tokyo 113–8656, Japan;
[b] Department of Bioengineering, School of Engineering,
The University of Tokyo, 7-3-1, Hongo, Bunkyo-ku, Tokyo 113–8656, Japan
*Email: ishihara@mpc.t.u-tokyo.ac.jp

5.1 Molecular Design of 2-Methacryloyloxyethyl Phosphorylcholine Polymers

5.1.1 Molecular Design Concept for 2-Methacryloyloxyethyl Phosphorylcholine Polymers

Cells are an integral component in the construction of biological systems. They act both independently and in response to chemical and physical signaling between groups of cells, which is often controlled by aspects of the cell membrane. This membrane is a molecular hybrid that is mainly composed of three different types of biomolecules: phospholipids, proteins, and polysaccharides.[1] The cell membrane has the mechanical strength to sustain the cell morphology and the ability to maintain the concentration of specific chemicals in the cytoplasm. In addition, cell–cell junctions that are formed between neighboring cells provide an important means of communication.

RSC Polymer Chemistry Series No. 11
Phosphorus-Based Polymers: From Synthesis to Applications
Edited by Sophie Monge and Ghislain David
© The Royal Society of Chemistry 2014
Published by the Royal Society of Chemistry, www.rsc.org

The high diversity of functions made possible by the structure of the cell membrane makes it an attractive candidate in the fabrication of nanostructured biomimetic materials for bio-, nano-, and information technology.

The role of phospholipids in the cell membrane is to provide efficient separation between the intracellular cytoplasm and the extracellular environment. They also provide a scaffold for the presentation of glycoprotein receptors and membrane proteins on the cell surface (Figure 5.1). Phospholipids consist of hydrophobic alkyl chains with a hydrophilic polar head group, enabling them to assemble spontaneously as a continuous bilayer membrane in aqueous solution. One of the major phospholipid polar groups contained within the cell membrane is phosphorylcholine, which is an electrically neutral, zwitterionic head group. In the field of biomimetic chemistry, phospholipid molecules have been utilized for the preparation of cell membrane-like structures, namely liposomes and Langmuir–Blodgett membranes. However, a major disadvantage of a molecular assembly of this kind is its inadequate chemical and/or physical stability. Stabilization of the phospholipid assembly is therefore an important topic of focus in the construction of interfaces between living and artificial systems. One approach to solve this issue is the design of a new type of polymer with phospholipid polar groups. Such a structure would have good chemical and physical stability owing to the bonding between the phospholipid-like moieties. The bioinspired development of the polymer proceeds *via* selection of the particular phospholipid polar group, followed by the introduction of this group into the polymer chain. Phosphorylcholine is a suitable polar group for obtaining a bioinert polymer. In addition, a conventional polymerization process can be used to synthesize a wide variety of molecular structures and material architectures. A methacrylate bearing a phosphorylcholine group in the side chain, 2-methacryloyloxyethyl phosphorylcholine (MPC), is widely recognized as being the most suitable monomer (Figure 5.2).[2–6]

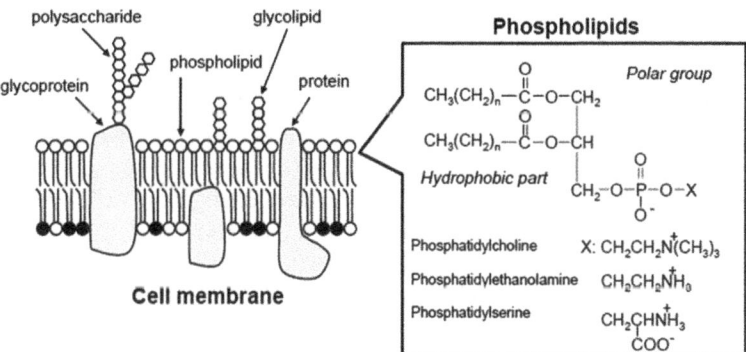

Figure 5.1 Schematic representation of the cell membrane and the chemical structures of phospholipids comprising the cell membrane.

2-Methacryloyloxyethyl phosphorylcholine (MPC)

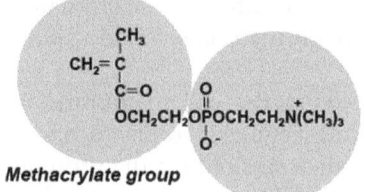

Methacrylate group

Phosphorylcholine group

Figure 5.2   Chemical structure of 2-methacryloyloxyethyl phosphorylcholine.

5.1.2   Synthesis of MPC

The total synthetic route to produce MPC is shown in Figure 5.3; in order to widen the scope of MPC polymer science, improved and elegant syntheses need to be developed. Ishihara identified a number of points for improvement of the classical synthetic route that were reported before 1986.[7,8] One such factor was control of the amount of water present during the procedure. Owing to the hygroscopic property of MPC, it was found that even trace amounts of water had a significant influence on the purity of the resulting compound.

The first step in the preparation of MPC is the reaction between the cyclic phosphoryl chloride, 2-chloro-2-oxo-1,3,2-dioxaphospholane (COP), with 2-hydroxyethyl methacrylate (HEMA) to form 2-(2-oxo-1,3,2-dioxaphospholan-2-yloxy)ethyl methacrylate (OPEMA).[9] Triethylamine is required during the reaction in order to trap hydrogen chloride as triethylammonium chloride. The purity of the OPEMA is dependent on the presence of water; therefore, a high purity, dry solvent is necessary, with effective elimination of the triethylammonium chloride required. The next step is the ring opening reaction of OPEMA with trimethylamine. As this reaction proceeds at 60 °C for 16 h, spontaneous polymerization induced heat needs to be suppressed. In addition, the boiling point of trimethylamine is 4 °C, resulting in high vapor pressure caused by the heating, and producing the necessity for an effectively sealed reaction vessel. On cooling of the reaction mixture, a white precipitate is formed. Crystallization of the MPC from the reaction mixture is performed by keeping the mixture in a freezer for 15 h. Subsequent filtration of the precipitate is carried out under dry argon gas, followed by vacuum drying. The filtrate is recrystallized again from acetonitrile, then the final product is obtained as a white powder with a melting point of over 140 °C. Characterization of the white powder with ^1H NMR, ^{13}C NMR, and IR spectroscopic techniques gives the following values: ^1H NMR (CDCl$_3$): $\delta = 1.90$ (–CH$_3$, 3H), 3.27–3.36 [–N(CH$_3$)$_3$, 9H], 3.70–3.80 (–CH$_2$N, 2H), 4.00–4.10 (POCH$_2$–, 2H), 4.21–4.31 (OCH$_2$–CH$_2$OP, 4H), 5.60 (CH=, 1H), and 6.10 (CH=, 1H); ^{13}C NMR (CDCl$_3$): $\delta = 18.59$ (–CH$_3$), 54.25 [N(CH$_3$)$_3$], 59.44 (CH$_2$N), 63.34 (POCH$_2$–), 64.41 (–CH$_2$OP), 66.20 (OCH$_2$), 126.09 (=CH$_2$), 136.22 (=C=), and 167.36 (–CH$_3$).

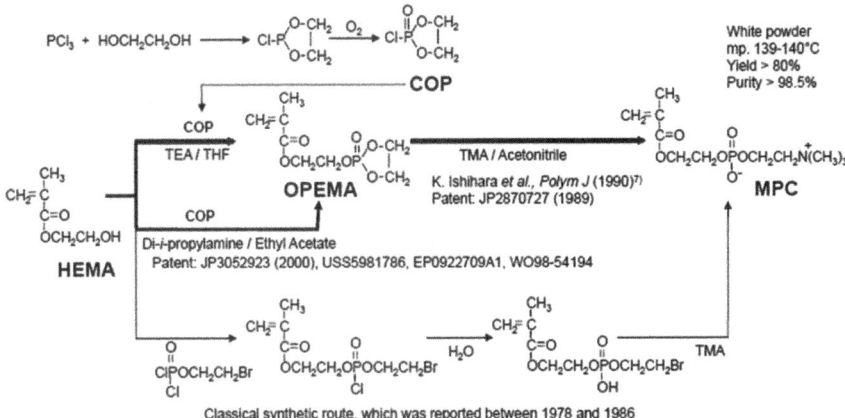

Figure 5.3 Synthetic route to MPC.

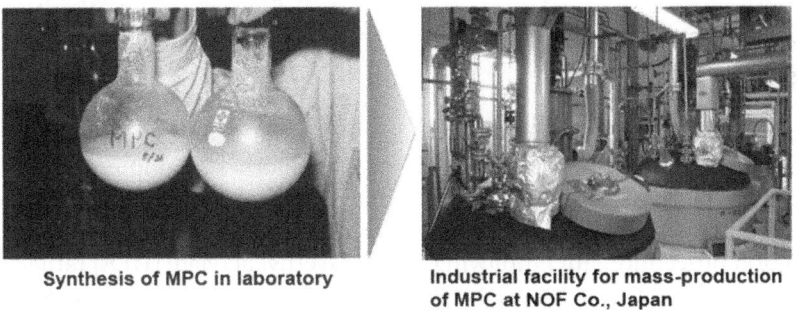

Synthesis of MPC in laboratory   Industrial facility for mass-production
                                         of MPC at NOF Co., Japan

Figure 5.4 MPC production.

This successful synthetic procedure for MPC was transferred to a Japanese chemical company, NOF Co. Ltd., with support from the Japan Science and Technology Agency. An industrial facility for large-scale production of MPC was set up in 1994 and finalized in 1999 (Figure 5.4).[10] During this period, small modifications were made to the synthetic procedure to allow for regulations involving the use of certain dangerous compounds. The most effective of these modifications was the use of diisopropylamine instead of triethylamine, which resulted in a significant increase in the purity of the OPEMA, and enabled the yield of MPC to reach 80%. The industrial plant currently runs well, producing almost 10 tonnes of MPC each year. A worldwide distribution network for MPC as a reagent means that researchers can obtain it easily.

The establishment of the fine synthetic route has encouraged further development of monomers bearing the phosphorylcholine group in the side chain, and a number of MPC derivatives have been synthesized (Figure 5.5).[11–18] These often contain different polymerizable groups, a variety of spacer units between these groups, and a phosphorylcholine

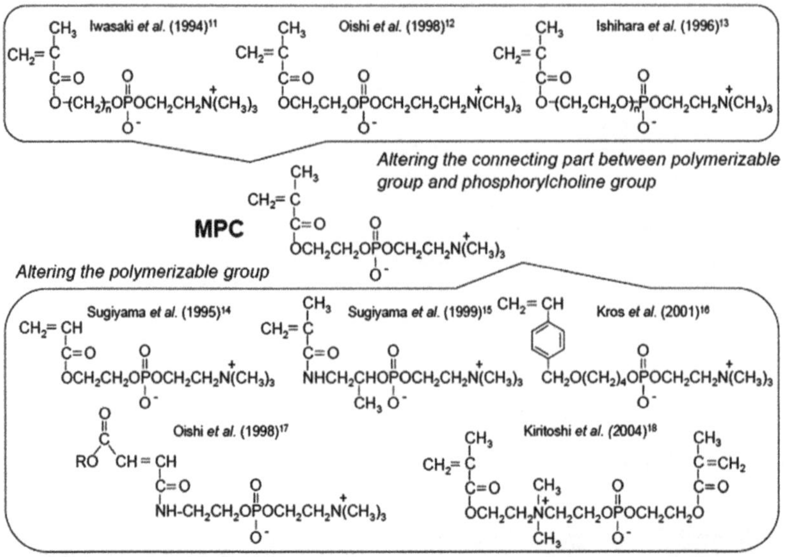

Figure 5.5 MPC derivatives.

group. From the viewpoint of synthetic cost, the chemical stability of the monomer, and its ability to be polymerized with other monomers, makes the methacrylate unit superior to other groups such as acrylate, fumarate, and styrene derivatives. The acrylate derivative can be synthesized *via* the same reaction procedure, but employing 2-hydroxyethyl acrylate (HEA) rather than HEMA as the starting material. Whilst HEA is commercially available owing to its use as a raw material for the production of paint, and the fact that it is more easily polymerized than methacrylate, the acrylate ester bond is easily hydrolyzed under physiological conditions.

5.1.3 Parameters for the Polymerization of MPC

As it is possible to obtain MPC with high purity, the polymerization parameters can be evaluated in detail. Some reports have been published on the solution polymerization of MPC. Sato *et al.* investigated the kinetics of the homogeneous polymerization of MPC in ethanol using dimethyl 2,2′-azo-bisisobutyrate (MAIB) as an initiator.[19] They calculated that the overall activation energy of the polymerization (E_a) was approximately 71 kJ mol^{-1}. This value is a little lower than the value of 84 kJ mol^{-1} calculated for the polymerization of methyl methacrylate (MMA) with 2,2′-azobisisobutyronitrile (AIBN). They also determined that the polymerization rate (R_p) could be expressed by $R_p = k[\text{MAIB}]^{0.54 \pm 0.05}[\text{MPC}]^{1.8 \pm 0.1}$. In addition, it was observed that R_p was highly dependent on the concentration of MPC in the feed. They speculated that this was due to acceleration of propagation caused by MPC aggregation, and also from retardation of termination due to a viscosity effect of the MPC. The rate constants of propagation (k_p) and termination (k_t) of

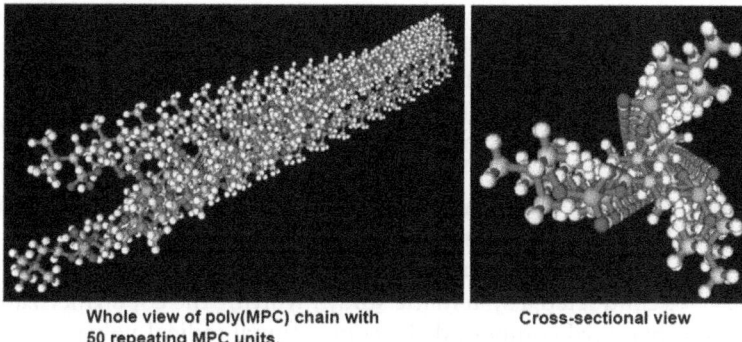

Whole view of poly(MPC) chain with 50 repeating MPC units.

Cross-sectional view

Figure 5.6 Computer illustration of the poly(MPC) chain in water.

MPC were calculated to be $k_p = 180$ L mol^{-1} s^{-1} and $k_t = 2.8 \times 10^4$ L mol^{-1} s^{-1} at 60 °C. This latter value is eight times larger than that of MMA in benzene owing to the slow termination process. Polymerization of MPC with MAIB in ethanol was seen to be accelerated by the presence of water and retarded by the presence of benzene or acetonitrile.

Sato *et al.* investigated the polymerization of MPC in water with potassium peroxydisulfate (KPS) as an initiator.[20] The E_a was calculated to be 53.8 kJ mol^{-1}, and the R_p at 40 °C was expressed by $R_p = k[\text{KPS}]^{0.98}[\text{MPC}]^{1.9}$. A kinetic study was also performed on the polymerization of MPC with KPS in water in the presence of 2.5 mol L^{-1} sodium chloride. In this case, the R_p at 40 °C was expressed by $R_p = k[\text{KPS}]^{0.6}[\text{MPC}]^{1.6}$, and an extremely low value of 19.7 kJ mol^{-1} was obtained as the overall E_a of the polymerization.

Recently, a computer simulation of poly(MPC) demonstrated that the poly(MPC) adopts a rod-like stretched form in water (Figure 5.6). The volume of the hydrophilic phosphorylcholine group in the side chain of MPC is relatively large, and its hydrophobic methacrylate units can aggregate through hydrophobic interactions. Polymerization proceeds readily between the methacrylate units due to their high local density in the aggregate. Thus, the E_a is lower under aqueous conditions compared to when the reaction is carried out in ethanol. The computer simulation of poly(MPC) suggested certain important characteristics, the most significant of which was that the hydrophobic methacrylate main chain has very little contact with water as it is covered completely with hydrophilic phosphorylcholine groups. It is therefore expected that the polymer would have extremely high solubility in aqueous solution, with super-hydrophilicity at the interface.

5.2 Variations in MPC Polymers

5.2.1 Control of Chemical Structure

The solubility of the MPC polymer strongly depends on its molecular structure, the particular composition of the MPC and other monomer units, and its molecular weight. Poly(MPC) has been shown to be easily dissolved in water and alcohol, but insoluble in acetone, acetonitrile, tetrahydrofuran, and a specific composition of an aqueous solution of ethanol (60–92 vol% ethanol).[21-24] However, introduction of other monomer units can change the solubility. Controlling the monomer feed ratio regulates the MPC unit composition in the polymer.[25,26] MPC polymers have been synthesized with other polymer architectures, including block-type copolymers and graft-type copolymers, using a conventional radical polymerization technique or a novel living radical polymerization technique.[27-40]

Ueda *et al.* conducted a systematic study involving the copolymerization of MPC with other vinyl compounds by conventional radical polymerization.[25] Typically, when the MPC was copolymerized in solution with styrene (St) and various alkyl methacrylates such as *n*-butyl methacrylate (BMA, forming PMB), *t*-butyl methacrylate (t-BMA), *n*-hexyl methacrylate (HMA, forming PMH), *n*-dodecyl methacrylate (DMA, forming PMD), or *n*-stearyl methacrylate (SMA, forming PMS), the polymerization progressed very well, and a statistically random sequence was obtained. The radical copolymerization of MPC (M_1) and St (M_2) in ethanol resulted in the following copolymerization parameters: monomer reactivity ratio for M_1 and M_2 are $r_1 = 0.39$, $r_2 = 0.46$, respectively. Also, the Q_1 and e_1 values of the MPC are calculated as $Q_1 = 0.76$ and $e_1 = +0.51$, respectively. In addition, for an MPC (M_1) and MMA (M_2) system, the monomer reactive ratio values are $r_1 = 1.61$, $r_2 = 0.66$.

The solubility of the MPC polymers was found to be dependent upon the proportion of MPC, the hydrophobicity of the alkyl methacrylate, and the molecular weight of the polymer. In Table 5.1, representative examples of copolymerization between MPC and alkyl methacrylate are summarized.

The synthesis of amphiphilic copolymers *via* a one-step reaction without additives is highly desirable, and is especially important for industrial applications. Soap-free heterogeneous polymerization, which copolymerizes hydrophilic and hydrophobic monomers in two phases, meets these requirements. Hatanaka *et al.* investigated the synthesis of various MPC copolymers with HMA (PMH), DMA (PMD), and SMA (PMS) using such a soap-free heterogeneous polymerization.[41] The reactions were carried out using the same monomer concentration but in solvents consisting of different compositions of aqueous dimethyl sulfoxide (DMSO). This control over the solvent enabled manipulation of the polymer composition, even though the monomer ratio was not varied. The polymer composition also depended on the specific alkyl methacrylate that was used. It appeared that changing the solvent polarity caused a change in the properties of the interface between the solution phase and the alkyl methacrylate phase. This phenomenon could be attributed to two main factors. Firstly, an increase in the solvent polarity could result in an increase in methacryloyl groups oriented towards the solution phase as such groups are more polar than alkyl chains. Secondly, the tendency for the alkyl methacrylate phase to disperse in the solution phase would decrease with increasing solvent polarity. The alkyl methacrylate content of PMH increased up to a point and then

Table 5.1 Free radical copolymerization of MPC with various alkyl methacrylates.

Abb.	Alkyl methacrylate	MPC mole fraction in feed	MPC mole fraction in copolymer	Solvent	Conversion %	Mw (10⁴)	Mw/Mn	Tg °C	H^c	MPC mole fraction coated on poly(BMA)	Solvent SP^e (10^{-3} J$^{1/2}$ m$^{-3/2}$) H$_2$O 47.9	EtOH 26.0	CHCl$_3$ 19.0	THF 18.6	Et$_2$O 15.1	HFIP^d –
PMB-20	BMA	0.20	0.13	EtOH/THF = 8/2	78	–	–	31	–	0.21	–^{f}	+^{f}	+	±^{f}	±	–
PMB-40	BMA	0.40	0.26	EtOH	70	–	–	–	–	0.34	–	+	±	–	–	–
PMB-60	BMA	0.60	0.40	EtOH	76	1.63	1.42	63	–	–	+	+	–	–	–	
PMB-80	BMA	0.80	0.58	EtOH	74	1.79	1.47	–	–	–	+	+	–	–	–	–
PMtB-20	tBMA	0.20	0.18	EtOH/THF = 8/2	81	3.63	1.69	–	0.39	0.21	–	+	+	+	±	–
PMtB-40	tBMA	0.40	0.31	EtOH	56	–	–	–	0.72	0.32	+	+	–	–	–	
PMtB-60	tBMA	0.60	0.48	EtOH	68	–	–	–	–	–	+	+	–	–	–	
PMtB-80	tBMA	0.80	0.66	EtOH	80	–	–	–	–	–	±	+	±	+	±	+
PMH-20	HMA	0.20	0.11	EtOH/THF = 8/2	39	6.51	1.04	5	0.65	0.22	±	+	±	+	±	
PMH-40	HMA	0.40	0.24	EtOH	71	–	–	–	–	0.34	±	+	±	–	–	+
PMH-60	HMA	0.60	0.31	EtOH	23	–	–	–	–	–	+	+	–	–	–	
PMH-80	HMA	0.80	0.35	EtOH	78	–	–	52	–	–	+	–	±	±	–	±
PMD-20	DMA	0.20	0.14	EtOH/THF = 5/5	66	–	–	–30	–	–	±	+	+	+	–	±
PMD-40	DMA	0.40	0.20	EtOH/THF = 5/5	31	7.71	1.05	–16	–	0.35	–	±	+	+	–	±
PMD-60^a	DMA	0.60	0.38	EtOH/THF = 9/1	44	–	–	–11	–	0.47	–	±	±	±	–	±
PMS-20	SMA	0.20	0.05	EtOH/THF = 3/7	83	–	–	–	0.17	–	–	–	±	±	–	
PMS-40	SMA	0.40	0.28	EtOH/THF = 6/4	62	22.9	1.30	39^a	0.37	–	–	–	+	+	–	
PMS-60	SMA	0.60	0.49	EtOH/THF = 9/1	46	–	–	39^b	0.58	–	±	±	±	–	–	
PMS-80	SMA	0.80	0.69	EtOH/THF = 9/1	60	–	–	–	–	–	±	+	±	–	–	

Polymerization time = 1.5 h, [Monomer] = 1 mol/L, [AIBN] = 1 mmol/L, Polymerization temp. : 60 °C.

^{a}Polymerization for 6 h.
^{b}Melting point.
^{c}Equilibrium water content.
^{d}Hexafluoro-iso-propanol.
^{e}Solubility parameter.
^{f}Solubility +: soluble, –: insoluble, ±: swelling.

remained constant as the water content of the solvent increased, whereas HMA and MPC were copolymerized even in a large amount of water. On the other hand, the alkyl methacrylate content of PMD decreased with increasing water content, with only MPC being polymerized in a large quantity of water. The alkyl methacrylate content of PMS behaved similarly to PMD. The investigation also compared copolymer compositions and molecular weights produced *via* heterogeneous and homogeneous polymerizations of MPC and HMA (polymerization was carried out in 90 vol% DMSO, [MPC]/[HMA] = 50 : 50 mol/mol). The MPC composition in the heterogeneously synthesized PMH (38 unit mol%) was found to be lower than that in the homogeneous PMH (44 unit mol%). This was attributed to the heterogeneous dispersion of HMA in the polymerization system, with the hydrophobic part of the MPC entrapping the hydrophobic HMA, producing a different local composition to that of the initial monomer feed. It was speculated that two kinds of polymer chains were formed: an HMA-rich high molecular weight copolymer and an MPC-rich low molecular weight copolymer. The monomer sequence of these polymer chains was arranged as a block-type copolymer. As mentioned above, in the case of the PMB system the water solubility of the polymer clearly depended on its molecular weight and the proportion of MPC. A representative example contained 30 unit mol% of MPC copolymer (PMB-30).[42] The random sequence PMB-X (X represents MPC unit mole proportion in the copolymer) could be easily dissolved in ethanol; however, if the molecular weight was above 5.0×10^5, it was insoluble in water. On the other hand, when the molecular weight becomes below 5.0×10^4, it is soluble in water. Figure 5.7 demonstrates the concentration dependence of the surface tension of an aqueous solution containing poly(MPC) and PMB-X. It can be seen that the surface tension did not change in the case of the poly(MPC) in the concentration range of 10^{-5} to 10^{-1} g dL^{-1}, remaining the same as that of pure water. On the other hand,

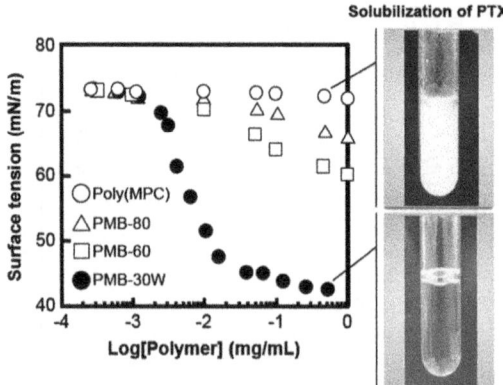

Figure 5.7 Concentration dependence of surface tension in poly(MPC) and PMB-X aqueous solution. The pictures represent the solubility of paclitaxel (PTX) in an aqueous solution of poly(MPC) and PMB-30W.

the surface tension of the PMB-X aqueous solution strongly depended on the polymer concentration. In particular, in the case of PMB-30W (W represents a water soluble polymer), it decreased from the 10^{-3} g dL^{-1} level and became constant above 10^{-2} g dL^{-1}. This result indicated that PMB-30W in water could aggregate and form a micelle-like structure when the concentration was above 10^{-2} g dL^{-1}. The diameter of the PMB-30W aggregate was determined by dynamic light scattering to be 23 ± 5 nm at a concentration of 0.01 g dL^{-1}.

It is possible that the PMB-30W aggregates may solubilize hydrophobic compounds. For example, paclitaxel (PTX), which is a commonly used anticancer drug, can be dissolved in an aqueous solution of PMB-30W.[43–47] PTX has a unique way of preventing the growth of cancer cells: it affects the microtubules in the cell, which play an important role in cellular functions. The microtubules become clogged, preventing the growth and division of the cancer cells. PTX is a highly hydrophobic drug that is barely soluble in water (water solubility <0.3 µg mL^{-1}). Because of this poor solubility in both water and many other acceptable pharmaceutical solvents, emulsifiers such as Cremophor EL® (CR-EL) are used to formulate PTX in commercial injected solutions. However, serious hypersensitivity reactions have been reported in some individuals as the content of CR-EL used in the PTX formulation is significantly higher than in any other marketed drug. Therefore, alternative dosage forms for PTX administration need to be developed to reduce the undesirable side effects induced by the CR-EL. The water solubility of PTX is enhanced dramatically by the amphiphilic nature of the PMB-30W. Figure 5.7 shows images of PTX in a solution of MPC polymer.[43] For aqueous solutions of poly(MPC) and PMB-80, it was not possible to dissolve 1.0 mg of PTX in 1.0 mL of a solution containing 45 mg mL^{-1} of MPC polymer. On the other hand, PMB-50 and PMB-30W, which have a much higher proportion of BMA units than PMB-80, were able to effectively dissolve PTX. The concentration of the PMB-30W also affected the PTX solubility. When the concentration of PMB-30W was higher than 9.0 mg mL^{-1}, the solubility of PTX dramatically increased, with 2.0 mg completely dissolving in the 90 mg mL^{-1} PMB-30W solution, which is 6.7×10^3 times higher than that found for pure water. The solubility of PTX in the PMB-30W aggregate was retained for at least a month. Subcutaneous injection of PTX/PMB-30W was found to cause no macroscopic or microscopic changes to the skin.[44] In addition, our results have shown that the cytotoxicity and antitumor activity of PTX/PMB-30W are similar to those of PTX/CR-EL. The PMB-30W alone did not demonstrate toxicity or antitumor activity. These results suggest controlled and targeted release of PTX following administration of the PTX/PMB-30W combination. This may occur because the size of the PMB-30W aggregate following loading of PTX is approximately 50 nm, and aggregates of this size have an enhanced permeation and retention effect, *i.e.* the molecules predominantly accumulate in tumors and inflamed regions where vessel wall permeability is enhanced. In addition, since the charge of the aggregates is neutral, there are no electrostatic interactions between biomolecules and the

PTX/PMB-30W, leading to good drug retention in the blood. Release of PTX from PMB-30W is likely to be diffusion controlled and dependent on the concentration gradient; hence, PTX/PMB-30W may release PTX in a time-dependent manner. Other *in vivo* experiments regarding anticancer effects using PTX/PMB-30W have been reported.[45] PMB-30W is currently commercially available as "PUREBRIGHT®" solubilizing reagent from NOF Co. Other water-soluble MPC polymers have been prepared, most of which have the function of enabling the dissolution of the hydrophobic groups in the side chain of the polymer. These functional groups include a *p*-nitrophenyl ester unit as an active ester group,[48–51] a phenyl boronic acid unit,[52,53] catechol groups as hydroxyl-containing reactive moieties,[54,55] and a ferrocene group as an electron transfer mediator.[56–58] In particular, a PMB-based polymer with phenylboronic acid units has been shown to spontaneously form a hydrogel on reaction with poly(vinyl alcohol) (PVA) in aqueous solution at room temperature. This gelation property can be applied for immobilizing cells inside the material, with their viability retained for a week. This polymer system is available from Sumitomo Bakelite Co. as a "Cell Cradle®" system. Moreover, another functional unit can be introduced into the PMBV, *i.e.* a vinylferrocene unit (PMBVF), to obtain an electron transfer hydrogel by crosslinking with PVA in aqueous solution.[58] The PMBVF/PVA hydrogel can entrap electron-generated bacteria and make a new-type "cell-based biofuel cell".

5.2.2 Surface Characteristics of MPC Polymers

Owing to the solubility of MPC polymers, PMB-30 is useful for preparing thin polymer layers on a substrate *via* a simple solvent casting method. Although the chemical structure of the PMB-30 is the same as that of PMB-30W, when the molecular weight of the PMB-30 exceeds 5×10^5 the PMB-30 is not soluble in aqueous media and a very stable polymer layer on a substrate is obtained. The modification of a surface with PMB-30 is an extremely easy process, with 0.2–0.5 wt% ethanolic solutions used for casting. On evaporating the solvent at room temperature under an ethanol atmosphere, a homogeneous thin polymer layer is formed on the substrate with a thickness of approximately 0.05–0.1 μm. The use of a spin-coating technique enables the formation of a much flatter membrane of approximately 0.02 μm in thickness. This coating process can be adapted for a wide variety of substrates, including polymers, metals, and ceramics that are wettable with ethanol. The PMB-30 layer is only attached to the substrate through noncovalent physical interactions; however, owing to its large molecular weight, the formed layer is stable even when the coated substrate is immersed in aqueous media. Characterization of MPC polymer-coated surfaces has been carried out using X-ray photoelectron spectroscopy (XPS), atomic force microscopy (AFM), contact angle measurement, and surface ζ-potential measurement. Representative XPS spectra of a poly(MMA) substrate coated with the PMB-30 are shown in Figure 5.8. Carbon peaks attributed to the main chain and alkyl groups in the BMA units can be observed at 285.0 eV, with that due to an ether bond

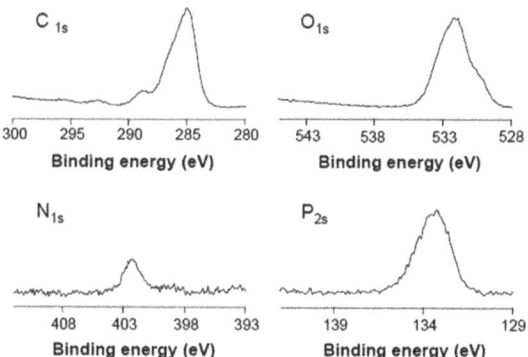

Figure 5.8 XPS spectra of PMB-30 coated on poly(MMA) substrate.

seen at 286.6 eV and an ester bond at 288.4 eV. A signal from an ammonium nitrogen atom is observed at 403.0 eV, and a phosphorus peak is evident at 133.5 eV due to the phosphate group. These XPS signals accurately correspond to the chemical structure of PMB-30.

The ability of a surface to be wetted with water is an important parameter that needs to be taken into account for biological applications. The contact angle that is formed between water and the substrate depends on the mobility of the molecules on the surface. Thus, the dynamic contact angle loop of an MPC surface has a significant hysteresis caused by movement of the phosphorylcholine groups in response to the environmental conditions.[25,59] AFM observation has indicated that the surface of a PMB-30 coating is uniformly flat when it is in aqueous solution. The surface ζ-potential of the PMB-30 reflects the state of the electrical charge of the MPC units. The values of the surface ζ-potential of glass, poly(BMA), and poly(HEMA) were identified to be −60, −40, and −16 mV, respectively. In contrast, coating with the PMB-30 gave a value of almost zero (−0.4 mV) due to hydration of the hydrophilic polymer and internal salt formation between trimethylammonium cations and phosphate anions. In general, inorganic salts affect the electronic charge of the polymer; however, in the case of the MPC polymers, no salt effects were observed.[60] This is further evidence for internal salt formation. When such inner salt formation occurs, the trimethylammonium cation and phosphate anion are situated close to one another, with the three methyl groups bound to the nitrogen atom of the MPC located outside of the polymer chains. This provides a site that can interact favorably with water through hydrophobic hydrations, inducing a more ordered water structure similar to that of free water in the bulk phase. Some biological aspects regarding the hydration state at a phospholipid bilayer have been reported. The water molecules make a "clathrate" around the trimethylammonium group of a phosphatidylcholine.[61–63] This is considered to be due to the unique solubility of poly(MPC) in an aqueous ethanol solution and the interfacial properties of PMB-30 in aqueous solution. Raman and IR spectroscopies carried out by Kitano *et al.*[64,65] and NMR spin–spin relaxation time analysis undertaken by

Morisaku *et al.*[66] support this hypothesis. The phosphorylcholine groups were found to possess a hydrophobic hydration layer that did not disturb the hydrogen bonding between the water molecules. PMB-30 is a promising polymer for use in the modification of a wide variety of substrates, providing a stable polymer layer that is resistant to protein adsorption and cell/bacteria adhesion, and can initiate an anti-inflammatory response.[67–71] The polymer has been filed in the Master Access File of the FDA, USA (# MAF2103, 2012) as LIPIDURE-CM5206, NOF Co., Japan.

5.2.3 Protein Adsorption Resistance on MPC Polymers

On initial contact of a biomaterial with bodily fluids such as blood, a layer of protein adsorbs to its surface. The amount of protein that adsorbs is one of the main factors that determines the biocompatibility and blood compatibility of the material. Therefore, surface modification is an important technique for improving the compatibility. Poly(MPC) layers have been demonstrated to provide a biocompatible interface, with many investigations into reducing protein adsorption having been carried out. Ishihara *et al.* hypothesized that the mechanism of protein resistance involved the structure of water on the poly(MPC) surface,[72] while Lu *et al.* similarly used the differences between bound and free water to explain the level of protein adsorption.[73] In general, adsorbed proteins and polymer surfaces have a fraction of bound water shared between them. The tightly adsorbed proteins, owing to these shared water molecules, become denatured by conformational change. The fraction of water present at the polymer surface has been evaluated using differential scanning calorimetry (DSC). The free water gave an exothermic peak at around 0 °C, and the free water fraction was calculated to give the water content in the equilibrium state. In the case of 36% water content, the PMB-30 contained a 0.69 fraction of free water, whereas poly(HEMA) had only 0.28 (Table 5.2). This result indicates that the hydrated PMB-30 had a large amount of free water.

Furthermore, the highly free water would provide stabilization for biomolecules such as enzymes and antibodies, even when adsorbed on the

Table 5.2 Relationship between hydration state and protein adsorption on the polymer surfaces.

Polymer	Water content[a] (%)	Free water fraction[b]	Amount of protein adsorbed on the surface ($\mu g/cm^2$) BSA	Fibrinogen	α-helix content of adsorbed protein (%)[d] BSA	Fibrinogen
Poly(HEMA)	40	0.34 (0.28)[c]	1.7 ± 0.7	3.4 ± 0.1	10	3
PMB-30	84	0.84 (0.69)	0.22 ± 0.1	1.1 ± 0.2	52	19
PMD-30	70	0.70 (0.62)	0.35 ± 0.1	1.2 ± 0.3	30	15

[a]Equilibrium hydration at 25°C.
[b]Free water content is measured equilibrium hydration.
[c]Free water content is at 36% water content.
[d]The α-helix contents of native BSA and fibrinogen are 54 % and 19 %, respectively.

surface. The stabilization of bovine serum albumin (BSA) was examined using circular dichroism (CD) spectroscopic measurements. The molar ellipticity of the free protein at 208 and 222 nm indicated α-helix and β-sheet secondary structures, respectively. In the case of BSA adsorbed on the PMB-30 surface, an almost identical spectrum was achieved, indicating that the conformation of the BSA was not significantly altered by the adsorption process. A poly(HEMA) coated surface, which exhibited a lower free water fraction, gave a significantly different spectrum, with the α-helix and β-sheet contributions decreasing with the amount of BSA that adsorbed. The level of BSA adsorption on the poly(HEMA) surface (1.7 μg cm^{-2}) was found to be eight times more than that on the PMB-30 surface (0.22 μg cm^{-2}). These results demonstrate that the PMB-30 surface provides a higher fraction of free water, resulting in suppression of protein adsorption and stabilization of the protein structure. Recently, a new type of MPC polymer surface has been prepared.[74–80] The use of surface-initiated atom transfer polymerization enabled a polymer brush layer to be formed at the surface. The amount of protein that adsorbed on such a surface was found to be less than 5 ng cm^{-2}. This value is below that which is necessary for platelet adhesion and subsequent thrombogenesis. This demonstrates that the added effect of the brush structure further improved the protein resistance of the MPC.

5.3 Functionalization of MPC Polymers

Research has been conducted with the aim of adding more specialized functionality to the MPC polymers. As discussed in Section 5.2.3, PMB-30 provides an excellent platform for reducing protein adsorption whilst maintaining protein secondary structure. Thus, if bioactive molecules such as enzymes, antibodies, and DNA could be incorporated with the PMB, the production of a bioactive surface could be possible. In Figure 5.9, systematic functionalization of the MPC polymers is demonstrated. Based on the chemical structure of PMB, new polymers have been designed to enable the conjugation of such biomolecules to a surface under mild conditions. In general, reactions involving biomolecules must be carried out in aqueous media at physiological pH and temperature owing to the poor stability of such structures. Therefore, active ester groups and the phenylboronic acid group would be useful for reacting with amino groups and sugar units in the biomolecules, respectively.

5.3.1 MPC Polymers for Bioconjugation

A water insoluble PMB-based polymer containing *p*-nitrophenyloxycarbonyl oligo(oxyethylene)methacrylate (MEONP) units (denoted as PMBN) has been designed and prepared with the aim of achieving an excellent biointerface. The incorporation of an active *p*-nitrophenyl ester group in the MEONP units enables a facile reaction with amino groups in biological molecules under mild conditions, forming a urethane bond that is connected to the

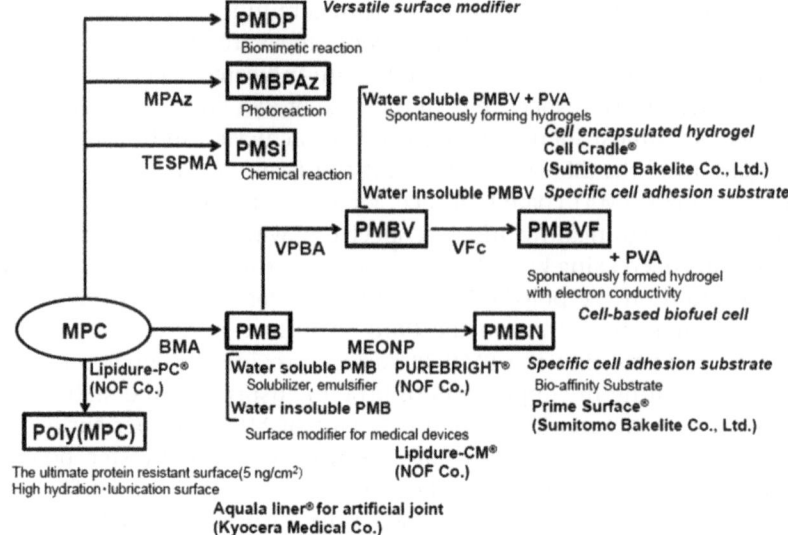

Figure 5.9   Chart of the molecular design and functionalization of MPC polymers.

methacrylate *via* the oxyethylene spacer.[81–83] The length of this spacer could be varied by changing the number of repeating units of the oxyethylene group in MEONP.[84] By increasing the length from 0.5 to 2 nm, the location of the immobilized biomolecules was altered. The phosphorylcholine groups were found to be at a position of 0.4 nm from the substrate, which was calculated by the molecular structure of the MPC unit. The spacer length was also found to play a key role in reducing background signals in γ-globulin (IgG)-immobilized biosensing applications. The bioconjugate reaction was accelerated under neutral or weak alkaline conditions (pH 7.4–8.0). By means of this PMBN polymer, biomolecules could be immobilized on a phosphorylcholine-enriched surface as a biomimic of the cell membrane.

The first report using PMBN involved the immobilization of an enzyme within an electrophoresis capillary.[81] The effect of the PMBN coating on the formation and dissociation of an enzyme–substrate complex of horseradish peroxidase (HRP) and 4-aminoantipyrine (AAP) or 3-(*p*-hydroxyphenyl)propionic acid (HPPA) was investigated, and regulation of the enzymatic activity of the conjugate was assessed. Using capillary electrophoresis, it was shown that the PMBN–AAP conjugate could form a complex with HRP, but that the conjugation decreased the quality of complex formation. From fluorescence studies, a complex between the PMBN–AAP conjugate and HRP could be seen to be easily dissociated by the addition of HPPA as an alternative substrate, while the complex between the non-conjugated AAP and HRP did not dissociate. Temporary blocking of the active site of the enzyme was realized by conjugating the substrate with PMBN, which provides a potential approach to reversibly blocking cellular receptors. It was speculated that ligands conjugated to PMBN could form a complex with the cellular receptor, but that this complex would easily dissociate. Additional research using polymer particles coated with PMBN revealed that the biological activity of biomolecules immobilized on the surface was retained, indicating that denaturation may have been suppressed. Owing to these useful characteristics, a PMBN-coated surface is now commercially available as a plastic plate for conjugation of biomolecules, named S-BIO PromeSurface® from Sumitomo Bakelite Co.

PMBN-coated surfaces are extremely thermally stable. Oligonucleotides immobilized on this substrate have been shown to be robust in boiling water, with no significant loss of hybridization activity during dissociation treatment.[85–87] This has allowed hybridization of the templates, extending the 3′-end of the immobilized DNA primers on the surface by DNA polymerase using deoxynucleotidyl triphosphates (dNTP) as extender units, releasing the templates by denaturalization, and then using the same templates for a second round of reactions, similar to that of the polymerase chain reaction (PCR) method. By repeating this cycle, a picomolar concentration range of the template oligonucleotide could be detected as a stable signal *via* the incorporation of labeled dUTP into the primers. This "multiple primer extension" (MPEX) method could be further extended as an alternative route for producing DNA microarrays for single nucleotide polymorphism (SNP) analyses *via* simple template preparation such as reverse transcription cDNA or restriction enzyme treatment of genomic DNA. The method utilizing the PMBN-coated plastic substrate has the potential to become a widely applicable tool for laboratories performing large-scale analyses, and for use as a DNA microarray platform. A microarray preparation kit for *in situ* PCR based on this technology is currently available from Sumitomo Bakelite Co.

It has been shown that nanometer-scale PMBN structures can be formed on the surface of an electrode using an electrospray deposition (ESD) method, enlarging the surface area available for conjugation of biomolecules. A much larger quantity of biomolecules could be immobilized on the ESD surface compared with a planar spin-coated surface. This resulted in a significant increase in the signal from an enzyme-linked immunosorbent assay (ELISA) carried out using immobilized enzymes.[88,89]

5.3.2 Surface Reactive MPC Polymers for Formation of a Durable Polymer Layer

The coating of a surface with PMB from solution is relatively simple, and the resulting layer can become tightly bound to the substrate on drying. A prehydration period over 5 h is needed to activate the surface functionalities of the PMB, although the amount of time depends on the thickness of the PMB layer.[90] However, such long prehydration processes cannot be applied to medical devices such as cardiovascular stents and blood separation devices. Thus, for practical applications it is desirable to use a more convenient and versatile method for immobilizing MPC polymers onto a surface. One suggested approach is the reduction of the time it takes to equilibrate the MPC and water, which could be achieved by increasing the proportion of MPC units in the polymer. However, owing to the resulting enhancement of the water solubility of the MPC polymer, a method by which to improve its fixation to the substrate would be necessary. MPC polymers have been designed to enable their effective attachment to pre-modified surfaces. For example, in order to achieve immobilization onto a metal oxide, poly[MPC-*co*-3-triethoxysilylpropyl methacrylate (TESPMA)] (PMSi) has been synthesized to enable a silane coupling reaction.[91] The PMSi was shown to react with metal oxide substrates, including glass, under mild conditions. An ethanolic solution of the PMSi could be cast on the substrate by simple dipping or spreading. The solvent was then evaporated and the substrate was heat-treated at 70 °C for 4 h. The triethoxysilyl group bound covalently to the oxide surface, forming a stably attached polymer layer. It was found that 10 unit mol% of triethoxysilylpropyl methacrylate was required for effective binding to the surface, with the remainder of the polymer made up of MPC. Thus, the surface was completely covered with MPC units, even when the thickness of the polymer layer was extremely thin (*ca.* 10 nm). In addition, the hydrophilic character of the poly(MPC) was relatively well retained, with the PMSi surface exhibiting a highly hydrophilic nature when in contact with water. Protein adsorption on the surface of the PMSi was significantly reduced in comparison to glass or other metal oxides, and cell adhesion was completely suppressed.

Using this methodology, Xu *et al.* treated the inner surface of microfluidic channels within a device made from glass for specific sensing of proteins.[92] Okahata *et al.* used PMSi for treatment of the surface of a quartz crystal microbalance (QCM) sensor to enhance the signal and reduce the noise level by lowering the level of non-specific protein binding.[93] Takeuchi *et al.* realized "cell-origami (cell force induced paper craft)", whereby the adhesion force of the cells on the substrate facilitated the fabrication of micrometer-sized devices spontaneously.[94] To attach cells to specific areas, the substrate was placed on the PMSi-treated surface. The cells then migrated to the specific substrate and adhered to it. The stability of the modification enabled the formation of covalent bonds and the excellent cell and protein resistance of the MPC units are promising for increasing sensing specificity.

An alternative approach to immobilize organic compounds such as MPC to metal substrates is the use of mussel-inspired chemistry. Mussels can rapidly and permanently adhere to all types of inorganic and organic surfaces in aqueous environments.[95] Such adhesive properties rely on repeats of the 3,4-dihydroxy-L-phenylalanine (DOPA) motif found in the foot protein of mussels. Although the exact mechanism of adhesion is not fully understood, it has been widely speculated that the 3,4-dihydroxyphenyl (DHP) group of DOPA is responsible for the adhesion. When a polymer with DHP groups was placed in contact with a metal substrate, a thin polymer film was observed to spontaneously deposit on the surface.[96] Functionalization of such a polymer was then able to impart new characteristics to the metal substrate.

Water-soluble MPC polymers containing DHP groups in the side chain (PMDP) have therefore been synthesized.[54,55,97] Surface modification of metal oxide substrates, *e.g.* titanium oxide, has been carried out using an aqueous solution of the polymer. The titanium alloy substrate was immersed in an aqueous solution of PMDP for different periods of time.[54] After immersion for 10 s, the substrate was dried under vacuum for observation using AFM. It was shown that such a polymer coating methodology was able to form an extremely smooth surface. The amount of polymer on the alloy substrate, as measured using QCM, was 354 ng cm^{-2}. These results showed that PMDP covered and adhered to the titanium alloy substrate immediately from the aqueous solution, forming a uniform coating *via* a simple dipping procedure. The hydrophilicity of the water-soluble PMDP was reflected by the dramatic change in water contact angle of the titanium dioxide from 54° to 0° on coating with the polymer. This low value was maintained even after the substrate had been immersed in water for 2 days, indicating that the PMDP remained on the substrate. These results strongly suggest that the DHP group in the PMDP played an important role in producing such a stable coating.

5.3.3 Photoreactive MPC Polymers for Simple Substrate Modification

It has been shown that immobilization of MPC polymers on substrates can be achieved *via* photoirradiation. Photoreaction is a powerful methodology for surface modification, enabling selective areas of substrates to be modified. Photoinduced chemical reactions are well known for painting and printing processes, and are a major component of the microfabrication methods used in semiconducting device production. It is essential that the energy for the photoreaction is provided in a rapid and convenient fashion, enabling effective modification under relatively dry conditions. There are many photoreactive molecules that can be incorporated into polymers to facilitate such surface modifications. Among these, the phenylazido group is a good candidate because the absorption wavelength for the reaction is 300 nm, and a high quantum yield can be achieved. The phenylazido group forms a nitrene and extracts a hydrogen radical from the methylene chain. Introduction of the phenylazido group into the MPC polymer has been carried out using two methods. The first of these is the chemical reaction between a carboxyl group of the MPC polymer and 4-azidoaniline. Konno *et al.* prepared poly(MPC-*co*-methacrylic acid) and reacted it with 4-azidoaniline in the presence of a water-soluble carbodiimide.[98,99] It was found that only 6 unit mol% of the phenylazido group was sufficient for binding the polymer to the surface of a polyalkene substrate by photoirradiation. They subsequently successfully achieved control of cell attachment to the substrates.

A second method involved the polymerization of a methacrylate with a phenylazido group in the side chain with MPC and BMA.[100,101] This procedure was found to be better for controlling the molecular structure, and also adding further functionality *via* copolymerization. In particular, the solubility of the obtained MPC polymer could be improved by changing the proportion of the MPC and BMA. Prior to the photoreaction, the polymer solution was spread on the surface of the substrate and was allowed to dry slightly. Thus, the wettability of the substrate was identified as being an important factor for producing a uniform coating. Taking this into account, Fukazawa *et al.* prepared a PMB-based polymer with 10 unit mol%.[102] A methacrylate monomer, 2-methacryloyloxyethyl 4-azidobenzoate (MPAz), was synthesized using a Schotten–Baumann reaction between HEMA and 4-azidobenzoyl chloride in the presence of triethylamine. The MPAz was then copolymerized with MPC and BMA using conventional radical polymerization. The resulting PMBPAz polymer was found to be soluble in ethanol and stable in the absence of light. In order to carry out the modification of the surface with the polymer, it was first dissolved in ethanol to a concentration of 0.20 wt%. The various substrates were washed using ethanol before surface modification. The various substrates were then immersed in the polymer solution for just 10 s, and then the solvent was evaporated at room temperature in air. Subsequently, the surfaces were irradiated with a UV lamp for 1.0 min (light wavelength: 254 nm, photo-intensity: 500 μJ cm^{-2}). The substrates included not only organic polymers, but also ceramics and metals. In the case of the ceramics and metals, silane-coupling chemistry was applied as a pretreatment for introduction of alkyl groups on the surface. The phenylazido groups reacted well with the surfaces, and the polymers were effectively immobilized. Accompanying this reaction, the wettability of the surface with water was dramatically increased. For example, by using PMBPAz with proportions of MPC, BMA, and MPAz of 79, 19, and 9 unit mol%, respectively, the water contact angle of a poly-ethylene surface decreased from 100° to 40°. The contact angle of a bubble of air in water with the substrate changed significantly from 100° to 170°, with the air bubble unable to attach to the surface after the photoreaction of the polyethylene with the PMBPAz. Protein adsorption and cell adhesion on the surface after the immobilization were seen to be suppressed. In combination with ink-jet printing technology, a cell-based array with a precise patterning was prepared using this methodology (Figure 5.10). Fukazawa *et al.* also prepared a molecularly printed surface for selective capturing of proteins and cells on polyethylene substrates.[100,101]

5.4   Biomedical Applications of MPC Polymers

5.4.1   Antithrombogenic Surface Coatings on Implantable Artificial Organs

It has been extensively shown that modification of surfaces with MPC polymers is effective for improving blood compatibility by suppressing protein adsorption, platelet adhesion, and platelet activation at the blood-contacting surface. MPC polymers have therefore been recently applied to

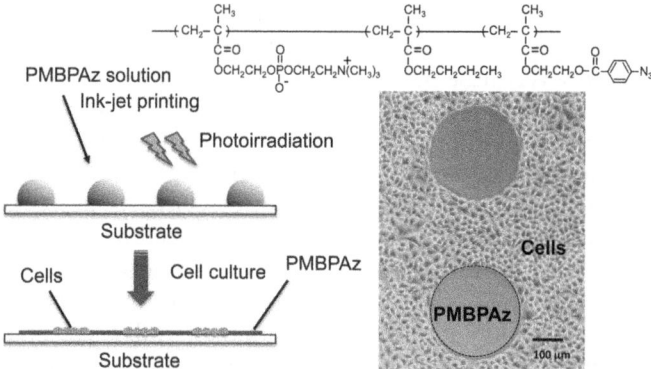

Figure 5.10 Fabrication of a patterned surface with ink-jet technology using a photoreactive MPC polymer (PMBAz) for regulation of cell attachment.

reduce the thrombogenicity of blood contacting devices and cardiovascular devices, including oxygenators (Sorin PrimO$_2$X®, Synthesis®), vascular prostheses (in development), stents (BiodivYsio AS®, Endeavor®), cardiopulmonary bypass devices, and ventricular assist devices (EVAHEART®). The treatment of acute closure of blood vessels increasingly involves the use of a coronary stent. To reduce the thrombogenicity of the metal stent frame, MPC copolymers have been coated onto their surfaces, with such devices being in use since 1995. The particular polymer used is based on PMD with a 2-hydroxypropyl methacrylate unit and a TESPMA unit incorporated as additional reactive moieties.[103] Reaction between triethoxysilyl groups and hydroxyl groups in the polymer facilitates stabilization of the polymer layer on the metal stent frame, even when the device is in its expanded state in the blood vessel and exposed to blood flow. Termed BiodivYsio AS® (Biocompatible Ltd.), this stent has been shown to be effective in preventing initial thrombus formation, and was approved by the FDA in 2000.[104,105] To date, over 850 000 such stents have been implanted in patients.

After the success of this particular device, a drug-eluting stent (DES) employing an MPC polymer (Endeavor® Sprint) has been developed.[106–108] This has been found to be even more efficient for reducing in-stent restenosis and target lesion revascularization. Because MPC polymers composed of alkyl methacrylates exhibit not only antithrombogenicity but also solute permeability, the polymer is highly effective as a coating material for DES. The bioactive reagent, Zotarolimus, which can suppress excessive cellular ingrowth, has been incorporated into the device. More than 22 000 Endeavor® Sprint stents having been implanted up to 2012.

The blood-contacting surface of oxygenators has also been modified with the MPC polymer, PMD, with the aim of reducing platelet activation during blood circulation (Sorin PrimO$_2$X®, Synthesis®). The porous hollow fiber polypropylene membranes that contact the blood were coated with the PMD.[109] As the affinity of the dodecyl group on the substrate was strong, and the solubility of the polymer in the aqueous medium was low, a stable coating on the substrate was achieved. Heparinized blood circulating within the hollow fibers gave very little platelet adhesion and activation, even though the membrane surface area was high. Sorin Synthesis® appeared to effectively reduce inflammation and prevent the blood coagulation cascade, resulting in lower transfusion and post-cardiopulmonary bypass inflammatory responses.

Implantation of a ventricular assist device (VAD) is associated with a number of complications, including infection, bleeding, and thromboembolism, with these biocompatibility concerns being a major reason why VADs are arguably underutilized for treating heart failure patients. In particular, platelet adhesion and thrombus formation still occur on the surface of the titanium from which the device is constructed, resulting in thromboembolism or an increased risk of bleeding due to the necessary use of anticoagulants. For the latest VADs, EVAHEART® VAD (Sun Medical Technology Research Co.) titanium alloy has been mainly used as the base material. The EVAHEART® is a centrifugal blood pump, with a pump size of 55×64 mm, made from pure titanium, and weighing 420 g. VADs provide circulatory support to those in end-stage heart failure *via* pulsatile or rotary actuation of blood. To improve the blood compatibility of the titanium alloy surface, one of the MPC polymers, PMB-30, has been applied as a surface treatment.[110,111] In May 2005, the first clinical implantation of EVAHEART® was successfully performed in a pilot study consisting of 18 patients. After obtaining permission for production in Japan, more than 100 patients have been supplied with the EVAHEART® as of December 2012. In Japan, the EVAHEART® has been covered by National Health Insurance since March 2011, and a clinical trial of the device is underway in the USA.

5.4.2 Prevention of Bioresponses at the Interface of Biomedical Devices

Soft contact lenses (SCLs) are one of the major products constructed from poly(HEMA) hydrogels. Since their introduction, a variety of such poly-(HEMA)-based hydrogels have been developed to improve lens properties. Both excellent protein adsorption resistance and wettability are required for SCLs, making MPC a good candidate as a suitable monomer.[112] Biocompatible Co. has produced MPC polymer-based SCLs that contain 20% MPC, 80% HEMA, and a small amount of crosslinker (Proclear®, omafilcon A),[113] with such lenses now commercially available from CooperVision Co. The Proclear® is the only contact lens for reduction of dry-eye syndrome that has been approved by the FDA in the USA.

5.4.3 Highly Lubricated Surfaces for Orthopedic Devices

The human body maintains the required lubrication of joints by the presence of synovia, which consists of many sugars, proteins, and lipids. It is important when restoring joint function that the same level of lubrication is achieved artificially. One approach to this is the development of neutral phosphatide artificial synovia that contains the same ingredients as the native synovial fluid and is adsorbed on the surface of an artificial joint. By employing the established procedure of surface-initiated graft polymerization of MPC, super low friction surfaces have been obtained. Ishihara *et al.* explored new methodology involving photoinduced polymerization of MPC on a polyethylene surface.[114] A water-insoluble photoinitiator, benzophenone (BP), was coated onto the surface of the substrate from acetone solution, and it was then exposed to UV light in an aqueous solution of MPC. Polymerization of the MPC was initiated from the surface of the polyethylene, and the molecular weight of the resulting poly(MPC) was observed to correlate with the photoirradiation period. Ishihara and Moro *et al.* applied this technique to the surface modification of a crosslinked ultrahigh molecular weight polyethylene (CLPE) cup of an artificial hip joint.[115–117] The surface was found to have high lubricity, also termed fluidic friction, and the wear of the poly(MPC)-grafted CLPE was significantly reduced compared with bare CLPE and that grafted with other hydrophilic polymers. Furthermore, the poly(MPC)-modified polyethylene particles did not stimulate osteoclast cells, which cause bone resorption, whereas normal polyethylene particles generated from friction wear of the artificial joint induced osteolysis. Very recently, a novel hip joint system with a poly(MPC)-grafted CLPE cup, Aquala® (Kyocera Medical Co.) has been approved in Japan, with the joint expected to be suitable as a long-term hip replacement, in comparison with conventional alternatives.[118] More than 6500 hip joints installed with this lining were implanted from November 2011 to March 2013.

Kyomoto and Ishihara found that self-initiated surface grafting of poly(MPC) could be achieved on poly(ether-ether-ketone) (PEEK), which is a group of super-engineering plastics with excellent mechanical properties and chemical stability (Figure 5.11).[119,120] The incorporation of a BP unit in the PEEK molecule enabled radicals to be generated by UV irradiation.

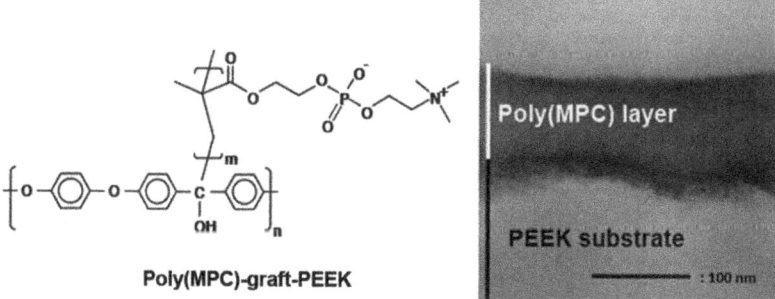

Poly(MPC)-graft-PEEK

Poly(MPC) layer

PEEK substrate

: 100 nm

Figure 5.11   Photoinduced and self-initiated graft polymerization of MPC on a PEEK substrate.

The PEEK specimens were soaked in an aqueous solution of MPC, and the photoinduced graft polymerization was subsequently carried out. Poly(MPC) graft chains were successfully generated on the PEEK surface without any reduction in bulk mechanical properties. PEEK has recently emerged as a leading high-performance thermoplastic candidate for replacing metal implant components, especially in the field of orthopedics and trauma. Improvements in the nonfouling, hydrophilic, and lubricant properties of PEEK by grafting of poly(MPC) may widen the scope of its potential for biomedical applications.[121]

5.5 Concluding Remarks

The fundamental properties of MPC polymers and their applications have here been summarized. These properties can be easily controlled in a systematic manner using conventional polymerization techniques. Many functions can be provided by the polymer molecules, resulting in their increasing use for biomedical and clinical medicine applications. A number of new implantable medical devices that employ MPC polymers have already been realized. The first synthesis of the MPC was reported in a 1978 Japanese article by a research group within the Tokyo Medical and Dental University (see Figure 5.3).[122] They used a condensation reaction between HEMA and 2-bromoethylphosphoryl dichloride, followed by quaternization with trimethylamine, to form the phosphorylcholine group. At that time, water was needed to convert the phosphoryl chloride group to the zwitterionic phosphorylcholine group, producing difficulties in crystallization of the product and a consequent low yield of pure material. Thus, a misunderstanding of the physical character and biological performance of MPC polymers was induced.[122–125] Subsequently, although improved processes have been reported, the purity of the product has remained fairly low due to hydration of the product.[123] The melting temperature of the MPC obtained by these procedures was low; the literature indicated that the temperature was 15 °C.[123,125] An excellent synthetic route with well-defined reaction conditions was first developed in 1987 and then reported in 1990.[7] This opened up the field of MPC polymer science for application in the biomedical and clinical medicine fields after high-purity MPC was successfully obtained. The science of MPC polymers has also initiated interest in zwitterionic polymer science, with attention recently being paid to the chemistry of carboxybetaine and sulfobetaine polymers.[126–130] The number of articles concerning these polymers increases year by year, and it is expected that biomedical devices incorporating them will soon be developed in a similar manner to the MPC polymers.

Acknowledgements

The authors thank Dr. Yuuki Inoue, Dr. Masayuki Kyomoto, and Dr. Tomohiro Konno (The University of Tokyo) for their kind help in the preparation of the manuscript.

References

1. S. J. Singer and G. L. Nicolson, *Science*, 1972, **175**, 720.
2. K. Ishihara, in *Encyclopedia of Polymer Science and Technology*, ed. H. F. Mark, Wiley, New York, 4th ed., 2014, in press; DOI: 10.1002/0471440264.pst574.
3. Y. Iwasaki and K. Ishihara, *Sci. Technol. Adv. Mater.*, 2012, **13**, 064101.
4. A. J. Lewis and A. W. Lloyd, in *Biomimetic, Bioresponsive, and Bioactive Materials: An Introduction to Integrating Materials with Tissues*, ed. M. Santin and G. Phillips, Wiley, New York, 2012, p. 95.
5. S. Monge, B. Canniccioni, A. Graillot and J. J. Robin, *Biomacromolecules*, 2011, **12**, 1973.
6. Y. Iwasaki and K. Ishihara, *Anal. Bioanal. Chem.*, 2005, **381**, 534.
7. K. Ishihara, T. Ueda and N. Nakabayashi, *Polym. J.*, 1990, **23**, 355.
8. K. Ishihara and N. Nakabayashi, *Jpn. Pat.*, 2 870 727, 1989.
9. R. S. Edmundson, *Chem. Ind. (London)*, 1962, 1828.
10. NOF Co. Ltd., *Jpn. Pat.*, 3 052 923, 2000.
11. Y. Iwasaki, K. Kurita, K. Ishihara and N. Nakabayashi, *J. Biomater. Sci., Polym. Ed.*, 1994, **6**, 447.
12. T. Oishi, H. Uchiyama, K. Onimura and H. Tsutsumi, *Polym. J.*, 1998, **30**, 17.
13. K. Ishihara, A. Fujiike, Y. Iwasaki, K. Kurita and N. Nakabayashi, *J. Polym. Sci., Part A: Polym. Chem.*, 1996, **34**, 199.
14. K. Sugiyama, K. Ohga and H. Aoki, *Makromol. Chem. Phys.*, 1995, **196**, 1907.
15. K. Sugiyama and K. Ohga, *Makromol. Chem. Phys.*, 1999, **200**, 1439.
16. A. Kros, M. Gerritsen, J. Murk, J. A. Jansen, N. A. J. M. Sommerdijk and R. J. M. Nolte, *J. Polym. Sci., Part A: Polym. Chem.*, 2001, **39**, 468.
17. T. Oishi, H. Uchiyama, K. Onimura and H. Tsutsumi, *Polym. J.*, 1998, **30**, 17.
18. Y. Kiritoshi and K. Ishihara, *Polymer*, 2004, **45**, 7449.
19. T. Sato, T. Miyoshi and M. Seno, *J. Polym. Sci., Part A: Polym. Chem.*, 2000, **38**, 509.
20. H. Wang, A. Miyamoto, T. Hirano, M. Seno and T. Sato, *Eur. Polym. J.*, 2004, **40**, 2287.
21. Y. Kiritoshi and K. Ishihara, *J. Biomater. Sci., Polym. Ed.*, 2002, **13**, 213.
22. Y. Kiritoshi and K. Ishihara, *Sci. Technol. Adv. Mater.*, 2003, **4**, 93.
23. Y. Matsuda, M. Kobayashi, M. Annaka, K. Ishihara and A. Takahara, *Polym. J.*, 2008, **40**, 479.
24. S. Edomondson, N. T. Nguyen, A. L. Lewis and S. P. Armes, *Langmuir*, 2010, **26**, 7216.
25. T. Ueda, H. Oshida, K. Kurita, K. Ishihara and N. Nakabayashi, *Polym. J.*, 1992, **24**, 1543.
26. A. L. Lewis, *Colloids Surf., B*, 2000, **18**, 261, and references cited therein.
27. K. Ishihara, T. Tsuji, Y. Sakai and N. Nakabayashi, *J. Polym. Sci., Part A: Polym. Chem*, 1994, **32**, 859.

28. K. Sugiyama, K. Shiraishi, K. Okada and O. Matsuo, *Polym. J.*, 1999, **31**, 883.
29. T. Uchida, T. Furuzono, K. Ishihara, N. Nakabayashi and M. Akashi, *J. Polym. Sci., Part A: Polym. Chem.*, 2000, **38**, 3052.
30. S. Yusa, K. Fukuda, T. Yamamoto, K. Ishihara and Y. Morishima, *Biomacromolecules*, 2005, **6**, 663.
31. B. Yu, A. B. Lowe and K. Ishihara, *Biomacromolecules*, 2009, **10**, 950.
32. N. Bhuchar, Z. Deng, K. Ishihara and R. Narain, *Polym. Chem.*, 2011, **2**, 632.
33. M. Ahmed, N. Bhuchar, K. Ishihara and R. Narain, *Bioconjugate Chem.*, 2011, **22**, 1228.
34. N. Bhuchar, R. Sunasee, K. Ishihara, T. Thundat and R. Narain, *Bioconjugate Chem.*, 2012, **23**, 75.
35. Y. Inoue, J. Watanabe, M. Takai and K. Ishihara, *J. Polym. Sci., Part A: Polym. Chem.*, 2005, **43**, 6073.
36. I. Y. Ma, E. J. Lobb, N. C. Billingham, S. P. Armes, A. L. Lewis, A. W. Lloyd and J. Salvage, *Macromolecules*, 2002, **35**, 9306.
37. Y. Ma, Y. Tang, N. C. Billingham, S. P. Armes, A. L. Lewis, A. W. Lloyd and J. P. Salvage, *Macromolecules*, 2003, **36**, 3475.
38. Y. Ma, Y. Tang, N. C. Billingham, S. P. Armes and A. L. Lewis, *Biomacromolecules*, 2003, **4**, 864.
39. Y. Li, Y. Tang, R. Narain, A. L. Lewis and S. P. Armes, *Langmuir*, 2005, **21**, 9946.
40. K. L. Thompson, I. Bannister, S. P. Armes and A. L. Lewis, *Langmuir*, 2010, **26**, 4693.
41. R. Kojima, M. C. Z. Kasuya, K. Ishihara and K. Hatanaka, *Polym. J.*, 2009, **41**, 370.
42. K. Ishihara, Y. Iwasaki and N. Nakabayashi, *Polym. J.*, 1999, **33**, 1231.
43. T. Konno, J. Watanabe and K. Ishihara, *J. Biomed. Mater. Res.*, 2003, **65A**, 210.
44. M. Wada, H. Jinno, M. Ueda, T. Ikeda, M. Kitajima, T. Konno, J. Watanabe and K. Ishihara, *Anticancer Res.*, 2007, **27**, 1431.
45. D. Soma, J. Kitayama, T. Konno, K. Ishihara, J. Yamada, T. Kamei, H. Ishigami, S. Kaisaki and H. Nagawa, *Cancer Sci.*, 2009, **100**, 1979.
46. J. Y. Kim, S. Kim, R. Pinal and K. Park, *J. Controlled Release*, 2011, **152**, 13.
47. T. Kano, C. Kakinuma, S. Wada, K. Morimoto and T. Ogihara, *Drug Metab. Pharmacokinet.*, 2011, **26**, 79.
48. T. Konno, J. Watanabe and K. Ishihara, *Biomacromolecules*, 2004, **5**, 342.
49. J. Watanabe and K. Ishihara, *Biomacromolecules*, 2006, **7**, 171.
50. Y. Goto, R. Matsuno, T. Konno, M. Takai and K. Ishihara, *Biomacromolecules*, 2008, **9**, 3252.
51. K. Ishihara, Y. Goto, R. Matsuno, Y. Inoue and T. Konno, *Biochim. Biophys. Acta, Gen*, 2011, **1810**, 268.
52. T. Konno and K. Ishihara, *Biomaterials*, 2007, **28**, 1770.
53. K. Ishihara, Y. Xu and T. Konno, *Adv. Polym. Sci.*, 2012, **247**, 141.

54. Y. Yao, K. Fukazawa, N. Huang and K. Ishihara, *Colloids Surf., B*, 2011, **88**, 215.
55. Y.-K. Gong, L.-P. Liu and P. B. Messersmith, *Macromol. Biosci.*, 2012, **12**, 979.
56. K. Nishio, R. Nakamura, X. Lin, T. Konno, K. Ishihara, S. Nakanishi and K. Hashimoto, *ChemPhysChem*, 2013, **14**.
57. Z. Li, T. Konno, M. Takai and K. Ishihara, *Biosens. Bioelectron.*, 2012, **34**, 191.
58. X. Lin, K. Nishio, T. Konno and K. Ishihara, *Biomaterials*, 2012, **33**, 8221.
59. T. Ueda, K. Ishihara and N. Nakabayashi, *J. Biomed. Mater. Res.*, 1995, **29**, 381.
60. M. Kikuchi, Y. Terayama, T. Ishikawa, T. Hoshino, M. Kobayashi, H. Ogawa, H. Masunaga, J.-I. Koike, M. Horigome, K. Ishihara and A. Takahara, *Polym. J.*, 2012, **44**, 121.
61. K. V. Damodaran and M. M. Kenneth Jr., *Langmuir*, 1993, **9**, 1179.
62. E. G. Finer and A. Darke, *Chem. Phys. Lipids*, 1974, **12**, 1.
63. K. V. Damodaran and M. M. Kenneth Jr., *Biophys. J.*, 1994, **65**, 1076.
64. H. Kitano, K. Sudo, K. Ichikawa, M. Ide and K. Ishihara, *J. Phys. Chem.*, 2000, **104**, 11425.
65. H. Kitano, M. Imai, T. Mori, M. Gemmei-Ide, Y. Yokoyama and K. Ishihara, *Langmuir*, 2003, **19**, 10260.
66. T. Morisaku, J. Watanabe, T. Konno, M. Takai and K. Ishihara, *Polymer*, 2008, **49**, 4652.
67. K. Ishihara, R. Aragaki, T. Ueda, A. Watanabe and N. Nakabayashi, *J. Biomed. Mater. Res.*, 1990, **24**, 1069.
68. K. Ishihara, N. P. Ziats, B. P. Tierney, N. Nakabayashi and J. M. Anderson, *J. Biomed. Mater. Res.*, 1991, **25**, 1397.
69. K. Ishihara and Y. Iwasaki, *Biomater. Appl.*, 1998, **13**, 111.
70. K. Ishihara, E. Ishikawa, Y. Iwasaki and N. Nakabayashi, *J. Biomater. Sci., Polym. Ed.*, 1999, **10**, 1047.
71. S.-I. Sawada, S. Sakaki, Y. Iwasaki, N. Nakabayashi and K. Ishihara, *J. Biomed. Mater. Res.*, 2003, **64A**, 411.
72. K. Ishihara, H. Nomura, T. Mihara, K. Kurita, Y. Iwasaki and N. Nakabayashi, *J. Biomed. Mater. Res.*, 1998, **39**, 323.
73. D. R. Lu, S. J. Lee and K. Park, *J. Biomater. Sci., Polym. Ed.*, 1991, **3**, 127.
74. W. Feng, S. Zhu, K. Ishihara and J. L. Brash, *Langmuir*, 2005, **21**, 5980.
75. W. Feng, S. Zhu, K. Ishihara and J. L. Brash, *Biointerphases*, 2006, **1**, 50.
76. W. Feng, J. L. Brash and S. Zhu, *Biomaterials*, 2006, **27**, 847.
77. M. Kobayashi, Y. Terayama, N. Hosaka, M. Kaido, A. Suzuki, N. Yamada, N. Torikai, K. Ishihara and A. Takahara, *Soft Mater.*, 2007, **3**, 740.
78. M. Chen, M. H. Briscoe, S. P. Armes and J. Klein, *Science*, 2009, **323**, 1698.
79. Y. Inoue and K. Ishihara, *Colloids Surf., B*, 2010, **81**, 350.

80. Y. Inoue, T. Nakanishi and K. Ishihara, *React. Funct. Polym.*, 2011, **71**, 350.
81. K. Takei, T. Konno, J. Watanabe and K. Ishihara, *Biomacromolecules*, 2004, **5**, 858.
82. K. Sakai-Kato, M. Kato, K. Ishihara and T. Toyo'oka, *Lab Chip*, 2004, **4**, 4.
83. K. Nishizawa, T. Konno, M. Takai and K. Ishihara, *Biomacromolecules*, 2008, **9**, 403.
84. J. Watanabe and K. Ishihara, *Nanobiotechnology*, 2008, **3**, 76.
85. K. Imai, Y. Ogai, D. Nishizawa, S. Kasai, K. Ikeda and H. Koga, *Mol. Biosyst.*, 2007, **3**, 547.
86. K. Kinoshita, K. Fujimoto, T. Yakabe, S. Saito, Y. Hamaguchi, T. Kikuchi, K. Nonaka, S. Murata, D. Masuda, W. Takada, S. Funaoka, S. Arai, H. Nakanishi, K. Yokoyama, K. Fujiwara and K. Matsubara, *Nucleic Acids Res.*, 2007, **35**, e3.
87. Y. Anzai, S. Saito, K. Fujimoto, K. Kinoshita and F. Kato, *J. Health Sci.*, 2008, **54**, 229.
88. K. Nishizawa, M. Takai and K. Ishihara, *Colloids Surf., B*, 2010, **77**, 263.
89. K. Nishizawa, M. Takai and K. Ishihara, *Methods Enzymol.*, 2011, **751**, 491.
90. A. Yamasaki, Y. Imamura, K. Kurita, Y. Iwasaki, N. Nakabayashi and K. Ishihara, *Colloids Surf., B*, 2003, **28**, 53.
91. T. Kihara, N. Yoshida, S. Mieda, K. Fukazawa, C. Nakamura, K. Ishihara and J. Miyake, *Nanobiotechnology*, 2007, **3**, 127.
92. Y. Xu, M. Takai, T. Konno and K. Ishihara, *Lab Chip*, 2007, **7**, 199.
93. H. Yoshimine, T. Kojima, H. Furusawa and Y. Okahata, *Anal. Chem.*, 2011, **83**, 8741.
94. K. Kuribayashi-Shigetomi, H. Onoe and S. Takeuchi, *PLoS One*, 2012, **7**, e51085.
95. H. Yamamoto, Y. Sakai and K. Ohkawa, *Biomacromolecules*, 2000, **1**, 543.
96. H. Lee, B. P. Lee and P. B. Messersmith, *Nature*, 2007, **448**, 338.
97. Y. Yao, K. Fukazawa, W. Ma, K. Ishihara and N. Huang, *Appl. Surf. Sci.*, 2012, **258**, 5418.
98. T. Konno, H. Hasuda, K. Ishihara and Y. Ito, *Biomaterials*, 2005, **26**, 1381.
99. T. Konno, K. Akita, K. Kurita and Y. Ito, *Biosci. Bioeng.*, 2005, **100**, 88.
100. K. Fukazawa and K. Ishihara, *Biosens. Bioelectron.*, 2009, **25**, 609.
101. K. Fukazawa, Q. Li, S. Seeger and K. Ishihara, *Biosens. Bioelectron.*, 2013, **40**, 96.
102. K. Fukazawa and K. Ishihara, *ACS Appl. Mater. Interfaces*, 2013, **5**, 6832.
103. A. L. Lewis, L. A. Tolhurst and P. W. Stratford, *Biomaterials*, 2002, **23**, 1697.
104. M. Galli, A. Bartorelli, F. Bedogni, N. DeCesare, S. Klugmann, L. Maiello, F. Miccoli, T. Moccetti, M. Onofri, V. Paolillo, R. Pirisi, P. Presbitero, P. Sganzerla, M. Viecca, S. Zerboni and G. Lanteri, *J. Invasive Cardiol.*, 2000, **12**, 452.

105. A. Bakhai, J. Booth, N. Delahunty, F. Nugara, T. Clayton, J. McNeill, S. W. Davies, D. C. Cumberland and R. H. Stables, *Int. J. Cardiol.*, 2005, **102**, 95.

106. J. Fajadet, W. Wijns, G. J. Laarman, K. H. Kuck, J. Ormiston, T. Münzel, J. J. Popma, P. L. Fitzgerald, R. Bonan and R. E. Kuntz, *Circulation*, 2006, **114**, 798.

107. A. Abizaid, A. J. Lansky, P. J. Fitzgerald, L. F. Tanajura, F. Feres, R. Staico, L. Mattos, A. Abizaid, A. Chaves, M. Centemero, A. G. Sousa, J. E. Sousa, M. J. Zaugg and L. B. Schwartz, *Am. J. Cardiol.*, 2007, **99**, 1403.

108. M. B. Leon, L. Mauri, J. J. Popma, D. E. Cutlip, E. Nikolsky, C. O'Shaughnessy, P. A. Overlie, B. T McLaurin, S. L. Solomon, J. S. Douglas Jr., M. W. Ball, R. P. Caputo, A. Jain, T. R. Tolleson, B. M. Reen 3rd, A. J. Kirtane, P. J. Fitzgerald, K. Thompson and D. E. Kandzari, *J. Am. Coll. Cardiol.*, 2010, **55**, 543.

109. G. J. Myers, D. R. Johnstone, W. J. Swyer, S. McTeer, S. L. Maxwell, C. Squires, S. N. Ditmore, C. V. Power, L. B. Mitchell, J. E. Ditmore, L. D. Aniuk, G. M. Hirsch and K. J. Buth, *J. Extra-Corpor. Technol.*, 2003, **35**, 6.

110. S. Kihara, K. Yamazaki, K. N. Litwak, P. Litwak, M. V. Kameneva, H. Ushiyama, T. Tokuno, D. C. Borzelleca, M. Umezu, J. Tomioka, O. Tagusari, T. Akimoto, H. Koyanagi, H. Kurosawa, R. L. Kormos and B. P. Griffith, *Artif. Organs*, 2003, **27**, 188.

111. T. A. Snyder, H. Tsukui, S. Kihara, T. Akimoto, K. N. Litwak, M. V. Kameneva, K. Yamazaki and W. R. Wagner, *J. Biomed. Mater. Res.*, *A*, 2007, **81**, 85.

112. T. Goda and K. Ishihara, *Expert Rev. Med. Devices*, 2006, **3**, 167.

113. J. L. Court, P. R. Redman, J. H. Wang, S. W. Leppard, V. J. Obyrne, S. A. Small, A. L. Lewis, S. A. Jones and P. W. Stratford, *Biomaterials*, 2001, **22**, 3261.

114. K. Ishihara, Y. Iwasaki, S. Ebihara, Y. Shindo and N. Nakabayashi, *Colloids Surf., B*, 2000, **18**, 325.

115. T. Moro, Y. Takatori, K. Ishihara, T. Konno, Y. Takigawa, T. Matsushita, U. I. Chung, K. Nakamura and H. Kawaguchi, *Nat. Mater.*, 2004, **3**, 829.

116. T. Moro, Y. Takatori, K. Ishihara, Nakamura and H. Kawaguchi, *Clin. Orthop. Relat. Res.*, 2006, **453**, 58.

117. T. Moro, H. Kawaguchi, K. Ishihara, M. Kyomoto, T. Karita, H. Ito, K. Nakamura and Y. Takatori, *Biomaterials*, 2009, **30**, 2995.

118. Y. Takatori, T. Moro, M. Kamogawa, H. Oda, S. Morimoto, T. Umeyama, M. Minami, H. Sugimoto, S. Nakamura, T. Karita, J. Kim, Y. Koyama, H. Ito, H. Kawaguchi and K. Nakamura, *J Artif. Organs*, 2013, **16**, 170.

119. M. Kyomoto and K. Ishihara, *ACS Appl. Mater. Interfaces*, 2009, **1**, 537.

120. M. Kyomoto, T. Moro, Y. Takatori, H. Kawaguchi, K. Nakamura and K. Ishihara, *Biomaterials*, 2010, **31**, 1017.

121. Y. Tateishi, M. Kyomoto, S. Kakinoki, T. Yamaoka and K. Ishihara, *J. Biomed. Mater. Res.*, *A*, 2013, in press; DOI: 10.1002/jbm.a.34809.

122. Y. Kadoma, N. Nakabayashi, E. Masuhara and J. Yamauchi, *Kobunshi Ronbunshu (Jpn. J. Polym. Sci. Technol.)*, 1978, **35**, 423.
123. T. Umeda, T. Nakaya and M. Imoto, *Makromol. Chem. Rapid Commun.*, 1982, **3**, 475.
124. S. Fukushima, Y. Kadoma and N. Nakabayashi, *Kobunshi Ronbunshu (Jpn. J. Polym. Sci. Technol.)*, 1983, **40**, 785.
125. T. Nakaya, H. Toyoda and M. Imoto, *Polym. J.*, 1986, **18**, 88.
126. A. B. Lowe, M. Vamvakaki, M. A. Wassall, L. Wong, N. C. Billingham, S. P. Armes and A. W. Lloyd, *J. Biomed. Mater. Res.*, 2000, **52**, 88.
127. S. L. West, J. P. Salvage, E. J. Lobb, S. P. Armes, N. C. Billingham, A. L. Lewis, G. W. Hanlona and A. W. Lloyd, *Biomaterials*, 2004, **25**, 1195.
128. Z. Zhang, T. Chao, S. Chen and S. Jiang, *Langmuir*, 2006, **22**, 10072.
129. W. Yang, T. Bai, L. R. Carr, A. J. Keefe, J. Xu, H. Xue, C. A. Irvin, S. Chen, J. Wang and S. Jiang, *Biomaterials*, 2012, **33**, 7945.
130. A. J. Keefe, N. D. Brault and S. Jiang, *Biomacromolecules*, 2012, **13**, 1683.

CHAPTER 6

Polyphosphoesters

VÉRONIQUE MONTEMBAULT AND LAURENT FONTAINE*

Institut des Molécules et des Matériaux du Mans, Equipe Méthodologie et Synthèse des Polymères, UMR CNRS 6283, Université du Maine, Avenue Olivier Messiaen, 72085 Le Mans Cedex 9, France
*Email: laurent.fontaine@univ-lemans.fr

6.1   Introduction

Polyphosphoesters (PPEs) are phosphorus-containing polymers with repeating phosphoester (P–O) linkages in the backbone (Scheme 6.1). The pentavalent nature of phosphorus allows the introduction of various side chains such as bioactive entities through P–O, P–N, and P–C bonds.

PPEs represent promising materials for biomedical applications because of their biodegradability, biocompatibility, versatile functionalities, and similarity to biomacromolecules such as nucleic acids.[1–5] Under physiological conditions, PPEs can degrade into small molecular components by hydrolytic or enzymatic cleavage of the phosphoester bonds.[6,7] PPEs have thus been extensively studied in biomedical applications such as drug delivery, scaffolds for tissue engineering, and gene carriers.[8–12] Moreover, it was found that some of the PPEs show thermoresponsivity in aqueous solution, giving rise to a lower critical solution temperature (LCST) type coacervate with potential applications within the biomedical field.[13–15]

PPEs can be prepared using a variety of synthetic routes: polycondensation (including transesterification and polyaddition), ring-opening polymerization (ROP), and post-polymerization modification (Scheme 6.2). Each of these techniques has advantages and limitations, which make them suitable

RSC Polymer Chemistry Series No. 11
Phosphorus-Based Polymers: From Synthesis to Applications
Edited by Sophie Monge and Ghislain David
© The Royal Society of Chemistry 2014
Published by the Royal Society of Chemistry, www.rsc.org

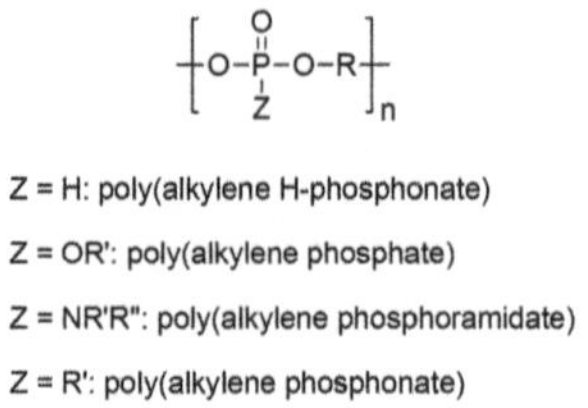

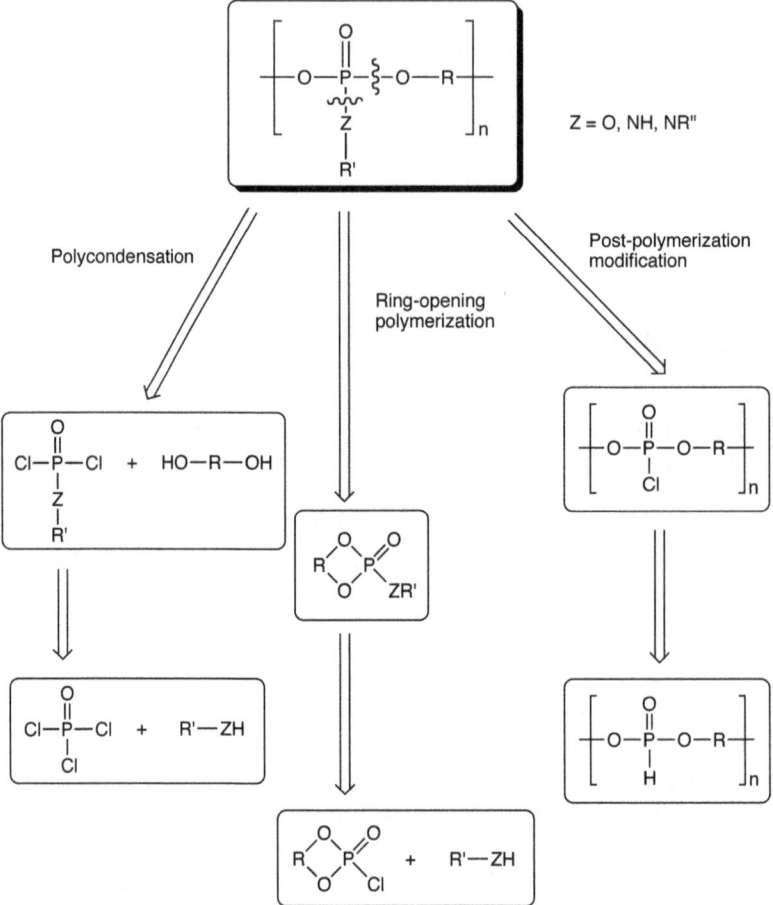

Z = H: poly(alkylene H-phosphonate)

Z = OR': poly(alkylene phosphate)

Z = NR'R": poly(alkylene phosphoramidate)

Z = R': poly(alkylene phosphonate)

Scheme 6.1   Polyphosphoesters: general structure.

Scheme 6.2   Retrosynthetic analysis of PPE synthesis.

for certain types of monomers and polymers. This chapter will describe synthesis of poly(alkylene H-phosphonate)s [P–H side chain, also named poly(alkylene hydrogenophosphonate)s], poly(alkylene phosphate)s (P–O side chain), poly(alkylene phosphoramidate)s (P–N side chain), and poly-(alkylene phosphonate)s (P–C side chain).

6.2 Synthesis by Polycondensation

Polycondensation between alkyl/aryl phosphoric dichlorides with diols or bisphenols (Scheme 6.2) is one of the oldest methods for obtaining PPEs. The earliest works about polycondensation were summarized more than 50 years ago in a book by Gefter.[16] Beside traditional melt and solution polycondensation in the presence of a tertiary amine as HCl scavenger,[6,17–25] this route also includes interfacial polycondensation.[8,26–31] The molecular weights of PPEs obtained through polycondensation between alkyl phosphoric dichlorides and aliphatic diols are often limited by side reactions (Scheme 6.3), such as chlorination of the diol, acidolysis of the side chain, and competitive formation of cyclic phosphorus esters in the case of 1,2- and 1,3-diols.[8,32–35]

Polycondensation of alkyl/aryl phosphoric dichlorides with hydroxy-telechelic oligomers [poly(glycolic acid), polylactide, and copolymers] has been used for the preparation of various PPEs with biomedical applications such as nerve guidance conduits[36] and microspheres for drug delivery systems.[37,38]

Polycondensation of phosphoric acid with ethylene glycol was investigated by Pretula *et al.*[39,40] in the presence of $Sc(OTf)_2$ or without any catalyst, in bulk or with added solvent (heptane or toluene). Low molecular weight polymers with different end-groups were formed, as found by NMR and MALDI-TOF MS ($n = 21$, Scheme 6.4).

The addition of phosphoric acid to epoxy groups of bisphenol A-based resins was studied by Penczek *et al.*[41] A series of oligomers with acidic end-groups was formed and the kinetics of the reaction was investigated using diglycidyl ether of bisphenol A as a model compound. Higher molecular weight linear or branched polyesters of phosphoric acid were prepared by the reaction of phosphoric acid with some diepoxides.[42–44] Liu *et al.*[45] employed the reaction of phosphoesters and epoxy groups to prepare poly(ethylene glycol) (PEG)-based hydrogels with degradable phosphoesters as cross-linked points. One of the main applications of polycondensation is related to the preparation of poly(alkylene H-phosphonate)s through polytransesterification of dialkyl phosphonates with diols (Scheme 6.5). The

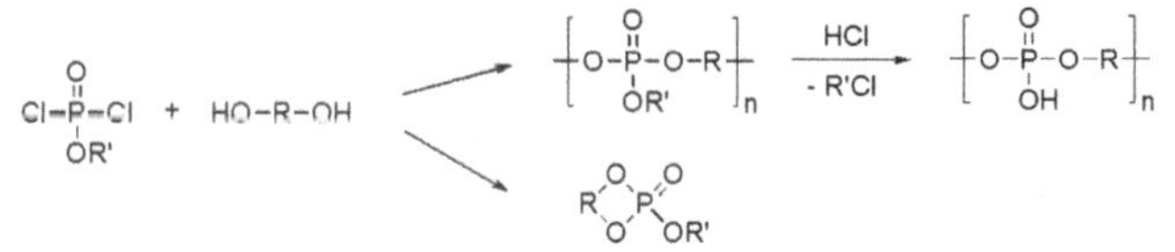

Scheme 6.3 Some side chain reactions during polycondensation of alkyl phosphoric dichlorides with aliphatic diols.

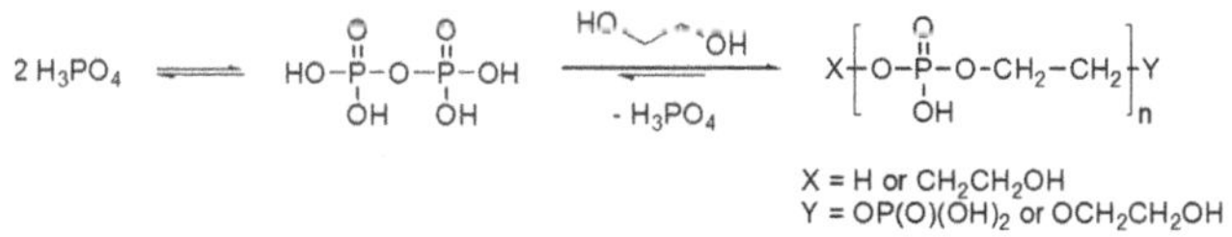

Scheme 6.4 Polycondensation of phosphoric acid with ethylene glycol.

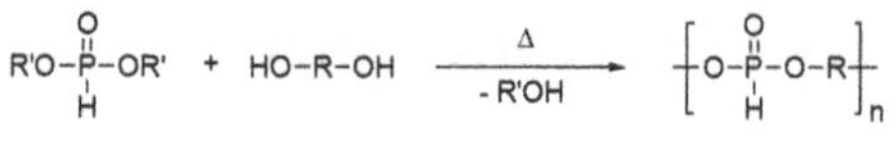

Scheme 6.5 Synthesis of poly(alkylene H-phosphonate)s through polytransesterification of dialkyl phosphonates with diols.

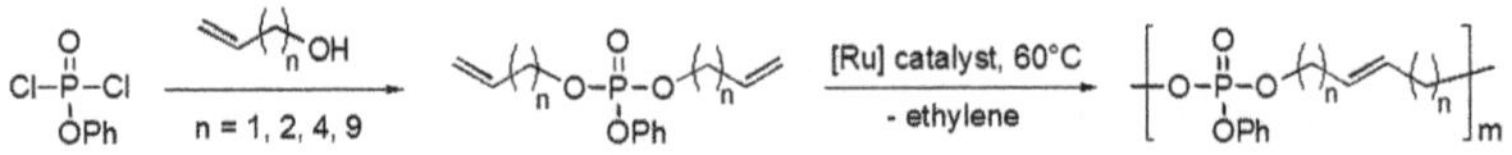

Scheme 6.6 Synthesis of unsaturated PPEs through ADMET polycondensation.

so-obtained poly(alkylene H-phosphonate)s are the starting points for further post-polymerization modification.[46–48] The reaction conducted by heating a mixture of dialkyl phosphonate and diol in the absence of a catalyst or in the presence of a catalytic amount of sodium metal (or sodium or potassium alcoholate) usually yields low molecular weight polymers (M_n about 10^3 g mol^{-1}).[49–51] By converting the phosphonic acid end-groups into methyl phosphonate groups using diazomethane, Branham *et al.*[52] could obtain molecular weights greater than 10^4 g mol^{-1}. A simpler alternative was found to carry out the transesterification in the presence of sodium carbonate, yielding polymers with M_n up to 4.5×10^4 g mol^{-1}.[53]

By using a two-stage polytransesterification process, Penczek *et al.* have found that higher molecular weights (M_n around 3×10^4 g mol^{-1}) can be obtained.[54–56] These authors also found that polycondensation of diphenyl phosphonate with diols gives high molecular weight polymers (M_n up to 3×10^4 g mol^{-1}) and allowed for elimination of the dealkylation side reaction.[57] Further transesterification of the phosphoester end-groups of such polymers with PEG monomethyl ether was used to prepare PEG-PPE-PEG triblock copolymers.[58] Phosphonic bis(dialkylamide)s have also been used as the comonomer in polycondensation with diols, leading to polymers with M_n up to 4×10^4 g mol^{-1}.[59]

Poly(oxyethylene H-phosphonate)s with various oligo(ethylene glycol) segment lengths have been prepared *via* the two-stage polytransesterification process between dimethyl phosphonate and PEGs of molecular weights 400–1000 g mol^{-1}, yielding polyphosphonates with molecular weights up to 10^4 g mol^{-1}.[60] Transesterification of PEG 400 with dimethyl phosphonate in a single step using microwave activation was reported, although yielding lower molecular weight polymers (M_n around 3×10^3 g mol^{-1}).[61] A new class of unsaturated PPEs were prepared through acyclic diene metathesis (ADMET)[62] polycondensation using ruthenium-based catalysts (Scheme 6.6). The so-obtained polymers have high molecular weights (7×10^3 to 5×10^4 g mol^{-1}) and polydispersities in the range 1.42–2.33.[63]

6.3 Synthesis by Ring-Opening Polymerization

Ring-opening polymerization (ROP) is the most efficient route to prepare PPEs with a controlled structure (Scheme 6.7). The first attempts to

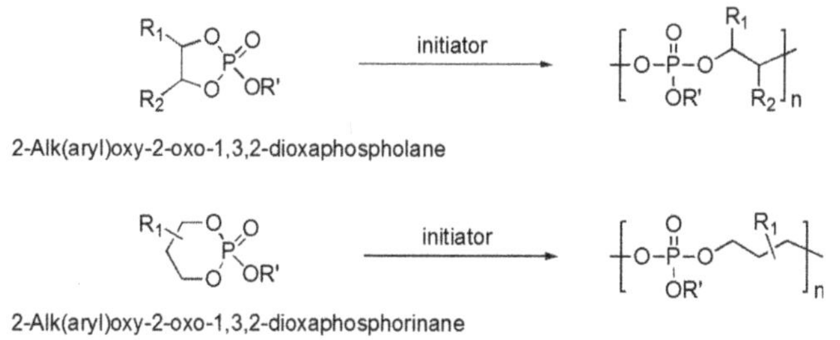

Scheme 6.7 Ring-opening polymerization of cyclic phosphorus esters.

polymerize cyclic phosphorus esters were made using water, sodium metal, ammonia, hydrogen chloride, or lithium aluminum hydride as catalysts and led to low molecular weight and rather ill-defined structures.[2] Although enzyme-catalyzed ROP using porcine pancreas lipase as a catalyst has been reported,[64] anionic initiators are the most useful systems to prepare PPEs with well-defined structures.[65] Most of the work devoted to the controlled/living ROP of cyclic phosphorus esters was pioneered by Penczek *et al.* in the 1970s and has been summarized in various publications.[2,5,65]

6.3.1 Monomer Synthesis

The monomers 2-alkoxy- or 2-aryloxy-2-oxo-1,3,2-dioxaphospholanes and 2-alkoxy- or 2-aryloxy-2-oxo-1,3,2-dioxaphosphorinanes are prepared through reaction of the appropriate 1,2- or 1,3-diol with an alkyl or aryl phosphoric dichloride[2] (Scheme 6.8, route A), or with phosphorus trichloride[66] followed by oxidation[67] and reaction with the corresponding alcohol[2,65,68] (Scheme 6.8, route B), or *via* transesterification with a dialkyl phosphonate[2,65,68–70] (Scheme 6.8, route C).

Cyclic alkylene phosphonates, namely 2-oxo-1,3,2-dioxaphospholanes and 2-oxo-1,3,2-dioxaphosphorinanes, can be obtained through either the controlled hydrolysis of the corresponding phosphorus(III) derivative[71,72] (Scheme 6.9, route A) or by a transesterification reaction between a diol and a dialkyl phosphonate[51,69,70,73–76] (Scheme 6.9, route B). An alternative method has been described, based on the reaction of cyclic ethers with phosphonic acid in the presence of acetic anhydride.[77]

6.3.2 ROP of Cyclic Alkylene Phosphates

Five-membered (2-alkoxy- or 2-aryloxy-2-oxo-1,3,2-dioxaphospholanes) and six-membered (2-alkoxy- or 2-aryloxy-2-oxo-1,3,2-dioxaphosphorinanes) cyclic alkylene phosphates have been the most studied monomers with the ROP process (Tables 6.1 and 6.2).[65]

ROP of monomers containing the THF ring[94] and deoxyribose and glucose rings[70] has also been described in the literature. The mechanisms,

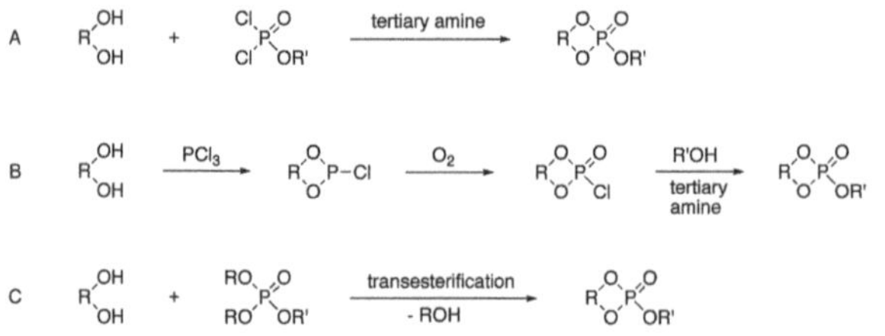

Scheme 6.8 Synthetic routes for cyclic phosphate monomer preparation.

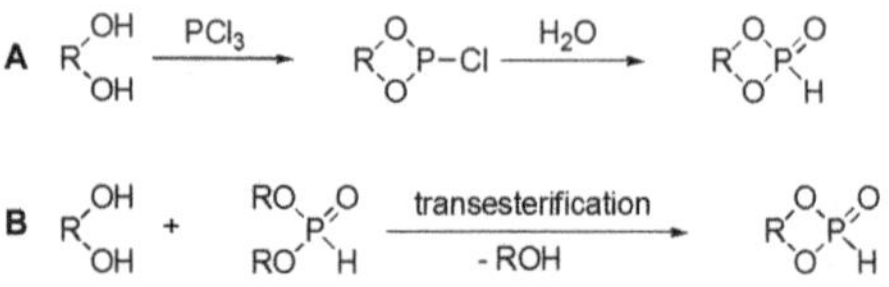

Scheme 6.9 Synthetic routes for cyclic phosphonate preparation.

kinetics, and thermodynamics of the ROP of most of those cyclic phosphorus monomers have been thoroughly investigated by Penczek et al.[2,65] Cationic polymerization of five-[79] and six-membered[91,93] cyclic alkylene phosphates initiated by triphenylcarbenium salts, CF_3SO_3H, CF_3SO_3Et, or $(CF_3SO_3)_2O$ gives low molecular weight (less than 10^3 g mol^{-1}) polymers. In contrast, higher polymers can usually be obtained using anionic initiators, which yield polymers with number-average molecular weights up to 1.5×10^4 g mol^{-1} (Tables 6.1 and 6.2). The high ring strain of five-membered rings (15–30 kJ mol^{-1})[95] makes the corresponding monomers much more reactive than their six-membered counterparts. However, the presence of substituents in the five-membered ring decreases the polymerizability of the corresponding monomer (see Table 6.1), as generally observed with most cyclic monomers through a decrease of the enthalpy of polymerization.[96] Tin(II) octoate, one of the most widely used initiators for cyclic ester polymerization, was first reported to induce polymerization of 2-ethoxy-2-oxo-1,3,2-dioxaphospholane in the presence of dodecanol by Xiao et al.[82] This ROP procedure yields PPEs with well-defined structures and controlled molecular weights (up 1.5×10^4 g mol^{-1} and polydispersity below 1.5).

To circumvent the potential problems associated with the use of metallic catalysts in the area of biomedical applications, organocatalysis was first investigated by Iwasaki et al.[15] using 1,8-diazabicyclo[5.4.0]undec-7-ene (DBU) and 1,5,7-triazabicyclo[4.4.0]dec-5-ene (TBD). This procedure was successfully used by other groups to elaborate complex PPE architectures.[89,90,97] Clément et al.[84] improved the procedure by using a combination of thiourea and DBU that was found to be an effective

Table 6.1 Ring-opening polymerization of 2-alk(aryl)oxy-2-oxo-1,3,2-dioxaphospholanes.

Monomer

$$\begin{array}{c} R_1 \\[-4pt] \diagdown \\[-6pt] {\rm O} \quad {\rm O} \\[-6pt] \diagup\kern-6pt\diagdown \\[-6pt] {\rm P} \\[-6pt] \diagup \quad \diagdown \\[-6pt] {\rm O} \qquad {\rm OR'} \\[-6pt] \diagup \\[-6pt] R_2 \end{array}$$

	R'	R_1	R_2	Initiator	T (°C)	Yield (%)	M_n (g mol^{-1})	Ref.
1	Me	H	H	Mg(C$_5$H$_5$)$_2$	−20	100	150 000	78,79
1	Me	H	H	C$_5$H$_5$N	100	65	50 000	80
2	Et	H	H	t-BuOK	3	61	12 600	81
2	Et	H	H	Al(i-Bu)$_3$	25	100	Up to 50 000	79
2	Et	H	H	Sn(Oct)$_2$/dodecanol	40	97[a]	Up to 15 000	82
3	n-Pr	H	H	Al(i-Bu)$_3$	25	100	100 000	79
4	i-Pr	H	H	Al(i-Bu)$_3$	25	100	35 000	79
4	i-Pr	H	H	DBU[b]	0	Up to 51[a]	Up to 8400	15
4	i-Pr	H	H	TBD[c]	0	Up to 50[a]	Up to 29 000	15
5	CH$_2$CCl$_3$	H	H	Al(i-Bu)$_3$	25	100	53 000	79
6	n-Bu	H	H	n-BuLi	25	100	54 000	79
7	t-Bu	H	H	MgEt$_2$	40	80	25 000	83
8	i-Bu	H	H	DBU + TU[d]/BnOH	50	98[a]	Up to 67 200	84
9	Ph	H	H	Al(i-Bu)$_3$	25	nd[e]	15 000	79,85
10	Me	Me	H	Et$_3$N	100	57	350	80
10	Me	Me	Me	Et$_3$N	100	90	310	80
11	Me	CH$_2$OAc	H	t-BuOK	25	nd[e]	3000	86
11	Me	CH$_2$OAc	H	Al(i-Bu)$_3$	−20	40	25 000	87
12	Et	CH$_2$OAc	H	t-BuOK	25	nd[e]	5000	86
13	Et	CH$_2$Cl	H	Al(i-Bu)$_3$	25	100	12 000	79
14	t-Bu	Me	H	MgEt$_2$	40	90	18 000	83
15	propargyl	H	H	Sn(Oct)$_2$/PCL-OH[f]	25	75	12 000	88
16	but-3-ynyl	H	H	DBU[b]/BnOH	0	80	Up to 18 000	89
17	2-ethylbutyl	H	H	TBD[c]/BnOH	0	99[e]	Up to 20 500	90

[a] Monomer conversion.
[b] DBU = 1,8-diazabicyclo[5.4.0]undec-7-ene.
[c] TBD = 1,5,7-triazabicyclo[4.4.0]dec-5-ene.
[d] DBU + TU = 1,8-diazabicyclo[5.4.0]undec-7-ene + thiourea.
[e] Not determined.
[f] PCL-OH = ω-hydroxypolyeaprolactone.

Table 6.2 Ring-opening polymerization of 2-alkoxy-2-oxo-1,3,2-dioxaphosphorinanes.

Monomer							
	R'	R_1	*Initiator*	T (°C)	*Yield (%)*	M_n (g mol^{-1})	*Ref.*
18	Me	H	Ph$_3$C$^+$PF$_6^-$	125	45	1200	91
18	Me	H	EtONa	135	10	900	92
19	Et	H	Ph$_3$C$^+$PF$_6^-$	105	51	800	93
19	Et	H	Al(Oi-Pr)$_3$	125	60	900	93
20	n-Pr	H	Ph$_3$C$^+$PF$_6^-$	140	77	800	93
21	t-Bu	4-Me	AlEt$_3$/0.9H$_2$O	40	nd^a	25 000	83

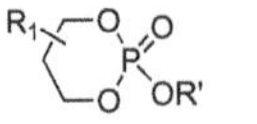

^{a}Not determined.

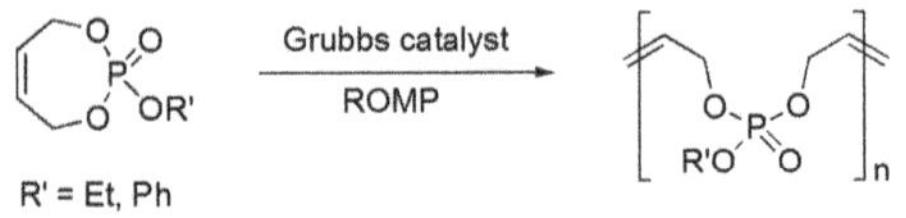

Scheme 6.10 Ring-opening metathesis polymerization of seven-membered cyclic phosphate monomers using a ruthenium-based (Grubbs) catalyst.

initiating system to prepare PPEs with predictable molecular weights (up to 70 000 g mol^{-1}) and narrow polydispersity (PDI < 1.10).

Besides the simple 2-alkoxy- and 2-aryloxy-2-oxo-1,3,2-dioxaphospholanes listed in Table 6.1, a variety of functional cyclic phosphorus monomers has also been described, including alkene-functionalized,[84,97,98] (meth)acrylate-functionalized,[99–103] PEG-ylated,[98,104] hydroxyl-functionalized,[105,106] protected hydroxyl-functionalized,[107] protected amino-functionalized,[108,109] protected thiol-functionalized,[110] and ATRP (atom transfer radical polymerization) initiator-functionalized.[111] A few reports have been devoted to the polymerization of seven- and eight-membered cyclic phosphates, bicyclic monomers and spirophosphoranes, which have been summarized in a book chapter by Lapienis and Penczek.[65] Random copolyphosphates have been synthesized by ring-opening copolymerization of cyclic phosphorus esters with a cyclic comonomer, including 2-alkoxy-2-oxo-1,3,2-dioxaphospholanes[13,14,100–102,111–114] D,L-lactide,[115,116] ε-caprolactone,[117] trimethylene carbonate,[118] and *p*-dioxanone.[119] The use of ring-opening metathesis polymerization (ROMP) to prepare PPEs was first reported[110] by Steinbach *et al.*[120] ROMP of novel seven-membered cyclic phosphate monomers and their copolymerization with cyclooctene was investigated using Grubbs ruthenium-based catalysts (Scheme 6.10). Using Grubbs third-generation catalyst, homopolymers and random copolymers having molecular weights up to 50 000 g mol^{-1} were prepared with polydispersity below 2.

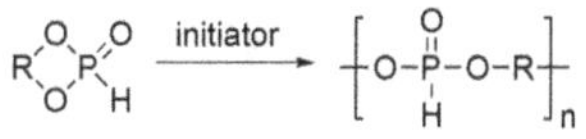

Scheme 6.11 ROP of cyclic alkylene phosphonates.

Table 6.3 Ring-opening polymerization of cyclic alkylene hydrogenophosphonates.

Monomer	Initiator	T (°C)	Yield (%)	Mn (g mol^{-1})	Ref.
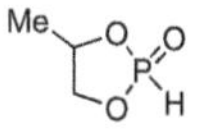 **22**	Al(i-Bu)$_3$	25	nd^a	nd^a	121,122
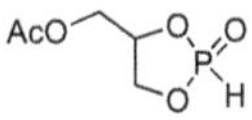 **23**	Al(i-Bu)$_3$	25	60	10 000	87
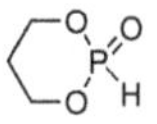 **24**	Al(i-Bu)$_3$	25	75	100 000	78,123,124

^{a}Not determined.

6.3.3 ROP of Cyclic Alkylene Phosphonates

Polymerization of cyclic alkylene phosphonates (Scheme 6.11) has been mostly investigated by Penczek *et al.* using triisobutylaluminum as the initiator (Table 6.3).

Investigation of the anionic polymerization of 2-oxo-1,3,2-dioxaphosphorinane (**24**, Table 6.3) showed that high molecular weight poly(alkylene H-phosphonate)s can be obtained (up to 10^5 g mol^{-1}) through a quasi-living process.[78,123] The microstructure of the polymers resulting from the anionic polymerization of racemic and optically active 4-methyl-2-oxo-1,3,2-dioxaphospholane (**22**, Table 6.3) was investigated by the same group on the basis of ^{1}H, ^{13}C, and ^{31}P NMR analyses.[122] Cationic polymerization of **22** was also achieved using a non-toxic proton-exchanged montmorillonite clay (maghnite-II$^+$) that is easily separated from the polymer product.[73]

6.3.4 Block, Graft, and Hyperbranched Copolymers *via* ROP

A variety of block, graft, and hyperbranched PPEs have been described that find applications mostly in the biomedical area,[125,126] including nanocarriers for drug delivery and PPE-based hydrogels.[127–129] PEG-based diblock[14,98,113] and triblock[130,131] copolymers have also been reported. PPE block copolymers based on degradable segments such as poly(ε-caprolactone) (PCL) and polylactide (PLA) have been thoroughly investigated for drug delivery, including PCL-based diblock[88,99,107,132–138] and triblock

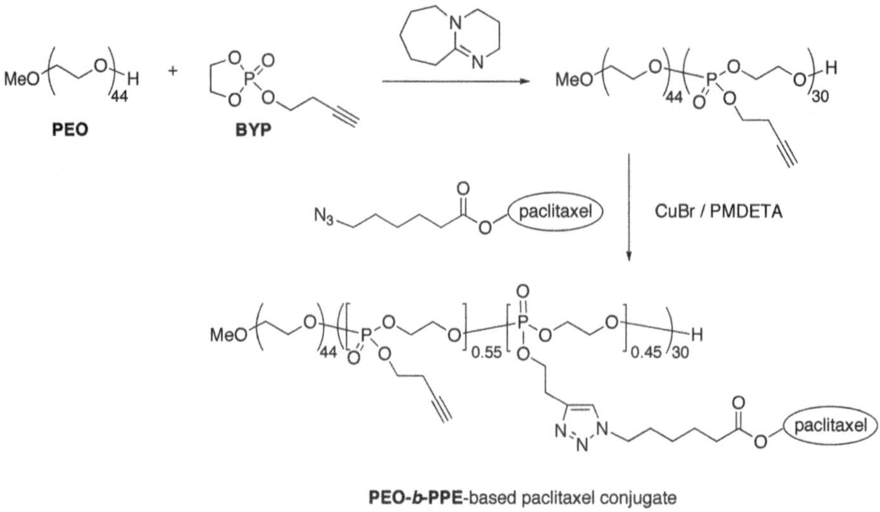

PEO-*b*-PPE-based paclitaxel conjugate

Scheme 6.12 Elaboration of poly(ethylene oxide)-*b*-PPE-based paclitaxel conjugates.[160]

PPEs,[103,108–110,139] PLA-*b*-PPEs,[140] and PLA-based triblock copolymers.[141,142] Double hydrophilic PPE-based block copolymers have been prepared by a combination of ROP and ATRP.[143,144]

Other macromolecular architectures include brush copolymers,[145–147] star-shaped block copolymers derived from PCL[148] or PLA,[149] miktoarm star polymers,[150,151] and hyperbranched PPE-derived polymers.[106,152–159]

The ability to prepare well-defined PPEs through metal-free ROP strategies has opened the way to highly complex PPE-based architectures for applications such as drug delivery. As an example, Scheme 6.12 shows a complex architecture obtained through the combination of organocatalyzed ROP of a "clickable" cyclic phosphorus monomer together (BYP, Scheme 6.12) with "click" chemistry to elaborate degradable nanoparticles made of poly-(ethylene oxide)-*b*-PPE-based paclitaxel conjugates for cancer treatment.[160]

6.4 Synthesis by Post-Polymerization Functionalization

The unique advantage of poly(alkylene H-phosphonate)s is the functionalization ability owing to the pentavalent nature of the phosphorus atoms, which makes it possible to conjugate many functional groups, as shown in Scheme 6.1. Thus, poly(alkylene phosphorothioate)s were prepared by sulfurization of the corresponding poly(alkylene H-phosphonate)s using a solution of sulfur in lutidine, and were shown to be able to act as inhibitors of viral diseases.[161] Similarly, a typical reaction of the P–H group with unsaturated C=O (Abramov reaction)[162,163] or C=C bonds (Pudovik

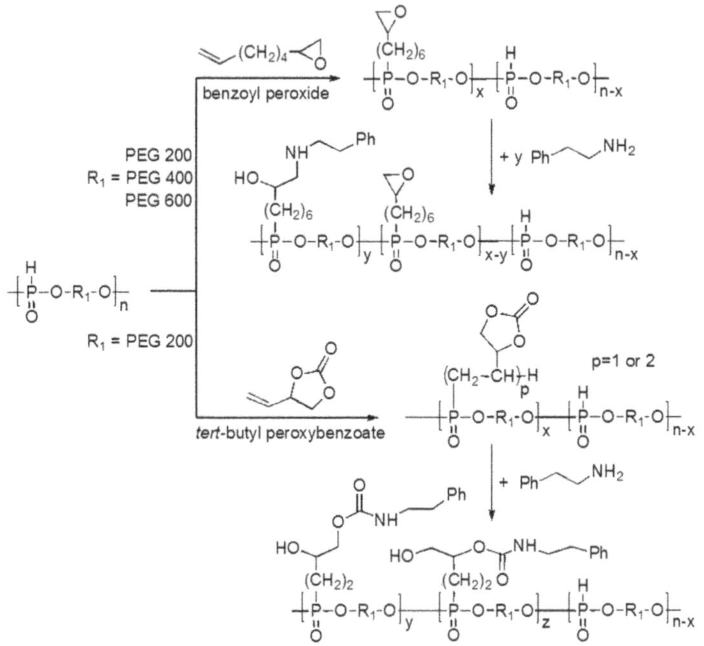

Scheme 6.13 Synthetic routes to poly(alkylene phosphonate)s.

reaction)[164,165], resulting into P–C bond formation, has been exploited to synthesize poly(alkylene phosphonate)s (Scheme 6.13). Reaction between poly(alkylene H-phosphonate)s and chloroacetone under phase-transfer catalysis conditions has been used for the preparation of poly(oxyethylene phosphonate)s bearing hydroxyl groups and oxirane moieties in the side chain through addition of the P–H bond to the carbonyl bond of chloroacetone. The resulting designed poly(oxyethylene phosphonate)s are of interest as drug carriers.[166] Reaction between poly(oxyethylene H-phosphonate)s and 1,2-epoxyoct-7-ene[167] or 4-vinyl-1,3-dioxolan-2-one[168] has resulted in poly(alkylene phosphonate)s bearing cyclic oxirane or carbonate side groups, respectively, that enable attachment of bioactive compounds, rending these PPEs as candidates for drug delivery applications.

Nevertheless, most of the work reported on the post-polymerization modification of poly(alkylene H-phosphonate)s concerns the functionalization with reactive pendant groups that include hydroxyl or amino groups, which make possible the introduction of a wide range of (bio)active molecules and leading to new reactive PPEs with tunable properties for biomedical applications.[3,126] Most of the post-polymerization functionalization methods involve the corresponding polymeric chlorophosphite, since it was early demonstrated that cyclic alkylene chlorophosphites such as 2-chloro-2-oxo-1,3,2-dioxaphospholane cannot be polymerized efficiently.[169]

6.4.1 Polyphosphates

Polyphosphates with P–O side chain functionalities are attractive bio-compatible and biodegradable biomaterials, as shown since the pioneering work of Penczek and co-workers at the end of 1970s.[2] They have investigated the synthesis of polyphosphates with repeating phosphoester units in the backbone of five or six atoms, resulting in a structural similarity to nucleic acid and teichoic acid.[5] Such biorelated polyphosphates can be obtained by oxidation of the corresponding poly(alkylene H-phosphonate)s (Scheme 6.14).[51,87,170] They were used as a result of their ability to bind cations and to actively transport cations of biological importance (Mg^{2+}, Ca^{2+}) through biomembranes,[171–175] mimicking the biomineralization process.[176]

Such polyphosphates have also been used to immobilize tertiary amines and tetraalkylammonium hydroxides (Scheme 6.15), leading to polysalts with long alkylene side groups.[55]

Biologically active substances have been linked to polyphosphates *via* ionic bonds between the substance and the polymer (Scheme 6.16). Troev and co-workers have reported the successful immobilization of drugs such as 2-phenylethylamine,[177] cysteamine,[178] and bendamustine hydro-chloride,[179] leading to enhanced stability and a lower cytotoxicity of the polymer-drug complexes than the pure analogs.

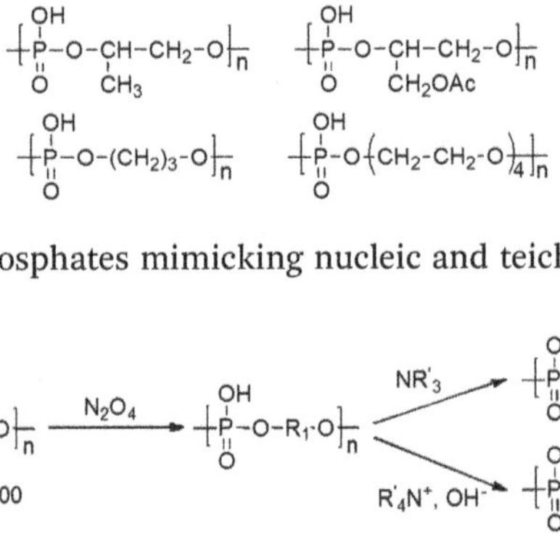

Scheme 6.14   Polyphosphates mimicking nucleic and teichoic acids.

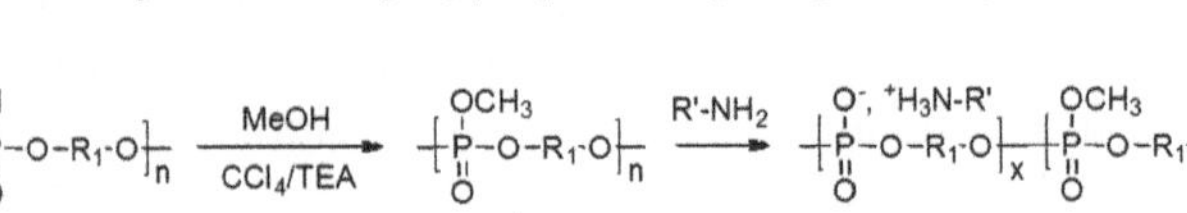

Scheme 6.15   Polysalts from poly(alkylene H-phosphonate)s.

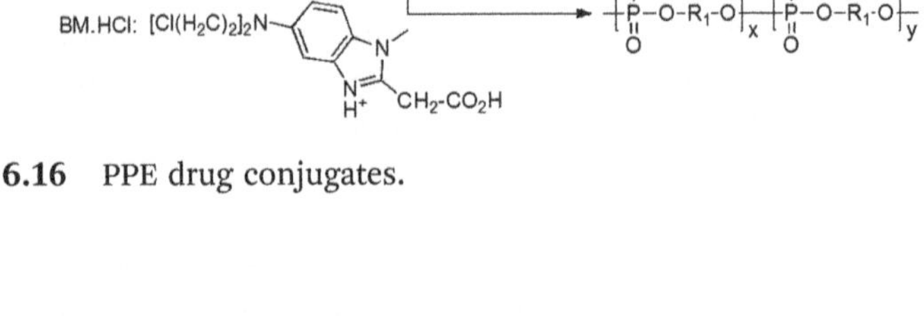

Scheme 6.16   PPE drug conjugates.

Since the 1980s, several groups have extensively investigated the biomedical applications of polyphosphates obtained by post-polymerization modification of poly(alkylene H-phosphonate)s to introduce reactive pendant groups, leading to new materials for controlled drug and gene delivery and tissue engineering applications.[3,126,180] Polyphosphates from poly(alkylene H-phosphonate)s as starting materials are mainly obtained according to a two-step process. The P–H bond is first converted into a P–Cl bond by chlorination. Poly(alkylene phosphate)s are then synthesized by reaction between the resulting poly(alkylene chlorophosphite) and hydroxyl compounds. Poly(alkylene chlorophosphate)s are commonly obtained *via* the Atherton-Todd reaction,[181,182] in which carbon tetrachloride or chlorine is used as an oxidizing reagent.[183] Troev *et al.*[184] have developed a novel and efficient method for the conversion of poly(alkylene H-phosphonate)s into the corresponding poly(alkylene chlorophosphate)s at room temperature in short reaction times using trichloroisocyanuric acid as chlorinating agent, without use of the carcinogenic carbon tetrachloride. Moreover, using different molar ratios between trichloroisocyanuric acid and poly(alkylene H-phosphonate), the degree of oxidation can be controlled and novel poly(alkylene H-phosphonate-*co*-chlorophosphate)s can be obtained.[184] Byrd *et al.*[185] reported a non-oxidative chlorination procedure using dichloro(2,4,6-tribromophenoxy)(2,2′-biphenoxy)phosphorane as the reactant to quantitatively convert poly(alkylene H-phosphonate)s into poly-(alkylene chlorophosphite)s in which the P atoms are in the +3 oxidation state. The resulting poly(alkylene chlorophosphite)s are versatile starting materials for a variety of new polymers, either *via* coordination to a transition metal through the unshared electron pair on the P and/or by nucleophilic substitution reactions of the chloride. Penczek and co-workers have converted a range of poly(alkylene H-phosphonate)s to poly(alkylene phosphate)s according to a two-step procedure by chlorination and subsequent reaction with alcohols in the presence of imidazole or pyridine (Table 6.4, entry 1), leading to amphiphilic polymers with a hydrophobic main chain and hydrophilic side chain, or *vice versa*, with self-assembling properties.[55,186] Poly(alkylene phosphate)s with N-containing bases in the side chain have been synthesized according to the same procedure by using bases with a hydroxyethyl spacer, since the P–N bond resulting from a direct coupling of N-containing bases to phosphorus is not stable enough for further applications of the poly(ester amide) in water (Table 6.4, entries 2–4).[183,187,188] These hydrophilic polymers with bases in the side chains, spaced similarly to their natural counterparts, can be used as models to mimic the functions of enzymes. The feasible modification at the P-center of poly-(alkylene H-phosphonate)s has been exploited to anchor molecules with specific properties. Fontaine *et al.*[189,190] have reported the synthesis of polyphosphates bearing chelating molecule of iron(III) according to an Atherton-Todd reaction between poly(propylene phosphonate) and 8-hydroxyquinoline (Table 6.4, entry 5). Poly(oxyethylene H-phosphonate)s have been converted into poly(oxyethylene H-phosphate)s with methoxy side chains, leading to a new type of

Table 6.4 Polyphosphates synthesized from poly(alkylene H-phosphonate)s by chlorination and subsequent reaction with an hydroxyl compound.

Entry	Structure of poly(alkylene H-phosphonate)	Reagents[a]	Structure of poly(alkylene phosphate)	Application	Ref.
1	$\left[\text{P-O-R}_1\text{-O}\right]_n$ with H and =O; R_1 = PEG, $(CH_2)x$, cyclohexyl, $-CH_2-\!\bigcirc\!-CH_2-$	1. Imidazole or pyridine 2. R_2OH	$\left[\text{P-O-R}_1\text{-O}\right]_n$ with OR_2 and =O; $R_2 = n\text{-}C_4H_9$, $n\text{-}C_9H_{19}$, $n\text{-}C_{16}H_{33}$, $(CH_2CH_2O)_{1-2}H$, $(CH_2CH(CH_3)O)_{1-2}H$, $(CH_2CH(C_2H_5)O)_{1-2}H$, CH_2CH_2OBn	Amphiphilic polymers	55,186
2	$\left[\text{P-O-(CH}_2)_3\text{-O}\right]_n$ with H and =O	1. Imidazole 2. N-Hydroxyethyl-imidazole	$\left[\text{P-O-(CH}_2)_3\text{-O}\right]_n$ with imidazolyl-ethyl-O and =O	–	183
3	$\left[\text{P-O-(CH}_2)_3\text{-O}\right]_n$ with H and =O	1. Imidazole 2. N-Hydroxyethyl-adenine	$\left[\text{P-O-(CH}_2)_3\text{-O}\right]_n$ with adenin-9-yl-ethyl-O and =O	Models of biopolymers containing nucleic and related bases to mimic the functions of enzymes	187
4	$\left[\text{P-O-(CH}_2)_3\text{-O}\right]_n$ with H and =O	1. Imidazole 2. N^1-Hydroxyethyl-uracil	$\left[\text{P-O-(CH}_2)_3\text{-O}\right]_n$ with uracil-1-yl-ethyl-O and =O	–	188

No.	Starting material	Reagents	Product	Application	Ref.
5	$\left[P(H)(=O)(O\text{-}CH\text{-}CH_2\text{-}O)\right]_n$, CH_3	quinolin-8-ol, base/CCl$_4$/DCM	quinolin-8-yl phosphoester	Chelating polymers	189,190
6	$\left[P(H)(=O)(O\text{-}CH_2\text{-}CH_2)_x O\right]_n$, x = 4; 9	1. EtN(i-Pr)$_2$/CCl$_4$ 2. MeOH	OCH$_3$ phosphoester	Organic biomarkers	191
7	$\left[P(H)(=O)(O\text{-}CH\text{-}CH_2\text{-}O)\right]_n$, CH_3	1. DMAP, CbzNR-(CH$_2$)$_m$OH 2. HCOOH, Pd/C, HCl	$(CH_2)_m$-NRH$_2^+$, Cl$^-$ phosphoester; R = H, m = 2; R = H, m = 6; R = CH$_3$, m = 2	Biodegradable gene carrier	9,11,192–194
8	$\left[P(H)(=O)(O\text{-}CH\text{-}CH_2\text{-}O)\right]_n$, CH_3	1. DMAP, PhCH$_2$-O(CH$_2$)$_2$OH 2. Pd/C, H$_2$/MeOH	OH phosphoester	Naked DNA-based gene therapy; Enhancement of gene expression in muscle	11,195
9	$\left[P(H)(=O)(O\text{-}CH_2\text{-}CH_2)_4 O\right]_n$	AZT, TEA/CCl$_4$/Cl(CH$_2$)$_2$Cl, MeCN	AZT phosphoester	Potent prodrug substance	196

[a]DCM = dichloromethane; DMAP = 4-(N,N-dimethylamino)pyridine; TEA = triethylamine.

luminescent polymer by coordination with ternary europium complexes, making them promising candidates for organic biomarkers (Table 6.4, entry 6).[191]

The biodegradable backbone of PPEs, their water solubility, and the possibility to anchor reactive pendant groups render these polymers attractive in biomedical applications. A series of polymers with a PPE backbone, containing different charge groups in the side chain connected to the backbone through a phosphate bond, have been synthesized by Leong and co-workers,[9,11,192,193] leading to new biodegradable gene carriers (Table 6.4, entry 7). The resulting PPE/DNA complexes release DNA in a sustained manner and enhance gene expression following intramuscular injection.[194] A water-soluble non-ionic PPE with hydroxyethyl side chains (Table 6.4, entry 8) has also been synthesized and has demonstrated intramuscular transfection efficiency of naked DNA co-delivered with this PPE.[11,195] Poly(alkylene H-phosphonate)s have shown to be promising candidates for an immobilization template. Indeed, Troev *et al.*[196] have reported the conjugation of 3′-azido-2′,3′-dideoxythymidine (AZT) with poly(oxyethylene H-phosphonate) under Atherton-Todd reaction conditions (Table 6.4, entry 9). This potent prodrug polymeric substance with high water solubility demonstrated significantly decreased toxicity in comparison with non-immobilized AZT.[196]

6.4.2   Polyphosphoramidates

Polyphosphoramidates with P–N side chain functionalities have been subjected to numerous studies since the pioneering work of Penczek *et al.*[183] Two approaches have been used to convert the P–H bond to the P–N bond, resulting in polyphosphoramidates with diverse amino groups in the polymer pendants. The first one is based on a two-step post-polymerization modification of poly(alkylene H-phosphonate)s consisting of a chlorination and a subsequent coupling with an amino-functionalized molecule. Such a strategy has been used to synthesize polyphosphoramidates with attached amino acids esters from glycine, L- and D,L-alanine, and valine (Table 6.5, entry 1).[186,197,198] These poly(alkylene phosphate) backbone polymers with phosphoramide-bonded side groups undergo selective hydrolysis at the amide bond in acidic and neutral conditions while retaining the backbone structure.[199] This highly pH-dependent hydrolysis behavior seems to support a potential application in the area of gene delivery and nerve guides. The second strategy, which has been widely used, consists of a direct conversion of P–H to a phosphoryl amide group through the Atherton–Todd reaction in the presence of CCl_4 as an oxidant. Fontaine *et al.* were the first to apply the Atherton–Todd reaction[181,182,200] for the post-polymerization modification of poly(alkylene H-phosphonate)s into polyphosphoramidates, and have anchored amino-functionalized active molecules with chelating properties[189,190] (Table 6.5, entry 2) or with therapeutic properties[201–203] (Table 6.5, entries 3 and 4).

Leong and co-workers have developed new polymeric gene carriers based on a polyphosphoramidate structure.[11,204–206] The pentavalency of the

Table 6.5 Polyphosphoramidates prepared from poly(alkylene H-phosphonate)s.

Entry	Structure of poly(alkylene H-phosphonate)	Reactants[a]	Structure of poly(alkylene phosphoramidate)	Application	Ref.
1	$\begin{array}{c}H\\|\\ -[P-O-R_1-O]_n\\|\\ O\end{array}$ $R_1 = (CH_2)_3$, $(CH_2CH_2O)_3CH_2CH_2$, $CH(CH_3)CH_2$	1. Cl_2/DCM 2. $H_2N-CHR'-COOR''$ $R' = H$, CH_3, $CH(CH_3)_2$ $R'' = CH_2CH_3$, CH_2Ph	$\begin{array}{c}CO_2R''\\|\\ HN-CHR'\\|\\ -[P-O-R_1-O]_n\\|\\ O\end{array}$	Gene delivery, nerve guides	186,197–199
2	$\begin{array}{c}H\\|\\ -[P-O-CH-CH_2-O]_n\\|\ \ \ \ \ \ \ \|\\ O\ \ \ \ \ CH_3\end{array}$	pyridyl-CH₂NH₂ base/CCl₄/DCM	pyridyl-CH₂-NH–$\begin{array}{c}|\\ -[P-O-CH-CH_2-O]_n\\|\ \ \ \ \ \ \ \|\\ O\ \ \ \ \ CH_3\end{array}$	Chelating polymers	189,190
3	$\begin{array}{c}H\\|\\ -[P-(O-CH-CH_2)_x O]_n\\|\ \ \ \ \ \ \ \ \ \ \|\\ O\ \ \ \ \ \ \ \ \ R_1\end{array}$ $R_1 = CH_3$; $x = 1$ $R_1 = H$; $x = 4$	$R'NH_2$/base; CCl_4/DCM	$\begin{array}{c}NHR'\\|\\ -[P-(O-CH-CH_2)_x O]_n\\|\ \ \ \ \ \ \ \ \ \|\\ O\ \ \ \ \ \ \ \ R_1\end{array}$ $R' = p\text{-}C_6H_4CO_2Et$, $(CH_2)_2Ph$	Drug carriers	202
4	$\begin{array}{c}H\\|\\ -[P-(O-CH_2-CH_2)_4-O]_n\\|\\ O\end{array}$	$HN(CH_2CH_2Cl)_2$, base/CCl₄/DCM	$(ClCH_2CH_2)_2N-\begin{array}{c}|\\ -[P-(O-CH_2-CH_2)_4-O]_n\\|\\ O\end{array}$	Antitumor drug	203
5	$\begin{array}{c}H\\|\\ -[P-O-CH-CH_2-O]_n\\|\ \ \ \ \ \ \ \|\\ O\ \ \ \ \ CH_3\end{array}$	1. $CbzNR(CH_2)_mNH_2/CCl_4$ 2. H_2, Pd/C	$\begin{array}{c}HN^{(CH_2)_m}NRR''\\|\\ -[P-O-CH-CH_2-O]_n\\|\ \ \ \ \ \ \ \|\\ O\ \ \ \ \ CH_3\end{array}$ $R', R'' = H$; $m = 2, 3, 4$ $R' = H$, $R'' = CH_3$; $m = 2$ $R', R'' = CH_3$; $m = 2$	Cationic polymeric gene carriers	11,204

Table 6.5 (*Continued*)

Entry	Structure of poly(alkylene H-phosphonate)	Reactants[a]	Structure of poly(alkylene phosphoramidate)	Application	Ref.
6	$\begin{matrix} H \\ \mid \\ \text{P-O-CH-CH}_2\text{-O} \\ \parallel \\ O \quad\;\; \mid \\ \;\;\;\; CH_3 \end{matrix}_n$	1. TFA-NH(CH$_2$)$_x$NH-(CH$_2$)$_y$NH-TFA 2. NH$_3$, H$_2$O	$\begin{matrix} (CH_2)_y NH_2 \\ \mid \\ N\text{-}(CH_2)_x NH_2 \\ \mid \\ \text{P-O-CH-CH}_2\text{-O} \\ \parallel \\ O \quad\; CH_3 \end{matrix}_n$ $x = y = 2,\ 3\ \ x = 2,\ 4;\ y = 3$		11,205–207
7	$\begin{matrix} H \\ \mid \\ \text{P}\left(\text{O-CH}_2\text{-CH}_2\right)_{/3}\text{O} \\ \parallel \\ O \end{matrix}_n$	H$_2$NR, 2H$_2$O; CCl$_4$/DMF R = (CH$_2$)$_3$NH(CH$_2$)$_2$-SP(O)(OH)$_2$, (CH$_2$)$_3$NH(CH$_2$)$_2$SH	$\begin{matrix} Z \\ \mid \\ \text{P}\left(\text{O-CH}_2\text{-CH}_2\right)_{/3}\text{O} \\ \parallel \\ O \end{matrix}_n$ Z = HNR; O$^-$ NH$_3$R$^+$; OH	Radioprotective effect	208

[a]DCM = dichloromethane; DMF = *N,N*-dimethylformamide; TFA = trifluoroacetic acid.

phosphorus atom in poly(alkylene H-phosphonate)s allowed incorporation of positive charges on the side chains by an Atherton–Todd reaction. The authors have designed polyphosphoramidates containing different types of amino groups in the pendant chains, with the aim of investigating the effect of different charge groups on the transfection abilities of the carriers (Table 6.5, entries 5 and 6). They have demonstrated that polyphosphoramidate carriers with primary amino groups were more efficient compared with other types of charged groups, based on assessing a series of polyphosphoramidates that have an identical backbone, the same side chain spacer, and a similar molecular weight.[11,204] Furthermore, polyphosphoramidates with branched side chains have proven to exhibit much higher gene transfer efficiency than linear ones.[11,205,206] A polyphosphoramidate/DNA complex with comparable transfection efficiency as the well-known poly(ethyleneimine)/DNA complex has been obtained by tuning the structure of the polyphosphoramidate.[207]

Troev *et al.* have reported the successful immobilization of aminothiol radioprotective agents used in cancer radiotherapy on poly(oxyethylene H-phosphonate)s through a covalent bond, *via* an Atherton–Todd reaction coupling, an ionic bond, and physical complexation (Table 6.5, entry 7).[208]

6.4.3 Complex Macromolecular Architectures From Poly(alkylene H-Phosphonate)s

Poly(alkylene H-phosphonate)s represent a wide range of biodegradable polymers with repeating phosphoester linkages in the backbone and have been inserted in a variety of complex macromolecular architectures (Scheme 6.17).

Penczek *et al.* have reported the synthesis of block and graft copolymers from poly(alkylene H-phosphonate)s. They have elaborated a triblock copolymer, PEG-*b*-poly(alkylene phosphate)-*b*-PEG, through transesterification of the methyl H-phosphonate end-groups of a poly(alkylene H-phosphonate) with PEG monomethyl ether and subsequent conversion of the H-phosphonate central blocks into the acidic blocks of diesters of phosphoric acid by oxidation (Scheme 6.17). These first ionic-nonionic block copolymers reported in the literature were successfully used as modifiers of the crystallization of $CaCO_3$.[5,58] They have also exploited the pendant functionality ability of poly(alkylene H-phosphonate)s. Introduction of biocompatible polymeric PEG pendant groups has been reported. Poly(alkylene H-phosphonate)s have first been functionalized into the corresponding poly(alkylene phosphate)s. The hydroxyl functionality has then been used to initiate the polymerization of ethylene oxide according to a "grafting from" strategy, leading to poly(alkylene phosphate)-*g*-PEGs with an average graft length of 1–2 units (Scheme 6.17).[55]

Zhuo and co-workers have synthesized PPE linkage-containing hydrogels based on PEG by photo-initial crosslinking polymerization of the

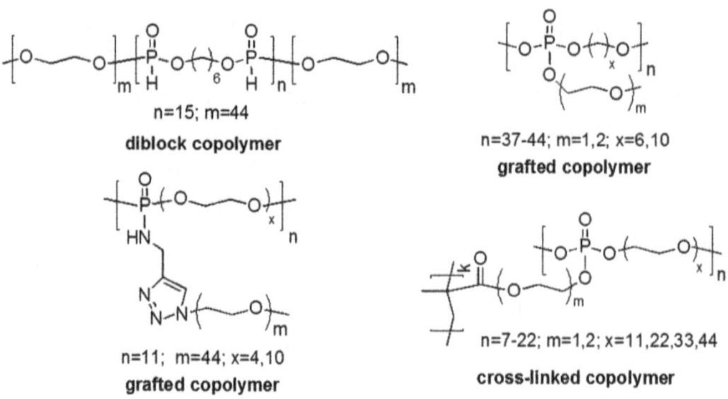

n=15; m=44
diblock copolymer

n=37-44; m=1,2; x=6,10
grafted copolymer

n=11; m=44; x=4,10
grafted copolymer

n=7-22; m=1,2; x=11,22,33,44
cross-linked copolymer

Scheme 6.17 Examples of macromolecular architectures from poly(alkylene H-phosphonate)s.

methacryloyl groups in the side chain of polyphosphates prepared by post-polymerization modification of the hydroxyl groups of the former poly(alkylene phosphate)-*g*-PEGs with methacryloyl chloride (Scheme 6.17).[145] These hydrogels were loaded with 5-fluorouracil for the investigation of drug release profiles. Fontaine and co-workers have synthesized poly(alkylene phosphate)-*g*-PEGs according to the versatile and tunable Cu(I)-catalyzed Huisgen 1,3-dipolar cycloaddition "click" reaction using the "grafting onto" approach.[209] First, a poly(alkylene phosphate) with pendant acetylene groups was obtained by the quantitative conversion of the P–H groups of a poly(alkylene H-phosphonate) through an Atherton-Todd reaction with *N*-propargylamine. The "click" functionalization of this acetylene-containing poly(alkylene phosphate) with an ω-azido PEG monomethyl ether 2000 resulted in a poly(alkylene phosphate)-*g*-PEG (Scheme 6.17).[209]

6.5 Conclusion

A variety of synthetic routes has been investigated up to now for the preparation of PPEs. With the advent of efficient and robust chemistries such as "click" chemistry and controlled polymerization processes using organocatalysis, PPEs represent nowadays a class of easily accessible and versatile biodegradable polymers with a wide range of properties and applications, including drug delivery systems and gene transfection. By combination of the various methodologies that are now available to the polymer chemists, PPEs will undoubtedly find new applications in the area of polymeric materials with specific and tunable properties, and with high added value.

References

1. P. Klosinski and S. Penczek, *Adv. Polym. Sci.*, 1986, **79**, 139.
2. S. Penczek and P. Klosinski, in *Models of Biopolymers by Ring-Opening Polymerization*, ed. S. Penczek, CRC Press, Boca Raton, 1990, p. 291.

3. Y. C. Wang, Y. Y. Yuan, J. Z. Du, X. Z. Yang and J. Wang, *Macromol. Biosci.*, 2009, **9**, 1154.
4. S. Monge, B. Canniccioni, A. Graillot and J. J. Robin, *Biomacromolecules*, 2011, **12**, 1973.
5. S. Penczek, J. B. Pretula, K. Kaluzynski and G. Lapienis, *Isr. J. Chem.*, 2012, **52**, 306.
6. M. L. Renier and D. H. Kohn, *J. Biomed. Mater. Res.*, 1997, **34**, 95.
7. J. Wen, H. Q. Mao, W. P. Li, K. Y. Lin and K. W. Leong, *J. Pharm. Sci.*, 2004, **93**, 2142.
8. B. I. Dahiyat, M. Richards and K. W. Leong, *J. Controlled Release*, 1995, **33**, 13.
9. Z. Zhao, J. Wang, H. Q. Mao and K. W. Leong, *Adv. Drug Delivery Rev.*, 2003, **35**, 483.
10. M. V. Chaubal, A. Sen Gupta, S. T. Lopina and D. F. Bruley, *Crit. Rev. Ther. Drug Carrier Syst.*, 2003, **20**, 295.
11. H. Q. Mao and K. W. Leong, *Adv. Genetics*, 2005, **53**, 275.
12. S. W. Huang and R. X. Zhuo, *Phosphorus, Sulfur Silicon Relat. Elem.*, 2008, **183**, 340.
13. Y. Iwasaki, C. Wachiralarpphaithoon and K. Akiyoshi, *Macromolecules*, 2007, **40**, 8136.
14. Y. C. Wang, L. Y. Tang, Y. Li and J. Wang, *Biomacromolecules*, 2009, **10**, 66.
15. Y. Iwasaki and E. Yamaguchi, *Macromolecules*, 2010, **43**, 2664.
16. E. L. Gefter, *Organophosphorus Monomers and Polymers*, Pergamon, Oxford, 1962.
17. D. Derouet, T. Piatti and J. C. Brosse, *Eur. Polym. J.*, 1987, **23**, 657.
18. D. J. Liaw and W. C. Shen, *Polymer*, 1993, **34**, 1336.
19. D. J. Liaw and D. W. Wang, *React. Funct. Polym.*, 1996, **30**, 309.
20. D. J. Liaw, *J. Appl. Polym. Sci.*, 1997, **65**, 59.
21. D. J. Liaw, *J. Polym. Sci., Part A: Polym. Chem.*, 1997, **35**, 2365.
22. M. V. Chaubal, B. Wang, G. Su and Z. Zhao, *J. Appl. Polym. Sci.*, 2003, **90**, 4021.
23. A. Sen Gupta and S. T. Lopina, *Polymer*, 2004, **45**, 4653.
24. A. Sen Gupta and S. T. Lopina, *Polymer*, 2005, **46**, 2133.
25. P. Sakthivel and P. Kannan, *Polymer*, 2005, **46**, 9821.
26. F. Millich and C. E. Carraher, *J. Polym. Sci., Part A1: Polym. Chem.*, 1969, **7**, 2669.
27. F. Millich and L. L. Lambing, *J. Polym. Sci., Part A: Polym. Chem.*, 1980, **18**, 2155.
28. M. Richards, B. I. Dahiyat, D. M. Arm, S. Lin and K. W. Leong, *J. Polym. Sci., Part A: Polym. Chem.*, 1991, **29**, 1157.
29. S. Iliescu, G. Ilia, A. Popa, G. Dehelean, L. Macarie, L. Pacureanu and N. Hurduc, *Polym. Bull.*, 2001, **46**, 165.
30. S. Iliescu, G. Ilia, N. Plesu, A. Popa and A. Pascariu, *Green Chem.*, 2006, **8**, 727.
31. S. Iliescu, A. Pascariu, N. Plesu, A. Popa, L. Macarie and G. Ilia, *Polym. Bull.*, 2009, **63**, 485.

32. A. Munoz, J. Navech and J. P. Vives, *Bull. Soc. Chim. Fr.*, 1966, 2350.
33. J. Navech, J. P. Vives and A. Munoz, *Bull. Soc. Chim. Fr.*, 1966, 2355.
34. A. Munoz, J. Navech, J. P. Vives and J. P. Majoral, *Bull. Soc. Chim. Fr.*, 1967, 3343.
35. D. Derouet, T. Piatti and J. C. Brosse, *Eur. Polym. J.*, 1986, **22**, 963.
36. A. C. A. Wan, H. Q. Mao, S. Wang, K. W. Leong, L. K. L. L. Ong and H. Yu, *Biomaterials*, 2001, **22**, 1147.
37. X. Y. Xu, H. Yu, S. J. Gao, H. Q. Mao, K. W. Leong and S. Wang, *Biomaterials*, 2002, **23**, 3765.
38. M. V. Chaubal, G. Su, E. Spicer, W. B. Dang, K. E. Branham, J. P. English and Z. Zhao, *J. Biomater. Sci., Polym. Ed.*, 2003, **14**, 45.
39. J. Pretula, K. Kaluzynski, B. Wisniewski, R. Szymanski, T. Loontjens and S. Penczek, *J. Polym. Sci., Part A: Polym. Chem.*, 2006, **44**, 2358.
40. J. Pretula, K. Kaluzynski, B. Wisniewski, R. Szymanski, T. Loontjens and S. Penczek, *J. Polym. Sci., Part A: Polym. Chem.*, 2008, **46**, 830.
41. S. Penczek, K. Kaluzynski and J. Pretula, *J. Appl. Polym. Sci.*, 2007, **105**, 246.
42. P. Klosinski and S. Penczek, *Makromol. Chem., Rapid Commun.*, 1988, **9**, 159.
43. A. Nyk, P. Klosinski and S. Penczek, *Makromol. Chem.*, 1991, **192**, 833.
44. T. Biedron, K. Kaluzynski, J. Pretula, P. Kubisa, S. Penczek and T. Loontjens, *J. Polym. Sci., Part A: Polym. Chem.*, 2001, **39**, 3024.
45. Z. X. Liu, L. Wang, C. Y. Bao, X. X. Li, L. Cao, K. R. Dai and L. Y. Zhu, *Biomacromolecules*, 2011, **12**, 2389.
46. J. Pretula, K. Kaluzynski, R. Szymanski and S. Penczek, *Macromol. Symp.*, 1997, **122**, 269.
47. K. D. Troev, *Chemistry and Application of H-Phosphonates*, Elsevier, Amsterdam, 2006.
48. K. D. Troev, *Polyphosphoesters*, Elsevier, Oxford, 2012, p. 1.
49. W. Vogt and S. Balasubramanian, *Makromol. Chem.*, 1973, **163**, 111.
50. K. Troev and M. Simeonov, *Phosphorus, Sulfur Silicon Relat. Elem.*, 1984, **19**, 363.
51. J. Pretula and S. Penczek, *Makromol. Chem., Rapid Commun.*, 1988, **9**, 731.
52. K. E. Branham, J. W. Mays, G. M. Gray, P. C. Bharara, H. Byrd, R. Bittinger and B. Farmer, *Polymer*, 2000, **41**, 3371.
53. R. D. Myrex, B. Farmer, G. M. Gray, Y. J. Wright, J. Dees, P. C. Bharara, H. Byrd and K. E. Branham, *Eur. Polym. J.*, 2003, **39**, 1105.
54. J. Pretula and S. Penczek, *Makromol. Chem.*, 1990, **191**, 671.
55. S. Penczek and J. Pretula, *Macromolecules*, 1993, **26**, 2228.
56. J. Pretula, K. Kaluzynski, R. Szymanski and S. Penczek, *J. Polym. Sci., Part A: Polym. Chem.*, 1999, **37**, 1365.
57. J. Pretula, K. Kaluzynski, R. Szymanski and S. Penczek, *Macromolecules*, 1997, **30**, 8172.
58. S. Penczek, J. Pretula and K. Kaluzynski, *J. Polym. Sci., Part A: Polym. Chem..*, 2005, **43**, 650.

59. J. Baran, P. Klosinski and S. Penczek, *Makromol. Chem.*, 1989, **190**, 1903.
60. I. Gitsov and F. E. Johnson, *J. Polym. Sci., Part A: Polym. Chem.*, 2008, **46**, 4130.
61. E. Bezdushna, H. Ritter and K. Troev, *Macromol. Rapid Commun.*, 2005, **26**, 471.
62. K. B. Wagener, J. M. Boncella and J. G. Nel, *Macromolecules*, 1991, **24**, 2649.
63. F. Marsico, M. Wagner, K. Landfester and F. R. Wurm, *Macromolecules*, 2012, **45**, 8511.
64. J. Wen and R. X. Zhuo, *Macromol. Rapid Commun.*, 1998, **19**, 641.
65. G. Lapienis and S. Penczek, in *Ring-Opening Polymerization*, ed. K. J. Ivin and T. Saegusa, Elsevier Applied Science, London, 1984, vol. 2, p. 919.
66. H. J. Lucas, F. W. Mitchell and C. N. Scully, *J. Am. Chem. Soc.*, 1950, **72**, 5491.
67. R. S. Edmundson, *Chem. Ind. (London)*, 1962, 1770.
68. L. Fontaine, D. Derouet and J. C. Brosse, *Eur. Polym. J.*, 1990, **26**, 857.
69. A. A. Oswald, *Can. J. Chem.*, 1959, **37**, 1498.
70. G. Lapienis, J. Pretula and S. Penczek, *Macromolecules*, 1983, **16**, 153.
71. A. Zwierzak, *Can. J. Chem.*, 1967, **45**, 2501.
72. E. E. Nifant'ev, J. S. Nasonovskij and A. A. Borisenko, *J. Gen. Chem. USSR (Engl Transl.)*, 1971, **41**, 1885.
73. K. Oussadi, V. Montembault, M. Belbachir and L. Fontaine, *J. Appl. Polym. Sci.*, 2011, **122**, 891.
74. M. Maffei and G. Buono, *Tetrahedron*, 2003, **59**, 8821.
75. C. Negrell-Guirao and B. Boutevin, *Macromolecules*, 2009, **42**, 2446.
76. C. Negrell-Guirao, B. Boutevin, G. David, A. Fruchier, R. Sonnier and J. M. Lopez-Cuesta, *Polym. Chem.*, 2011, **2**, 236.
77. P. Klosinski, *Tetrahedron Lett.*, 1990, **31**, 2025.
78. K. Kaluzynski, J. Libiszowski and S. Penczek, *Macromolecules*, 1976, **9**, 365.
79. J. Libiszowski, K. Kaluzynski and S. Penczek, *J. Polym. Sci., Part A: Polym. Chem.*, 1978, **16**, 1275.
80. H. Yasuda, M. Sumitani, K. Lee, T. Araki and A. Nakamura, *Macromolecules*, 1982, **15**, 1231.
81. W. Vogt and R. Pfluger, *Makromol. Chem., Suppl.*, 1975, **1**, 97.
82. C. S. Xiao, Y. C. Wang, J. Z. Du, X. S. Chen and J. Wang, *Macromolecules*, 2006, **39**, 6825.
83. H. Yasuda, M. Sumitani and A. Nakamura, *Macromolecules*, 1981, **14**, 458.
84. B. Clément, B. Grignard, L. Koole, C. Jerome and P. Lecomte, *Macromolecules*, 2012, **45**, 4476.
85. W. Vogt and N. U. Ahmad, *Makromol. Chem.*, 1977, **178**, 1711.
86. T. Gehrmann and W. Vogt, *Makromol. Chem.*, 1981, **182**, 3069.
87. P. Klosinski and S. Penczek, *Macromolecules*, 1983, **16**, 316.
88. Y. C. Wang, Y. Y. Yuan, F. Wang and J. Wang, *J. Polym. Sci., Part A: Polym. Chem.*, 2011, **49**, 487.

89. S. Y. Zhang, A. Li, J. Zou, L. Y. Lin and K. L. Wooley, *ACS Macro Lett.*, 2012, **1**, 328.

90. S. Y. Zhang, J. Zou, F. W. Zhang, M. Elsabahy, S. E. Felder, J. H. Zhu, D. J. Pochan and K. L. Wooley, *J. Am. Chem. Soc.*, 2012, **134**, 18467.

91. G. Lapienis and S. Penczek, *Macromolecules*, 1974, 7, 166.

92. G. Lapienis and S. Penczek, *J. Polym. Sci., Part A: Polym. Chem.*, 1977, **15**, 371.

93. G. Lapienis and S. Penczek, *Macromolecules*, 1977, **10**, 1301.

94. G. Lapienis and S. Penczek, *J. Polym. Sci., Part A: Polym. Chem.*, 1990, **28**, 1743.

95. S. Sosnowski, J. Libiszowski, S. Slomkowski and S. Penczek, *Makromol. Rapid Commun.*, 1984, **5**, 239.

96. K. J. Ivin, in *Reactivity, Mechanism and Structure in Polymer Chemistry*, ed. A. D. Jenkins and A. Ledwith, Wiley, New York, 1974, p. 514.

97. Y. Y. Yuan, J. Z. Du and J. Wang, *Chem. Commun.*, 2012, **48**, 570.

98. J.-Z. Du, X.-J. Du, C.-Q. Mao and J. Wang, *J. Am. Chem. Soc.*, 2011, **133**, 17560.

99. H. Y. Shao, M. Z. Zhang, J. L. He and P. H. Ni, *Polymer*, 2012, **53**, 2854.

100. C. Wachiralarpphaithoon, Y. Iwasaki and K. Akiyoshi, *Biomaterials*, 2007, **28**, 984.

101. Y. Iwasaki and K. Akiyoshi, *Biomacromolecules*, 2006, **7**, 1433.

102. Y. Iwasaki, C. Nakagawa, M. Ohtomi, K. Ishihara and K. Akiyoshi, *Biomacromolecules*, 2004, **5**, 1110.

103. C. Q. Mao, J. Z. Du, T. M. Sun, Y. D. Yao, P. Z. Zhang, E. W. Song and J. Wang, *Biomaterials*, 2011, **32**, 3124.

104. W. P. Zhu, S. Sun, N. Xu, P. F. Gou and Z. Q. Shen, *J. Appl. Polym. Sci.*, 2012, **123**, 365.

105. J. Liu, W. Huang, Y. Pang, X. Zhu, Y. Zhou and D. Yan, *Biomaterials*, 2010, **31**, 5643.

106. J. Liu, Y. Pang, W. Huang, Z. Zhu, X. Zhu, Y. Zhou and D. Yan, *Biomacromolecules*, 2011, **12**, 2407.

107. W. J. Song, J. Z. Du, N. J. Liu, S. Dou, J. Cheng and J. Wang, *Macromolecules*, 2008, **41**, 6935.

108. T. M. Sun, J. Z. Du, L. F. Yan, H. Q. Mao and J. Wang, *Biomaterials*, 2008, **29**, 4348.

109. T. M. Sun, J. Z. Du, Y. D. Yao, C. Q. Mao, S. Dou, S. Y. Huang, P. Z. Zhang, K. W. Leong, E. W. Song and J. Wang, *ACS Nano*, 2011, **5**, 1483.

110. Y. C. Wang, Y. Li, T. M. Sun, M. H. Xiong, J. Wu, Y. Y. Yang and J. Wang, *Macromol. Rapid Commun.*, 2010, **31**, 1201.

111. Y. Iwasaki and K. Akiyoshi, *Macromolecules*, 2004, **37**, 7637.

112. L. Fontaine, D. Derouet and J. C. Brosse, *Eur. Polym. J.*, 1990, **26**, 865.

113. Y. Y. Yuan, X. Q. Liu, Y. C. Wang and J. Wang, *Langmuir*, 2009, **25**, 10298.

114. R. Ikeuchi and Y. Iwasaki, *J. Biomed. Mater. Res., A*, 2013, **101A**, 318.

115. J. Wen and R. X. Zhuo, *Polym. Int.*, 1998, **47**, 503.

116. J. Wen, G. J. A. Kim and K. W. Leong, *J. Controlled Release*, 2003, **92**, 39.
117. W. P. Zhu, S. Sun, N. Xu and Z. Q. Shen, *J. Polym. Sci., Part A: Polym. Chem.*, 2011, **49**, 4987.
118. X. L. Wang, R. X. Zhuo and L. J. Liu, *Polym. Int.*, 2001, **50**, 1175.
119. F. Li, J. Feng and R. Zhuo, *J. Appl. Polym. Sci.*, 2006, **102**, 5507.
120. T. Steinbach, E. M. Alexandrino and F. R. Wurm, *Polym. Chem.*, 2013, in press; DOI: 10.1039/c3py00437f.
121. T. Biela, S. Penczek, S. Slomkowski and O. Vogl, *Makromol. Chem. Rapid Commun.*, 1982, **3**, 667(erratum: *ibid.*, 1983, **4**, 443).
122. T. Biela, P. Klosinski and S. Penczek, *J. Polym. Sci., Part A: Polym. Chem.*, 1989, **27**, 763.
123. K. Kaluzynski, J. Libiszowski and S. Penczek, *Makromol. Chem.*, 1977, **178**, 2943.
124. K. Kaluzynski, R. Szymanski and S. Penczek, *J. Polym. Sci., Part A: Polym. Chem.*, 1991, **29**, 1825.
125. L. S. Nair and C. T. Laurencin, *Prog. Polym. Sci.*, 2007, **32**, 762.
126. H. Tian, Z. Tang, X. Zhuang, X. Chen and X. Jing, *Prog. Polym. Sci.*, 2012, **37**, 237.
127. Q. Li, J. Wang, S. Shahani, D. D. N. Sun, B. Sharma, J. H. Elisseeff and K. W. Leong, *Biomaterials*, 2006, **27**, 1027.
128. J. He, M. Zhang and P. Ni, *Soft Matter*, 2012, **8**, 6033.
129. L. Zhang, Y. I. Jeong, S. Zheng, S. I. Jang, H. Suh, D. H. Kang and I. Kim, *Polym. Chem.*, 2013, **4**, 1084.
130. Y. C. Wang, J. Wu, Y. Li, J. Z. Du, Y. Y. Yuan and J. Wang, *Chem. Commun.*, 2010, **46**, 3520.
131. J. Wu, X. Q. Liu, Y. C. Wang and J. Wang, *J. Mater. Chem.*, 2009, **19**, 7856.
132. D. P. Chen and J. Wang, *Macromolecules*, 2006, **39**, 473.
133. Y. C. Wang, S. Y. Shen, Q. P. Wu, D. P. Chen, J. Wang, G. Steinhoff and N. Ma, *Macromolecules*, 2006, **39**, 8992.
134. Y. C. Wang, Y. Li, X. Z. Yang, Y. Y. Yuan, L. F. Yan and J. Wang, *Macromolecules*, 2009, **42**, 3026.
135. F. Wang, Y. C. Wang, L. F. Yan and J. Wang, *Polymer*, 2009, **50**, 5048.
136. Y. C. Wang, X. Q. Liu, T. M. Sun, M. H. Xiong and J. Wang, *J. Controlled Release*, 2008, **128**, 32.
137. P. C. Zhang, L. J. Hu, Q. Yin, Z. W. Zhang, L. Y. Feng and Y. P. Li, *J. Controlled Release*, 2012, **159**, 429.
138. H. X. Wang, M. H. Xiong, Y. C. Wang, J. Zhu and J. Wang, *J. Controlled Release*, 2013, **166**, 106.
139. Y. C. Wang, L. Y. Tang, T. M. Sun, C. H. Li, M. H. Xiong and J. Wang, *Biomacromolecules*, 2008, **9**, 388.
140. X. Z. Yang, T. M. Sun, S. Dou, J. Wu, Y. C. Wang and J. Wang, *Biomacromolecules*, 2009, **10**, 2213.
141. X. Z. Yang, Y. C. Wang, L. Y. Tang, H. Xia and J. Wang, *J. Polym. Sci., Part A: Polym. Chem.*, 2008, **46**, 6425.

142. Q. H. Wu, C. Wang, D. Zhang, X. M. Song, F. Verpoort and G. L. Zhang, *React. Funct. Polym.*, 2011, **71**, 980.

143. X. Liu, P. H. Ni, J. L. He and M. Z. Zhang, *Macromolecules*, 2010, **43**, 4771.

144. J. Bian, M. Z. Zhang, J. L. He and P. H. Ni, *React. Funct. Polym.*, 2013, **73**, 579.

145. J. Wang and R. X. Zhuo, *Eur. Polym. J.*, 1999, **35**, 491.

146. J. Z. Du, D. P. Chen, Y. C. Wang, C. S. Xiao, Y. J. Lu, J. Wang and G. Z. Zhang, *Biomacromolecules*, 2006, **7**, 1898.

147. Y. Hao, J. He, M. Zhang, Y. Tao, J. Liu and P. Ni, *J. Polym. Sci., Part A: Polym. Chem.*, 2013, **51**, 2150.

148. J. Cheng, J. X. Ding, Y. C. Wang and J. Wang, *Polymer*, 2008, **49**, 4784.

149. Q. H. Wu, C. Wang, D. Zhang, X. M. Song, D. L. Liu, L. P. Wang and G. L. Zhang, *React. Funct. Polym.*, 2012, **72**, 372.

150. Y. Y. Yuan, Y. C. Wang, J. Z. Du and J. Wang, *Macromolecules*, 2008, **41**, 8620.

151. Y. Y. Yuan and J. Wang, *Colloids Surf., B*, 2011, **85**, 81.

152. J. Liu, Y. Pang, W. Huang, X. Zhai, X. Zhu, Y. Zhou and D. Yan, *Macromolecules*, 2010, **43**, 8416.

153. M. H. Xiong, J. Wu, Y. C. Wang, L. S. Li, X. B. Liu, G. Z. Zhang, L. F. Yan and J. Wang, *Macromolecules*, 2009, **42**, 893.

154. J. Liu, W. Huang, Y. Pang, X. Zhu, Y. Zhou and D. Yan, *Biomacromolecules*, 2010, **11**, 1564.

155. J. Liu, W. Huang, Y. Pang, X. Zhu, Y. Zhou and D. Yan, *Langmuir*, 2010, **26**, 10585.

156. J. Liu, Y. Pang, W. Huang, X. Huang, L. Meng, X. Zhu, Y. Zhou and D. Yan, *Biomacromolecules*, 2011, **12**, 1567.

157. M. H. Xiong, Y. Bao, X. Z. Yang, Y. C. Wang, B. L. Sun and J. Wang, *J. Am. Chem. Soc.*, 2012, **134**, 4355.

158. C. J. Chen, Q. Jin, G. Y. Liu, D. D. Li, J. L. Wang and J. Ji, *Polymer*, 2012, **53**, 3695.

159. M. H. Xiong, Y. J. Li, Y. Bao, X. Z. Yang, B. Hu and J. Wang, *Adv. Mater.*, 2012, **24**, 6175.

160. S. Y. Zhang, J. Zou, M. Elsabahy, A. Karwa, A. Li, D. A. Moore, R. B. Dorshow and K. L. Wooley, *Chem. Sci.*, 2013, **4**, 2122.

161. O. Kuznetsova, K. Kaluzynski, G. Lapienis, J. Pretula and S. Penczek, *J. Bioact. Compat. Polym.*, 1999, **14**, 232.

162. B. Abramov, *Dokl. Akad. Nauk SSSR*, 1950, 487; *Chem. Abstr.*, 1951, **45**, 2855.

163. K. D. Troev, *Chemistry and Application of H-Phosphonates*, Elsevier, Amsterdam, 2006, p. 59.

164. A. N. Pudovik and B. Arubzov, *Izv. Akad. Nauk SSSR, Otd. Khim. Nauk*, 1949, 522; *Chem. Abstr.*, 1953, **47**, 4300.

165. K. D. Troev, *Chemistry and Application of H-Phosphonates*, Elsevier, Amsterdam, 2006, p. 54.

166. K. Kossev, A. Vassilev, Y. Popova, I. Ivanov and K. Troev, *Polymer*, 2003, **44**, 1987.
167. R. Tzevi, P. Novakov, K. Troev and D. M. Roundhill, *J. Polym. Sci., Part A: Polym. Chem.*, 1997, **35**, 625.
168. N. Koseva, A. Bogomilova, K. Atkova and K. Troev, *React. Funct. Polym.*, 2008, **68**, 954.
169. W. Vogt, *Makromol. Chem.*, 1973, **163**, 89.
170. S. Penczek, P. Klosinski, A. Narebska and R. Wodzki, *Makromol. Chem., Macromol. Symp.*, 1991, **48/49**, 1.
171. J. Ostrowska-Czubenko and R. Wodzki, in *Biophosphates and Their Analogues: Synthesis, Structure, Metabolism and Activity*, ed. K. S. Bruzik and W. J. Stec, Elsevier, Amsterdam, 1987, p. 575.
172. J. Ostrowska-Czubenko, *Colloid Polym. Sci.*, 1994, **272**, 562.
173. J. Ostrowska-Czubenko, *Phys. Chem. Chem. Phys.*, 1999, **1**, 4371.
174. S. Penczek, J. Pretula and K. Kaluzynski, *Pol. J. Chem.*, 2001, **75**, 1171.
175. J. Ostrowska-Czubenko, *Colloid Polym. Sci.*, 2002, **280**, 1015.
176. S. Penczek, J. Pretula and K. Kaluzynski, *Biomacromolecules*, 2005, **6**, 547.
177. R. Tzevi, G. Todorova, K. Kossev, K. Troev, E. M. Georgiev and D. M. Roundhill, *Makromol. Chem.*, 1993, **194**, 3261.
178. R. Georgieva, R. Tsevi, K. Kossev, R. Kusheva, M. Balgjiska, R. Petrova, V. Tenchova, I. Gitsov and K. Troev, *J. Med. Chem.*, 2002, **45**, 5797.
179. I. Pencheva, A. Bogomilova, N. Koseva, D. Obreshkova and K. Troev, *J. Pharm. Biomed. Anal.*, 2008, **48**, 1143.
180. M. A. Mintzer and E. E. Simanek, *Chem. Rev.*, 2009, **109**, 259.
181. F. R. Atherton, H. T. Openshaw and A. R. Todd, *J. Chem. Soc.*, 1945, 382.
182. F. R. Atherton and A. R. Todd, *J. Chem. Soc.*, 1947, 674.
183. J. Pretula, K. Kaluzynski and S. Penczek, *Macromolecules*, 1986, **19**, 1797.
184. K. Troev, A. Naruoka, H. Terada, A. Kikuchi and K. Makino, *Macromolecules*, 2012, **45**, 5698.
185. H. Byrd, D. Bond-Garcia, G. M. Gray and K. E. Branham, *Inorg. Chim. Acta*, 2006, **359**, 4001.
186. S. Penczek, G. Lapienis, K. Kaluzynski and A. Nyk, *Pol. J. Chem.*, 1994, **68**, 2129.
187. G. Lapienis, S. Penczek, G. P. Aleksiuk and V. A. Kropachev, *J. Polym. Sci., Part A: Polym. Chem.*, 1987, **25**, 1729.
188. G. Lapienis and S. Penczek, *J. Polym. Sci., Part A: Polym. Chem.*, 1990, **28**, 1519.
189. L. Fontaine, D. Derouet and J. C. Brosse, *Eur. Polym. J.*, 1990, **26**, 865.
190. L. Fontaine, D. Derouet, S. Chairatanathavorn and J. C. Brosse, *React. Polym.*, 1993, **19**, 47.
191. S. Stanimirov, A. Vasilev, E. Haupt, I. Petkov and T. Deligeorgiev, *J. Fluoresc.*, 2009, **19**, 85.
192. J. Wang, H. Q. Mao and K. W. Leong, *J. Am. Chem. Soc.*, 2001, **123**, 9480.

193. J. Wang, S. W. Huang, P. C. Zhang, H. Q. Mao and K. W. Leong, *Int. J. Pharm.*, 2003, **265**, 75.
194. J. Wang, P. C. Zhang, H. Q. Mao and K. W. Leong, *Gene Ther.*, 2002, **9**, 1254.
195. S. W. Huang, J. Wang, P. C. Zhang, H. Q. Mao, R. X. Zhuo and K. W. Leong, *Biomacromolecules*, 2004, **5**, 306.
196. K. D. Troev, V. A. Mitova and I. G. Ivanov, *Tetrahedron Lett.*, 2010, **51**, 6123.
197. K. Kaluzynski and S. Penczek, *Macromol. Chem. Phys.*, 1994, **195**, 3855.
198. S. Penczek, K. Kaluzynski and J. Baran, in *Macromolecules 1992*, ed. J. Kahovec, VSP International, Utrecht, 1993, p. 231.
199. J. Baran, K. Kaluzynski, R. Szymanski and S. Penczek, *Biomacromolecules*, 2004, **5**, 1841.
200. F. R. Atherton, H. T. Openshaw and A. R. Todd, *J. Chem. Soc.*, 1945, 660.
201. J. C. Brosse, L. Fontaine, D. Derouet and S. Chairatanathavorn, *Makromol. Chem.*, 1989, **190**, 2329.
202. J. C. Brosse, D. Derouet, L. Fontaine and S. Chairatanathavorn, *Makromol. Chem.*, 1989, **190**, 2339.
203. L. Fontaine, C. Marboeuf, J. C. Brosse, M. Maingault and F. Dehaut, *Macromol. Chem. Phys.*, 1996, **197**, 3613.
204. J. Wang, S. J. Gao, P. C. Zhang, S. Wang, M. Q. Mao and K. W. Leong, *Gene Ther.*, 2004, **11**, 1001.
205. J. Wang, P. C. Zhang, H. F. Lu, N. Ma, S. Wang, H. Q. Mao and K. W. Leong, *J. Controlled Release*, 2002, **83**, 157.
206. X. Q. Zhang, X. L. Wang, P. C. Zhang, Z. L. Liu, R. X. Zhuo, H. Q. Mao and K. W. Leong, *J. Controlled Release*, 2005, **102**, 749.
207. Y. Ren, X. Jiang, D. Pan and H. Q. Mao, *Biomacromolecules*, 2010, **11**, 3432.
208. K. Troev, I. Tsatcheva, N. Koseva, R. Georgieva and I. Gitsov, *J. Polym. Sci., Part A: Polym. Chem.*, 2007, **45**, 1349.
209. K. Oussadi, V. Montembault and L. Fontaine, *J. Polym. Sci., Part A: Polym. Chem.*, 2011, **49**, 5124.

CHAPTER 7

Phosphazene High Polymers

HARRY R. ALLCOCK

Department of Chemistry, The Pennsylvania State University,
University Park, PA 16802, USA
Email: hra@chem.psu.edu

7.1 Introduction

The incorporation of phosphorus into polymer molecules has a long tradition, ranging from the study of mineralogical polyphosphates,[1,2] through organic polymers with phosphorus atoms in the side chains,[3,4] to the large and expanding family of polymers known as polyphosphazenes (Scheme 7.1).[5]

Small-molecule phosphazene chemistry has a long and distinguished history that extends back to the early 1800s. The first air-stable members of the high polymeric phosphazene series were synthesized in the author's laboratory in the mid-1960s[6–8] and the number of different polymers in this class has expanded year-by-year until several hundred different examples are now known.[5] This diversity of structure is a result of the unusual methods used for their synthesis and the enormous variety of different side groups, R, that have now been linked to the polyphosphazene chain. In turn, this has led to the identification of numerous diverse uses for these polymers, ranging from medical materials to aerospace polymers, membranes, polymer electrolytes, optical materials, and fire-resistant polymers. In the following sections a summary is given of the synthetic techniques that are involved, how the synthetic methodology is used to change the structure–property relationships, and the actual and prospective uses.

RSC Polymer Chemistry Series No. 11
Phosphorus-Based Polymers: From Synthesis to Applications
Edited by Sophie Monge and Ghislain David
© The Royal Society of Chemistry 2014
Published by the Royal Society of Chemistry, www.rsc.org

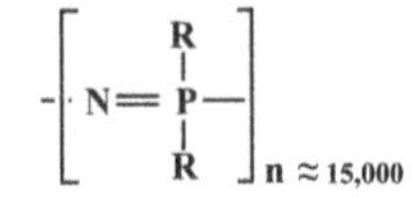

$$\left[-N = \underset{\underset{R}{|}}{\overset{\overset{R}{|}}{P}} - \right]_{n \approx 15,000}$$

Scheme 7.1   The fundamental polyphosphazene structure.

The chapter will conclude with some observations about the challenges to be met before this and any other pioneering polymer system can find its way into the commercial sector.

7.2   Synthesis

A number of routes have been developed for the synthesis of poly(organophosphazenes). The main methods are described in Sections 7.2.1–7.2.4.

7.2.1   Ring-Opening Polymerization of a Small-Molecule Cyclic Chlorophosphazene such as (NPCl$_2$)$_3$, followed by Replacement of the Chlorine Atoms in the Resultant Chlorophosphazene High Polymer by Organic Groups

This is the method pioneered in our laboratory[5–18] and it has been the vehicle for the synthesis of most of the known poly(organophosphazenes). It is summarized in Scheme 7.2. Thus, hexachlorocyclotriphosphazene (structure **2** in Scheme 7.2, see also Figure 7.1) (produced *via* the reaction of PCl$_5$ and NH$_4$Cl) is heated in the molten state at 250 °C to form high molecular weight, essentially linear, poly(dichlorophosphazene) (**3**). Depending on the polymerization conditions, the chain lengths of this polymer may equal or exceed 15 000 repeating units, with molecular weights that reach several million. This is a water-sensitive rubbery elastomer that dissolves in organic solvents such as benzene, toluene, or tetrahydrofuran. Stable polymers (**4–6**) are produced by replacement of the chlorine atoms by treatment of this polymer with nucleophiles such as alkoxides, aryloxides, or amines. More than 250 different nucleophiles have been shown to undergo this reaction. Two or more different side groups can be linked to the polymer chain by simultaneous or sequential addition, and the pattern of mixed substituent introduction can be controlled by the reaction conditions.

Similar but more restricted substitutions are possible if fluorophosphazene intermediates are used in place of the chloro derivatives, but the fluorophosphazenes are more appropriate for halogen replacement by organometallic nucleophiles. Replacement of the halogen atoms in the chloro or fluoro polymers by organic groups turns a hydrolytically sensitive polymer into water-stable derivatives, and this is the main reason for the emergence of this field as a serious component of polymer science. It should be noted that organic-substituted small-molecule cyclic

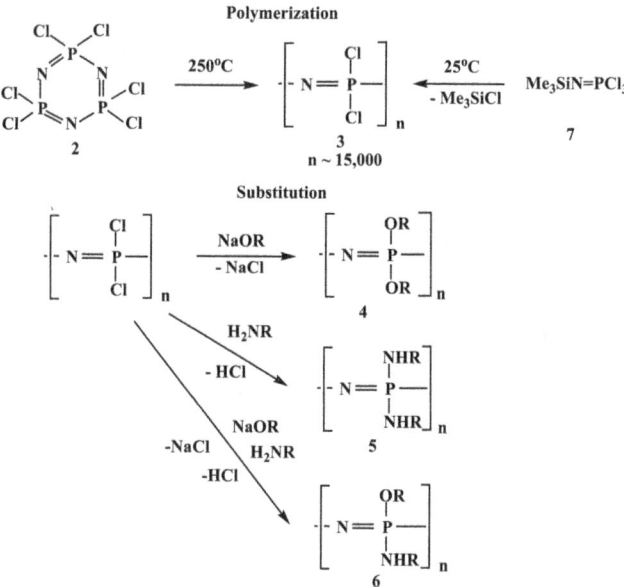

Scheme 7.2 The macromolecular substitution route for the synthesis of poly(organophosphazenes).

Figure 7.1 Crystals of hexachlorocyclotriphosphazene, the starting material for the macromolecular substitution route.

phosphazenes such as $(NPR_2)_3$ cannot be polymerized, although a few compounds with both halogen and organic side groups will polymerize at moderate temperatures.

One of the advantages of the macromolecular substitution route is the ease of introduction of two or more different substituents. In a sense this is a counterpart of the copolymerization of different monomers in conventional organic polymer chemistry, but with a special nuance. Thus, the distribution of different side groups along a phosphazene chain is controlled by steric factors, reagent nucleophilicity, the ratio of reactants in the reaction mixture, and whether the groups are introduced simultaneously or sequentially. This provides enormous scope for changing structures and properties.[18]

7.2.2 Living Cationic Condensation Polymerization of a Chlorophosphoranimine

This approach was discovered by Allcock, Manners, and co-workers in 1995. In this method the phosphoranimine, $Cl_3P{=}N{-}Si(CH_3)_3$ (7 in Scheme 7.2), undergoes a rapid conversion to poly(dichlorophosphazene) in solution *at room temperature* in the presence of a trace of Lewis acid initiator such as PCl_5.[19–44] The polymer chain length is controlled by the ratio of Lewis acid to the monomer (each initiator molecule yields one polymer chain), and very narrow molecular weight distributions are obtained. Moreover, the living nature of the polymerization allows coupling of the end groups to other polymers [phosphazenes, organic polymers, or poly(organosiloxanes)] to produce block copolymers, combs, stars, and dendrimers. An additional advantage is the fact that the chlorotrimethylsilane employed in the synthesis of the phosphoranimine is released as polymerization takes place, thus allowing this reactant to be recycled back into the monomer synthesis. A disadvantage is that the monomer must be stored at low temperatures, otherwise it will cyclize at room temperature to the cyclic trimer even in the absence of a Lewis acid.

7.2.3 Production of Poly(dichlorophosphazene) *via* the Solution Phase Thermal Condensation Polymerization of $Cl_3P{=}N{-}POCl_2$, itself Obtained by Reaction of PCl_5 and an Ammonium Salt

This process has been studied thoroughly in France by De Jaeger and co-workers.[45] It has the advantage of being relatively easy to scale-up in acid-resistant equipment (hydrogen chloride is evolved), but gives broad molecular weight poly(dichlorophosphazene) that is generally of lower average molecular weight than the polymer produced by the ring-opening method. Unlike the ring-opening polymerization, it lacks the ease of isolation, purification, and storage of the crystalline cyclic trimer (2). Like the ring-opening process, the chlorophosphazene polymer can be used as a substrate for the replacement of chlorine by organic groups.

7.2.4 Polymerization of Organophosphoranimines

This method was pioneered by Neilson and Wisian-Nielson,[46–48] with a catalytic component introduced by Matyjaszewski.[41] The idea behind this approach is to link the organic side groups to phosphorus at the condensation monomer synthesis stage, thereby avoiding the macromolecular substitution step. It is a method for the production of relatively low molecular weight polyphosphazenes with alkyl or aryl side groups. These polymers, with P–C linkages to the side groups rather than P–O or P–N linkages, are difficult to produce by the other methods. In general, alkoxy, aryloxy, or amino side groups cannot be utilized by this method, although Flindt and Rose[49] and Matyjaszewski and co-workers[50–52] have synthesized trifluoroethoxyphosphazene polymers by this method.

7.3 Properties Generated by the Phosphazene Skeleton

7.3.1 Vital Reactivity of the P–Cl Bond

The characteristics of polyphosphazenes depend on both the skeleton and the types of side groups. The feasibility of the macromolecular substitution process described above is a consequence of the polarity and high reactivity of P–Cl bonds. This characteristic is similar to the substitutive reactivity of PCl_3, PCl_5, and $OPCl_3$. Without this reactivity it would not be possible to replace the roughly 30 000 chlorine atoms along the skeleton of poly(dichlorophosphazene). The lack of such reactivity is the reason why organic polymers such as poly(vinyl chloride) or poly(vinylidene chloride) cannot be used as macromolecular intermediates.

7.3.2 Unusual Characteristics of the P–N Bond

The nature of the bonding in the phosphazene backbone has generated considerable debate. An organic polymer with seemingly alternating single and double bonds, such as polyacetylene, has a stiff backbone due to the π–π orbital overlap within the delocalized structure. This overlap is also a reason for the electronic conductivity, color, and metal-like reflectivity of conjugated organic polymers. None of these properties are found in polyphosphazenes. Unless the side groups are colored, the polymers are transparent throughout the visible region and down to the 220 nanometer region of the ultraviolet. No evidence of electronic conductivity along the backbone has yet been detected, although side-group-to-side-group semiconductivity can be generated by the use of specific substituents. Moreover, the barrier to torsion of the backbone bonds is extremely low, as evidenced by glass transition temperatures of some derivatives in the –100 °C region.

Various theories have been proposed to explain this anomaly,[5] but the most appealing qualitative idea from a chemistry point of view was suggested for small-molecule phosphazenes by Dewar, Lucken, and Whitehead in the 1960s: that the phosphorus atoms can utilize their 3d orbitals for overlap with the adjacent nitrogen 2p orbitals, with a node at each phosphorus.[53] This would allow delocalization over each P–N–P segment, interrupted at each phosphorus. Thus, the torsional mobility of the backbone would be very high because four of the five d-orbitals could switch into the overlap position as the skeletal bonds undergo torsion. Moreover, the orbital overlap would explain the P–N bond shortening and increased strength compared to single-bonded phosphorus–nitrogen units. Other theories invoke zwitterionic structures or *ab initio* interpretations.

The presence of phosphorus in the polymer backbone has a very practical consequence, quite apart from the structural issues. Phosphorus is one of the most important elements that prevent the combustion of organic materials. The presence of both phosphorus and nitrogen is synergistic. Thus, the phosphorus–nitrogen backbone in polyphosphazenes ensures that many poly(organophosphazenes) are not only nonflammable but also quench combustion of other compounds with which they are in contact. The mechanism of this fire suppression is believed to be both an interruption of the free radical processes that occur in a flame and the formation of an intumescent "char" that shields the material from the ingress of oxygen.

7.3.3   Hydrolytic Sensitivity on Demand

Another consequence of a backbone with both phosphorus and nitrogen is that, in the special cases where the polymer is designed to be hydrolytically unstable in a biological environment (see later), the backbone decomposes to phosphate and ammonia, a buffered combination that prevents necrosis of living tissues. Most organic side groups protect the backbone from hydrolysis, but amino acid esters, glyceryl units, sugar molecules, and ethoxy groups facilitate backbone hydrolysis.

7.3.4   Radiation Stability

Finally, the backbone in polyphosphazenes is more stable to high-energy radiation than the carbon-containing skeletons in most classical organic polymers. The UV/visible transparency means that the free radical bond-cleavage mechanisms that restrict the long-term stability of many organic polymers in sunlight do not exist for polyphosphazenes. Hence, provided the side groups are transparent, the longevity of many polyphosphazenes can exceed that of classical polymers. The phosphorus–nitrogen backbone also appears to be more stable than many organic counterparts when irradiated with X-rays or γ-rays, especially in an oxygen-containing atmosphere.

7.4   Influence by the Side Groups

Although the backbone imparts its own characteristics to these polymers, the side groups also play a critical role based on their size, hydrophilicity or

Table 7.1 Influence of different side groups on the properties of polyphosphazenes.

Side group	Examples	Properties/uses
Fluoroalkoxy[a]	CF_3CH_2O-, $CH_2H(CF_2)_xCH_2O-$	Hydrophobic, films, fibers, elastomers
Alkoxy[a]	CH_3CH_2O-, $CH_3(CH_2)_xO-$	Low T_g elastomers
Alkyl ether[a]	$CH_3O(CH_2CH_2)_xO-$	Ionic conductors, hydrogel formers
Crown ethers	18-crown-6-CH_2O-	For capture of cations
Cyclodextrins		Biology and clathration
Aryloxy and substituted aryloxy[a]	$PhO-$, p-O_2NC_6H4O-, FC_6H_4O-	PhO polymer is fire inhibitor; fluoro derivatives, potential dielectric materials
Biphenyl, arylazido, halogenoaryloxy derivatives[a]	$PhN{=}NC_6H_4CH_2CH_2O-$	Liquid crystalline and NLO polymers, high refractive index polymers
Carboxyphenoxy (PCPP)	p-$HO_2CC_6H_4O-$	Gel former with divalent or trivalent cations; microspheres for oral vaccine delivery
Amino acid esters[a]	$EtO_2CCH(R)NH-$	Bioerodible polymers for tissue regeneration
Coordination ligands for metals	$Ph_2PC_6H_4O-$	Immobilized catalysts
Metal carbonyls	$(CO)_2CpFeC_6H_4O-$	Immobilized catalysts
Iron porphyrins	Porph-Fe(II)-$NHC(O)(CH_2)_2C(O)NH$-$(CH_2)_3NMe-$	Experimental oxygen carriers
Metal phthalocyanines	Phth-Cu(II)-$OCH_2C_6H_4O-$	Electrochemical activity
Organosilicon units	$(CH_3)_3SiCH_2O-$	Elastomers
Borazine rings	$Me_5N_3B_3$-$NH-$	Preceramic polymers
Carboranes	*nido*-Carborane-$Rh(PPh_3)_3Cl$, $B_{10}C_2H_{12}R$	Hydrogenation catalysts; heat resistant polymers

[a]A significant number of polymers with these side groups have been studied.

hydrophobicity, ionic or covalent character, or their optical absorption characteristics. A few examples are summarized in Table 7.1.

7.4.1 Fluorinated Side Groups

Polyphosphazenes with fluorinated organic side groups have played a significant role in polyphosphazene science and technology. The polymer $[NP(OCH_2CF_3)_2]_n$ is a hydrophobic, film- and fiber-forming material. The contact angle to water is 100 °C,[54] which is a typical value for fluorinated organic polymers such as Teflon. However, unlike Teflon, this polymer is soluble in organic solvents and can be solution-cast into films or solution-spun into fibers. When electrospun to form nanofibers, the contact angle rises to 160 °C, which is in the superhydrophobic range.[55] Fluorinated aryloxy side groups also generate hydrophobicity. The influence of fluorine on the properties of polyphosphazenes has been discussed in a recent publication.[56] When fluorine alone constitutes the side group, as in $(NPF_2)_n$, the polymer is less soluble in organic solvents than the chloro derivative and is less reactive to nucleophiles. Therefore it is not as useful as a macromolecular intermediate. However, a polymer intermediate with both phenyl and fluorine side groups is soluble and is a good substrate for reactions with organometallic reagents.[57]

7.4.2 Aryloxy Side Groups

Aryloxy side groups raise the polymer glass transition temperatures due to the steric restrictions they impose on the torsional mobility of the backbone.[5-7,58-66] Polycyclic aryloxy groups have an even more dramatic effect. A second characteristic is that aryloxy groups protect the backbone against attack by aggressive reagents such as acids or bases, to a degree that substitution chemistry can be carried out on the aryl groups without affecting the backbone. For example, sulfonation of the aryl groups is possible to yield polymers that function as proton conductors. Some complex aryl units confer liquid crystallinity and nonlinear optical behavior to the polymers. However, preliminary evidence suggests that bulky polycyclic aryloxy groups such as naphthyloxy or anthracenyloxy hinder complete replacement of the chlorine atoms during the macromolecular substitution step and also lower the ceiling temperature, the point at which a polymer depolymerizes to form small molecule rings. These restrictions can sometimes be overcome by the use of cesium instead of sodium salts in the halogen replacement reactions[67] or by the use of phase transfer agents such as *tert*-butylammonium salts. The presence of substituents on the aryl rings, such as carboxylic acid, hydroxyl, or amino groups, can convert the surface properties from hydrophobic to hydrophilic, but these functional groups must usually be introduced *via* protection–deprotection chemistry to prevent crosslinking of the system during macromolecular substitution.

7.4.3 Alkylamino and Arylamino Side Groups

In general, these side groups have been studied to a lesser degree than their alkoxy or aryloxy counterparts.[8,68-72] For primary amino groups the N–H bonds can participate in hydrogen bonding, which raises the T_g of the polymer and often lowers its solubility in organic solvents. The basicity of alkylamino side groups favors salt formation with the hydrogen chloride released during macromolecular substitution, and stronger bases such as triethylamine are used to preferentially capture the acid during the substitution process. Steric hindrance problems become evident during the linkage of secondary amino groups to the phosphazene skeleton. For example, the reactions of poly(dichlorophosphazene) with diethylamine or diphenylamine result in the replacement of only one chlorine per phosphorus unless forcing reaction conditions are employed. However, by far the most extensive aminolysis studies have been carried out using amino acid esters, as described in the following section.

7.4.4 Amino Acid Ester Side Groups

Amino acid ethyl ester side groups generate an important set of properties in polyphosphazenes.[73–89] Specifically, these polymers are hydrolytically sensitive and hydrolyze to phosphate, ammonium ion, amino acid, and ethanol. The rate of hydrolysis depends on the specific amino acid employed, with substituents on the α-carbon slowing the rate of polymer breakdown. The importance of this behavior becomes manifest when these polymers are used as matrices for mammalian tissue regeneration in the process known as "tissue engineering". In this process a porous construct fabricated from the polymer is seeded with the patient's own cells, which multiply as the matrix hydrolyzes and dissolves. Because the hydrolysis products from the polyphosphazenes are not only biologically benign but also are nutrients for cell growth, the regeneration of tissues is encouraged. This process has been used by the author's collaborators at the University of Connecticut to speed the regeneration of living bone. We are also developing polyphosphazenes specifically designed to function as tissue engineering matrices for ligaments and tendons. The hydrolysis behavior of these polymers has also been used for experimental controlled drug release systems.[77–79]

7.4.5 Inorganic and Organometallic Side Groups

Although most of the high polymeric phosphazenes synthesized to date bear organic side groups, several examples have been synthesized and studied that bear inorganic or organometallic side group structures. Examples include polyphosphazenes with phosphazo, borazine, organosilicon, and transition metal carbonyl side groups, and other research teams have prepared side units with coordinated transition metals.[70,90–95] Scheme 7.3 illustrates the synthesis of polyphosphazenes with phosphazo side groups. The chlorine or fluorine atoms on the side groups and along the backbone can be replaced by organic units in the normal way. Scheme 7.4 shows a borazinyl derivative, which was prepared by the reaction of a chloroborazine with methylamino side groups linked to the backbone. This was studied as a preceramic polymer for pyrolysis to phosphorus–nitrogen–boron ceramic materials.[96] Also shown in Scheme 7.4 are two organosilicon derivatives, one at the small-molecule model compound level and the other at the high polymer level.[97–102] This is in addition to the block copolymers between phosphazenes and poly(dimethylsiloxane) to be described below. Platinum antitumor agents have been coordinated to amino side groups on polyphosphazenes, and immobilized cobalt and rhodium catalysts have also been incorporated into the side group structures. Polyphosphazenes with

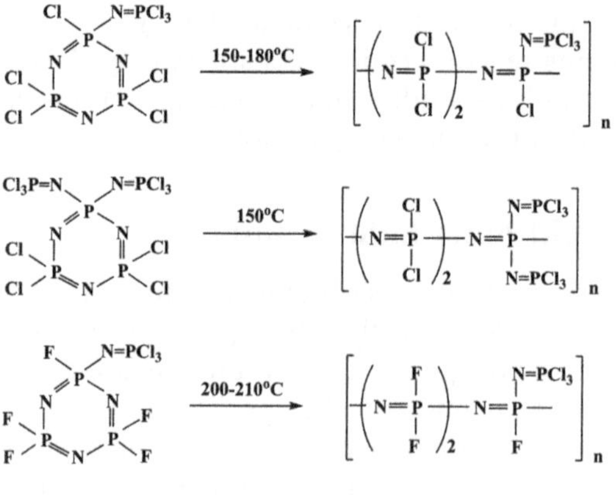

150-180°C

150°C

200-210°C

Scheme 7.3 Polyphosphazenes with phosphazo side groups.

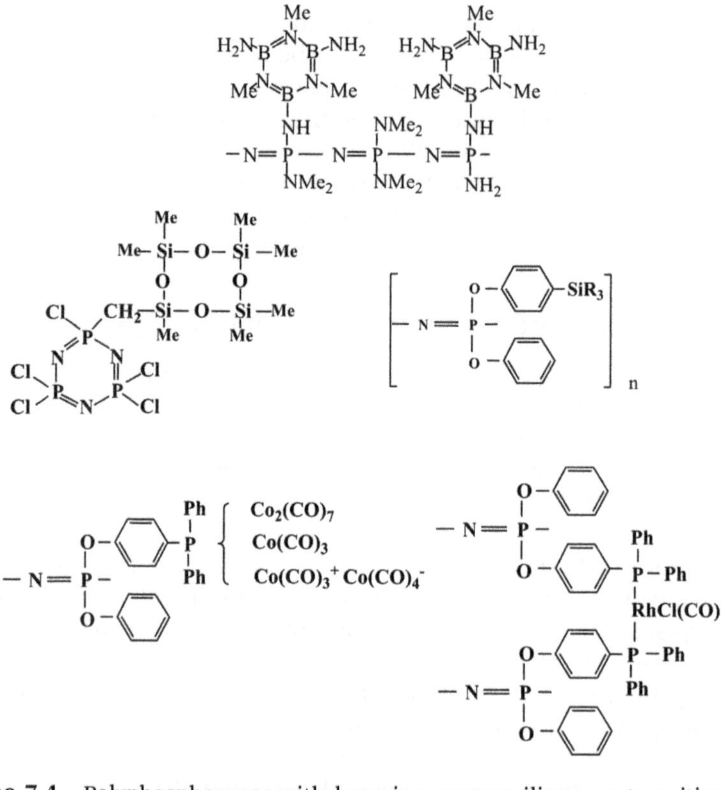

Me

Me

H₂N–B=N–B–NH₂   H₂N–B=N–B–NH₂

Me–N=B–N–Me   Me–N=B–N–Me

NH   NMe₂   NH

$-N=P-\ N=P-\ N=P-$

NMe₂   NMe₂   NH₂

Me   Me

Me–Si–O–Si–Me

O   O

Cl   CH₂–Si–O–Si–Me

Me   Me

$\begin{bmatrix} -N=P- \end{bmatrix}_n$ with O–⟨⟩–SiR₃

Co₂(CO)₇
Co(CO)₃
Co(CO)₃⁺ Co(CO)₄⁻

RhCl(CO)

Scheme 7.4 Polyphosphazenes with borazine, organosilicon, or transition metal organometallic side groups.

metallocene side groups are accessible by the reactions shown in Scheme 7.5. Here again, the fluorine atoms can be replaced in the usual way by organic co-substituents.

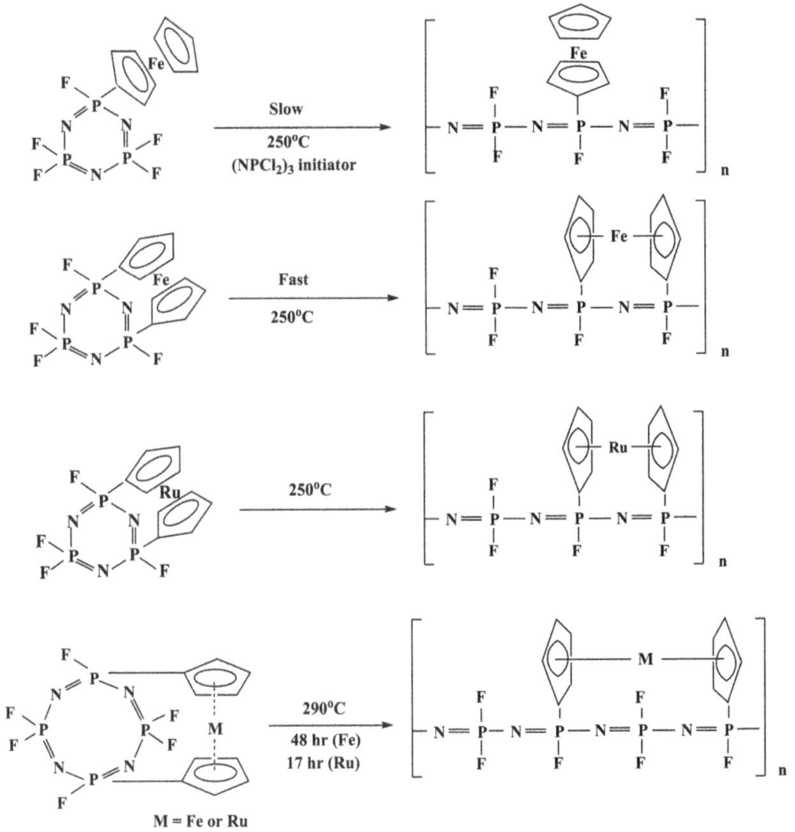

Scheme 7.5 Synthesis of polyphosphazenes with metallocene side units.

7.5 Different Architectures: Block Copolymers, Combs, Grafts, Stars, and Dendrimers

The polymer skeleton is not restricted to linear phosphazene macromolecules, but includes a number of other systems developed in our laboratory. These are shown in Scheme 7.6.

7.5.1 Block Copolymers

Access to the living cationic polymerization technique mentioned earlier has allowed the synthesis of a range of block copolymers, either phosphazene–phosphazene blocks, phosphazene–organic polymer blocks, or phosphazene–polyorganosiloxane blocks.[103–114] Both di- and triblock copolymers have been produced, and a summary of these structures is shown in Scheme 7.7.

There are three techniques to prepare block copolymers. In the first, a nonterminated living polyphosphazene is allowed to react with the

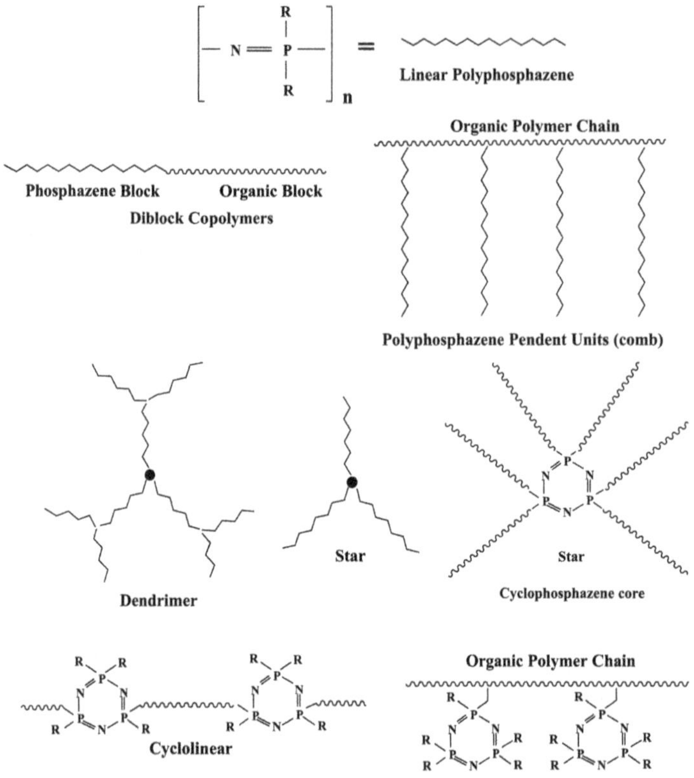

Scheme 7.6 Different structures obtained from polyphosphazenes.

functional end of an organic or organosilicon polymer, which acts as a ter-
minator. In the second method, the organic polymer terminus is functionalized with a phosphoranimine end group, and a polyphosphazene chain is
grown from that terminus. The third method is where a living polyphosphazene end unit is employed as a cationic initiator for the addition polymerization of an organic monomer. In practice, method 1 has been the most
widely used, with method 2 being the next most useful. Method 3 has not yet
been exploited in detail.

Several reasons exist for the synthesis of block copolymers. One reason is
to produce amphiphilic polymers that may self-assemble in a liquid medium
to form micelles. If the medium is aqueous, then the hydrophobic blocks
occupy the micelle interior. This arrangement facilitates the capture of
hydrophobic molecules such as hydrophobic drugs in the micelle interior,
thus providing a mechanism for the transport of hydrophobic molecules in
an aqueous medium. In organic media the opposite structure may be
formed, with the hydrophilic blocks occupying the interior of the micelle. In
the solid state, block copolymers may self-assemble to generate spherical
micelles, worms, layers, or toroidal structures.

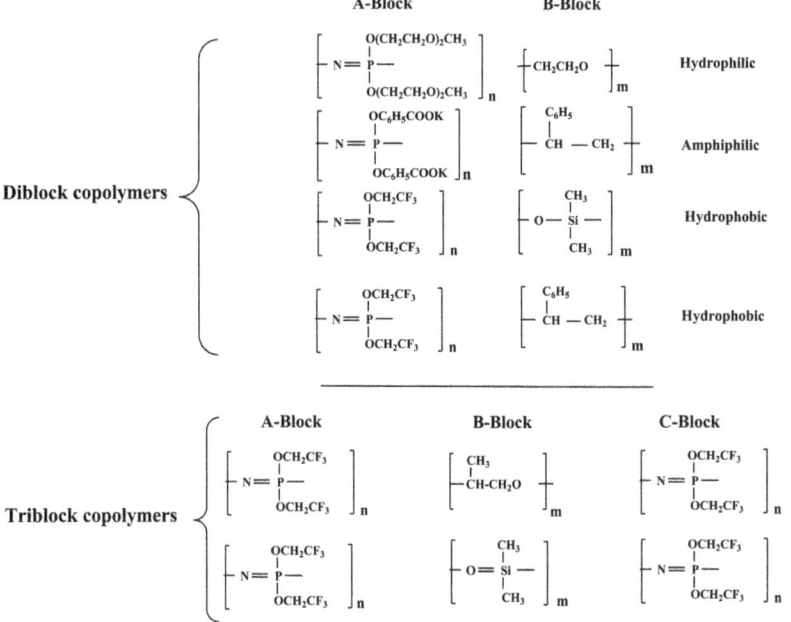

Scheme 7.7 Examples of phosphazene block copolymers prepared by the living cationic polymerization route.

7.5.2 Grafts, Combs, Stars, and Dendrimers

These structures are accessible by a variety of techniques that include atom transfer radical polymerization (ATRP) methods and the use of living cationic polymerization.[115–117] Tri-arm star structures are accessible by the living cationic growth of three chains from a triamino core structure[118] and dendrimers have been obtained by the growth of structures based on a diaminobutane polypropyleneimine core with polyphosphazene outer branches.[119]

7.5.3 Polymers with Pendent Cyclotriphosphazene Rings

The polymerization of cyclic organic monomers to which are attached cyclophosphazene rings yields linear organic polymers with pendent phosphazene units (Scheme 7.8). This has been accomplished in our program using the following method. Hexaorgano-substituted cyclotriphosphazenes were synthesized with one site bearing a norbornene or cyclooctene group. Ring-opening metathesis polymerization (ROMP) of the unsaturated cyclic organic units then yielded polynorbornene or polyoctenamer polymers with pendent phosphazene rings. The side groups on the rings may be hydrophobic units such as trifluoroethoxy or ion coordinative groups such as methoxyethoxyethoxy units, or both.[120] This is an alternative to the free

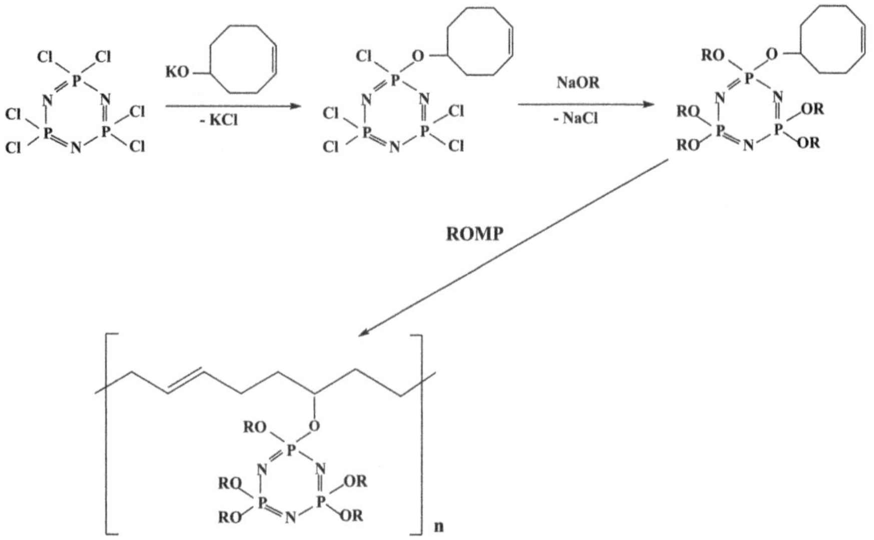

Scheme 7.8 An organic polymer with cyclophosphazene side groups formed by the polymerization of cyclooctene molecules pendent to a cyclophosphazene ring.

radical polymerization of vinyl or allyl groups linked to cyclophosphazene rings employed by Allen and co-workers.[121] Another architecture studied recently consists of a cyclotriphosphazene core with six organic polymer chains radiating from this core.

7.6 Polyphosphazenes with Other Elements in the Main Chain

In addition to the systems discussed above, an attempt has been made to expand the scope of phosphazene polymers by the incorporation of skeletal elements other than phosphorus and nitrogen. Those developed to date include examples with carbon, thio, or thionyl units in the skeleton together with phosphorus and nitrogen (Scheme 7.9).

These polymers were obtained by ring-opening polymerization of the appropriate cyclic monomers, with stabilization against hydrolysis subsequently accomplished by replacement of the chlorine atoms by organic groups. The introduction or carbon into the backbone has a stiffening effect and results in an increase of roughly 10–20 °C in the glass transition temperature. Presumably this effect is a consequence of the introduction of p_π-p_π bonding, with a consequent increase in the torsional barrier height.[122,123] Poly(thiophosphazenes) are also accessible by the polymerization of the cyclic trimer,[124,125] and the resultant polymer reacts with nucleophiles in the usual way. The S–Cl bonds are more reactive than the P–Cl bonds, and this presents scope for the controlled introduction of

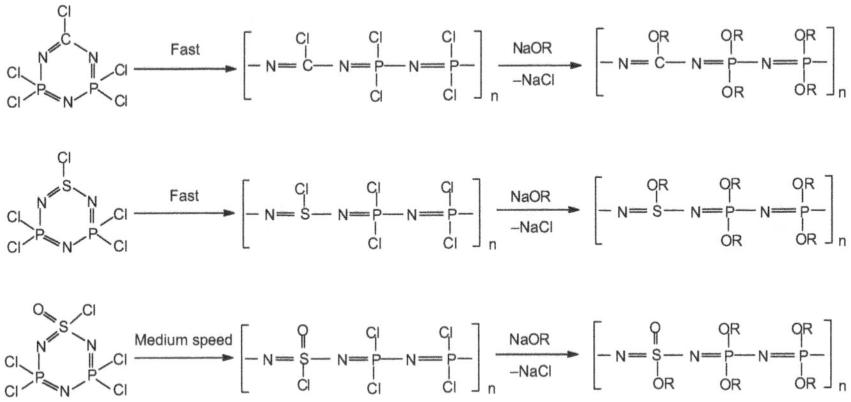

Scheme 7.9 Synthesis of polyphosphazenes with carbon or sulfur as the third element in the backbone.

different substituents at specific locations along the chain. Poly(thionylphosphazenes) were produced similarly by polymerization of the appropriate cyclothionylphosphazene.[126,127]

7.7 Actual and Potential Uses of Polyphosphazenes

7.7.1 Fluoroalkoxyphosphazene Polymers

The first polyphosphazenes to be developed commercially had trifluoroethoxy and a mixture of longer chain fluoroalkoxy groups distributed randomly along the polymer chains.[128–135] The mixture of different side groups introduces sufficient disorder that these derivatives do not crystallize, but form a series of elastomers. The difference between a single-substituent and a multiple substituent polymer is shown in Figures 7.2 and 7.3. The telomer alkoxy unit was chosen because the terminal CF_2H group gives polymers that are more soluble and easier to process than the counterparts with $-CF_3$ end groups. Polymers of this type, known as "PN-F" or "Eypel-F", have been employed to make fuel lines, gaskets, and O-rings for aerospace and land vehicles because of their flexibility at low temperatures, resistance to hydrocarbons or aggressive hydraulic fluids, fire resistance, and their biological compatibility. Cardiovascular components and dental devices[136] have also been made from this elastomer. The single-substituent polymer, $[NP(OCH_2CF_3)_2]_n$, has been reported to be a useful coating in cardiovascular devices. This same polymer and PN-F have recently shown promise as anti-icing coatings for aircraft and helicopters.

7.7.2 Polyphosphazenes with Oligoethylenoxy Side Chains

An example is the polymer MEEP (methoxyethoxyethoxyphosphazene), shown in Table 7.1. This polymer is an amorphous gum that has the ability

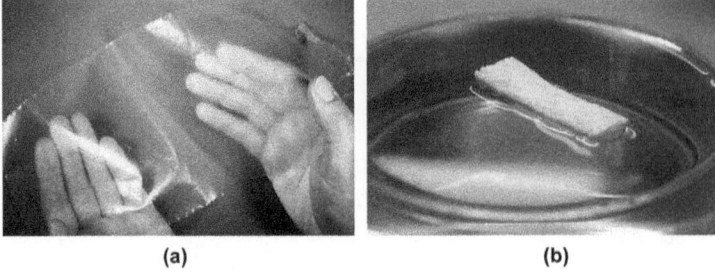

Figure 7.2 Poly[bis(2,2,2-trifluoroethoxy)phosphazene], [NP(OCH$_2$CF$_3$)$_2$]$_n$, can be solvent cast into (a) films or (b) rigid nonflammable expanded foam materials.

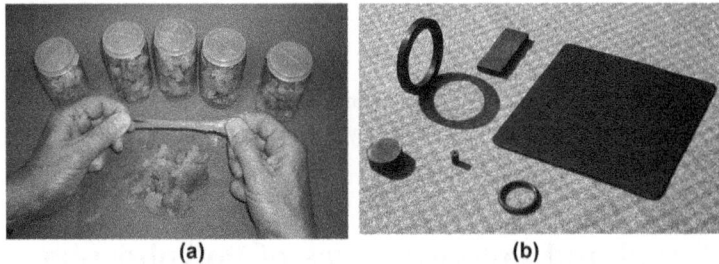

Figure 7.3 Mixed-substituent fluoroalkoxyphosphazenes such as PN-F are (a) elastomers that are compounded into (b) rubbery materials appropriate for aerospace and automotive applications.

to serve as a solvent for salts such as lithium triflate.[137-147] As such, it functions as the electrolyte in experimental rechargeable lithium batteries, to replace the organic solvent-based electrolytes currently in common use. Test batteries do not leak solvent when overcharged, and the electrolyte is nonflammable. The conductivity of this electrolyte is in the range of 10^{-4} S cm^{-1}, which is lower than liquid-based systems, but the use of a small amount of a plasticizer such as propylene carbonate raises the ionic conductivity to acceptable levels and has a minimal effect on the gel structure or the fire resistance. We have recently used a system of MEEP plasticized by phosphate esters, which provides some advantages over the MEEP/propylene carbonate system. MEEP has another important property: it is soluble in water and is infinitely stable in this medium. It can be crosslinked by UV or γ radiation, and the crosslinked system swells to form a hydrogel as it absorbs water, but it does not dissolve (Figure 7.4).[148-150] Both the aqueous solution and the hydrogel undergo an important transition as the temperature is raised. At room temperature the uncrosslinked polymer is soluble in water, and the hydrogel is fully expanded. As the temperature is raised a point is reached (the critical solution temperature) where the uncrosslinked polymer becomes insoluble and the gel contracts by exuding water. Such behavior is utilized in "intelligent" membranes in which the permeability of the membrane can be controlled by temperature changes. Use has been made of this

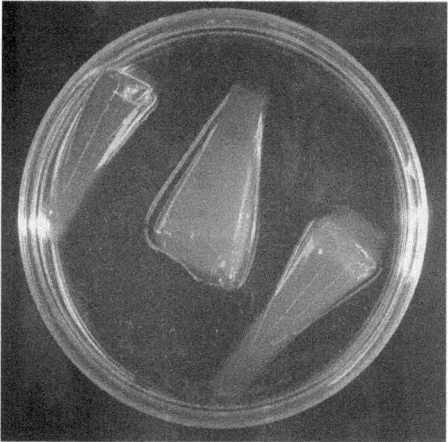

Figure 7.4 Hydrogels from lightly crosslinked MEEP absorb water but do not dissolve. The gels extrude water when the temperature is raised above the lower critical solution temperature (65 °C). Hydrogels derived from the ethoxyethoxyethoxy derivative have a lower critical solution temperature (LCST) of 38 °C (close to human body temperature).

phenomenon to trap enzymes in the gel and control enzyme activity by expanding or contracting the membrane. Polyphosphazenes with variety of different alkylenoxy side chains behave in the same way.

7.7.3 Poly(*p*-carboxyphenoxyphosphazene)

This polymer, abbreviated to PCPP, has the structure $[NP(OC_6H_4CO_2H)_2]_n$.[151–155] It is produced by the prior preparation of the ethyl ester of the polymer side group followed by removal of the ester protective group. PCPP has moderate solubility in water, but treatment with base generates the sodium salt, which is very soluble in water. Solutions of the sodium salt can be converted to a gel by the addition of a soluble calcium salt, which forms ionic crosslinks. This is used as a process for the immobilization of vaccines in the form of microspheres. Considerable development work has been carried out to utilize this system as an oral-delivery anti-influenza drug.[156] Polyphosphazenes that bear both methoxyethoxyethoxy and *p*-carboxyphenoxy side groups have properties that reflect the characteristics of both side groups. For example, they have been used as membranes that respond to pH, calcium salts, ion strength, and temperature to allow or restrict the transmission of small molecules, or to control access of small molecules to enzymes trapped in the membrane.[154,155]

7.7.4 Fire-Resistant Polymers

Poly(diphenoxyphosphazene), $[NP(OPh)_2]_n$, was developed commercially as an expanded foam material for fire-resistant heat and sound insulation, and PCPP (see above) has been coupled to polyurethane precursors to make fire-resistant foam rubber.[157–159] Relatively small amounts of the polyphosphazene (~10%) have a striking fire-suppression property.

7.7.5 Optical Materials

One of the main benefits of the polyphosphazene platform is the ability to link molecules that have specific properties as small molecules and immobilize them to a polymer chain. In many cases an improvement over the small-molecule properties can be anticipated, arising from a reduction in volatility or crystallinity and improvements in mechanical properties such as impact resistance or processability.[58–66] An example is the immobilization of chromophore molecules on the phosphazene skeleton. This can be accomplished readily by the use of azo dyes that bear hydroxyl or amino functional groups as nucleophiles for reaction with poly(dichlorophosphazene).[66] High refractive index (RI) glasses are accessible *via* the linkage of biphenylenoxy, naphthyloxy, or halogenated aryloxy side groups to the chain. The high electron density of the phosphazene backbone adds an increment to the refractive index component generated by the side groups. Nonlinear optical (NLO) and liquid crystalline (LC) polymers are accessible by the use of classical small-molecule NLO or LC molecules as side groups.[64]

Polyphosphazenes with photonic properties include the examples shown in Scheme 7.10.

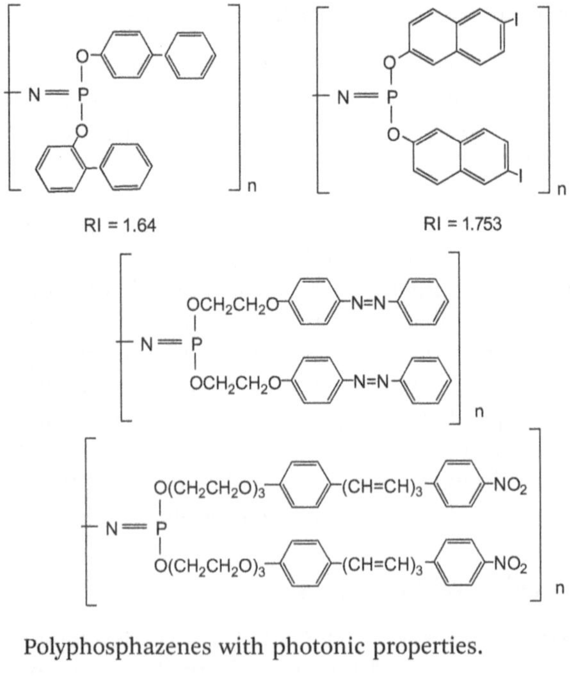

Scheme 7.10   Polyphosphazenes with photonic properties.

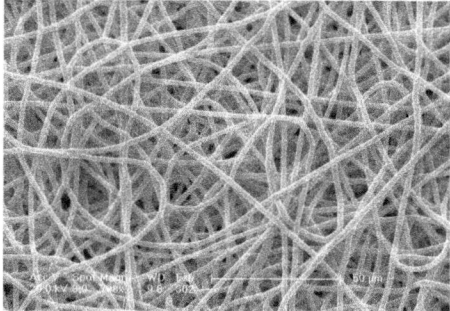

Figure 7.5 Fibers of poly(diphenoxyphosphazene) prepared by electrospinning.

7.7.6 Ion Conducting Membranes

The polymer known as MEEP is a good candidate for use as a non-volatile, nonflammable membrane electrolyte in rechargeable lithium batteries.[137–147] It is believed that lithium ions are transported across the membrane by "hopping" from one etheric side group oxygen site to another. However, this process can be amplified by the introduction of plasticizers such as propylene carbonate or nonflammable phosphate esters. Polymers designed for use in specialized batteries, such as lithium seawater primary cells, have also be synthesized.[160] Proton conducting membranes are important in low-temperature hydrogen or direct methanol fuel cells. Such membranes have been produced by the sulfonation or trifluoromethylsulfonation of aryloxy-substituted polyphosphazenes.[161–167]

7.7.7 Films, Fibers, and Nanofibers

Finally, many polyphosphazenes have been studied as thin films, solvent-extruded fibers, and nanofibers by solvent casting, spin casting, solvent extrusion, and electrospinning methods (Figures 7.2 and 7.5). These techniques have proved useful for semi-crystalline fluoroalkoxyphosphazene polymers and for aryloxy-substituted derivatives.

7.8 Future Challenges

From a fundamental scientific point of view, opportunities and challenges exist in the expansion of the polyphosphazene platform to incorporate elements other than phosphorus and nitrogen. As discussed, this has already been accomplished for skeletal systems that contain phosphorus, nitrogen, and carbon or sulfur, but numerous alternatives exist using other elements from both the main group and transition series. Especially intriguing is the possibility that transition metals can be incorporated into the skeleton with the possible development of electronic conductivity along the skeleton or the preparation of new types of immobilized, recoverable catalysts. The main challenges in the near future do not lie in the fundamental synthesis or molecular design aspects, which are well developed at the laboratory level. The main challenge is in the scale-up of these syntheses to allow the different polymers to be evaluated and developed for their industrial utility. This is the main barrier to the utilization of nearly all new polymers, and polyphosphazenes are no exception. The unusual synthesis methods, that involve the use of hydrolytically sensitive intermediates, have undoubtedly discouraged many organizations from participating in this field, as indeed also happened initially with the poly(organosiloxanes) (silicones). However, large-scale control of the ring-opening polymerization of $(NPCl_2)_3$ and the thermal condensation of $Cl_3P=N–POCl_2$ has been accomplished as part of the manufacture of fluoroalkoxyphosphazene elastomers and aryloxyphosphazene fire-retardant polymers, respectively. Thus, based on both laboratory and pilot plant studies, there appear to be no impediments to the scale-up of many other polyphosphazenes and of their commercialization. The range of properties accessible *via* the polyphosphazene platform is so wide that it seems inevitable that this will occur in the near future.

References

1. E. Thilo, *Adv. Inorg. Chem. Radiochem.*, 1962, **4**, 1.
2. J. R. Van Wazer, *Phosphorus and its Compounds*, Wiley-Interscience, New York, 1958.
3. H. R. Allcock and R. L. Kugel, *J. Polym. Sci., Part A: Gen. Pap.*, 1963, **1**, 3627.[†]
4. H. R. Allcock, *J. Polym. Sci., Part A: Gen. Pap.*, 1964, **2**, 4087.
5. H. R. Allcock, *Phosphorus-Nitrogen Compounds*, Academic Press, New York, 1972.
6. H. R. Allcock and R. L. Kugel, *J. Am. Chem. Soc.*, 1965, **87**, 4216.
7. H. R. Allcock, R. L. Kugel and K. J. Valan, *Inorg. Chem.*, 1966, **5**, 1709.
8. H. R. Allcock and R. L. Kugel, *Inorg. Chem.*, 1966, **5**, 1716.
9. H. R. Allcock, J. E. Gardner and K. M. Smeltz, *Macromolecules*, 1975, **8**, 36.
10. H. R. Allcock and G. Y. Moore, *Macromolecules*, 1975, **8**, 377.
11. H. R. Allcock, J. L. Schmutz and K. M. Kosydar, *Acc. Chem. Res.*, 1978, **11**, 179.
12. H. R. Allcock, D. B. Patterson and T. L. Evans, *Macromolecules*, 1979, **12**, 172.
13. R. J. Ritchie, P. J. Harris and H. R. Allcock, *Macromolecules*, 1979, **12**, 1014.
14. H. R. Allcock, T. L. Evans and D. B. Patterson, *Macromolecules*, 1989, **13**, 201.

[†]A complete list of 612 publications from the Penn State group can be found at www.research.chem.psu.edu/hragroup.

15. H. R. Allcock, R. J. Ritchie and P. J. Harris, *Macromolecules*, 1980, **13**, 1332.
16. P. E. Austin, G. H. Riding and H. R. Allcock, *Macromolecules*, 1983, **16**, 719.
17. H. R. Allcock and M. S. Connolly, *Macromolecules*, 1985, **18**, 1330.
18. H. R. Allcock, Generation of structural diversity in polyphosphazenes, in *Applied Organometallic Chemistry*, ed. R. Laine, Wiley Interscience, 2013, vol. 4, pp. 1826–1835.
19. C. H. Honeyman, I. Manners, C. T. Morrissey and H. R. Allcock, *J. Am. Chem. Soc.*, 1995, **117**, 7035.
20. H. R. Allcock, C. A. Crane, C. T. Morrissey, J. M. Nelson, S. D. Reeves, C. H. Honeyman and I. Manners, *Macromolecules*, 1996, **29**, 7740.
21. H. R. Allcock, J. M. Nelson, S. D. Reeves, C. H. Honeyman and I. Manners, *Macromolecules*, 1997, **30**, 50.
22. J. M. Nelson and H. R. Allcock, *Macromolecules*, 1997, **30**, 1854.
23. H. R. Allcock, S. D. Reeves, J. M. Nelson, C. A. Crane and I. Manners, *Macromolecules*, 1997, **30**, 2213.
24. J. M. Nelson, H. R. Allcock and I. Manners, *Macromolecules*, 1997, **30**, 3191.
25. J. M. Nelson, A. P. Primrose, T. J. Hartle, H. R. Allcock and I. Manners, *Macromolecules*, 1998, **31**, 947.
26. H. R. Allcock, C. A. Crane, C. T. Morrissey and M. A. Olshavsky, *Inorg. Chem.*, 1999, **38**, 280.
27. H. R. Allcock, J. M. Nelson, R. Prange, C. A. Crane and C. R. de Denus, *Macromolecules*, 1999, **32**, 5736.
28. R. Prange and H. R. Allcock, *Macromolecules*, 1999, **32**, 6390.
29. H. R. Allcock, C. R. de Denus, R. Prange and J. M. Nelson, *Macromolecules*, 1999, **32**, 7999.
30. H. R. Allcock, S. D. Reeves, J. M. Nelson and I. Manners, *Macromolecules*, 2000, **33**, 3999.
31. H. R. Allcock, S. D. Reeves, C. R. de Denus and C. A. Crane, *Macromolecules*, 2001, **34**, 748.
32. Y. Chang, S. C. Lee, C. Kim, S. D. Reeves and H. R. Allcock, *Macromolecules*, 2001, **34**, 269.
33. C. Kim, Y. Chang, S. C. Lee, H. R. Allcock and S. D. Reeves, *Polym. Prepr. (Am. Chem. Soc., Div. Polym. Chem.)*, 2000, **41**, 609.
34. R. Prange, S. D. Reeves and H. R. Allcock, *Macromolecules*, 2000, **33**, 5763.
35. H. R. Allcock, R. Prange and T. J. Hartle, *Macromolecules*, 2001, **34**, 5463.
36. H. R. Allcock and R. Prange, *Macromolecules*, 2001, **34**, 6858.
37. Y. Chang, E. S. Powell and H. R. Allcock, *J. Polym. Sci., Part A: Polym. Chem.*, 2005, **43**, 2912.
38. H. R. Allcock, E. S. Powell, Y. Chang and C. Kim, *Macromolecules*, 2004, **37**, 7163.

39. H. R. Allcock, S. Y. Cho and L. B. Steely, *Macromolecules*, 2006, **39**, 8334.
40. S. Y. Cho and H. R. Allcock, *Macromolecules*, 2007, **40**, 3115.
41. Y. Chang, S. Y. Cho, L. B. Steely and H. R. Allcock, *J. Polym. Sci., Part A: Polym. Chem.*, 2009, **47**, 692.
42. S. Y. Cho and H. R. Allcock, *Macromolecules*, 2009, **42**, 4484.
43. X. Liu, Z. Tian, C. Chen and H. R. Allcock, *Macromolecules*, 2012, **45**, 1417.
44. Z. Tian, X. Liu, C. Chen and H. R. Allcock, *Macromolecules*, 2012, **45**, 2502.
45. R. De Jaeger and P. Potin, *Synthesis and Characterization of Poly-(organophosphazenes)*, ed. M. Gleria and R. De Jaeger, Nova Science, New York, 2004, pp. 25–45.
46. R. H. Neilson and P. Wisian-Neilson, *J. Am. Chem. Soc.*, 1980, **102**, 2848.
47. R. H. Neilson and P. Wisian-Neilson, *Chem. Rev.*, 1988, **88**, 541.
48. P. Wisian-Neilson, *Synthesis and Characterization of Poly(organophos-phazenes)*, ed. M. Gleria and R. De Jaeger, Nova Science, New York, 2004, pp. 109–124.
49. E. P. Flindt and H. Z. Rose, *Z. Anorg. Allg. Chem.*, 1977, **428**, 204.
50. R. A. Montague and K. Matyjaszewski, *J. Am. Chem. Soc.*, 1990, **112**, 6721.
51. M. White and K. Matyjaszewski, *Macromol. Chem. Phys.*, 1997, **198**, 665.
52. K. Matyjaszewski and M. Cypryk, J. R. Montague and M. White, *Makromol. Chem., Macromol. Symp.*, 1992, **54/55**, 13.
53. M. J. S. Dewar, E. A. C. Lucken and M. A. Whitehead, *J. Chem. Soc.*, 1960, 2423.
54. A. Singh, N. Krogman, S. Sethuraman and L. S. Nair, J. L. Sturgeon, P. W. Brown, C. T. Laurencin and H. R. Allcock, *Polym. Prepr. (Am. Chem. Soc., Div. Polym. Chem.)*, 2005, **46**, 713.
55. H. R. Allcock, L. B. Steely and A. Singh, *Polym. Int.*, 2006, **55**, 621.
56. H. R. Allcock, in *Handbook of Fluoropolymer Science and Technology*, ed. D. W. Smith, S. S. Iyer and S. T. Iacano, Wiley, Hoboken, NJ, in press.
57. H. R. Allcock, T. L. Evans and D. B. Patterson, *Macromolecules*, 1980, **13**, 201.
58. H. R. Allcock and C. Kim, *Macromolecules*, 1990, **24**, 2841.
59. A. A. Dembek, C. Kim, H. R. Allcock, R. L. S. Devine, W. H. Steier and C. W. Spangler, *Chem. Mater.*, 1990, **2**, 97.
60. H. R. Allcock and C. Kim, *Macromolecules*, 1991, **113**, 2628.
61. H. R. Allcock, A. A. Dembek, C. Kim, R. L. S. Devine, Y. Shi, W. H. Steier and C. W. Spangler, *Macromolecules*, 1991, **24**, 1000.
62. H. R. Allcock and E. H. Klingenberg, *Macromolecules*, 1995, **28**, 4351.
63. M. Olshavsky and H. R. Allcock, *Macromolecules*, 1997, **30**, 4179.
64. H. R. Allcock, J. D. Bender, Y. Chang, M. McKenzie and M. M. Fone, *Chem. Mater.*, 2003, **15**, 473.
65. H. R. Allcock, M. N. Mang, A. A. Dembek and K. J. Wynne, *Macromolecules*, 1989, **22**, 4179.

66. H. R. Allcock, S. D. Wright and K. M. Kosydar, *Macromolecules*, 1978, **11**, 357.
67. G. A. Carriedo, F. J. Garcia Alonso and P. A. Gonzalez, *Macromol. Rapid Commun.*, 1997, **18**, 371.
68. H. R. Allcock, W. J. Cook and D. P. Mack, *Inorg. Chem.*, 1972, **11**, 2584.
69. H. R. Allcock, R. W. Allen and J. P. O'Brien, *J. Chem. Soc., Chem. Commun.*, 1976, 717.
70. H. R. Allcock, R. W. Allen and J. P. O'Brien, *J. Am. Chem. Soc.*, 1977, **99**, 3984.
71. H. R. Allcock, P. P. Greigger, J. E. Gardner and J. L. Schmutz, *J. Am. Chem. Soc.*, 1979, **101**, 606.
72. N. R. Krogman, M. D. Hindenlang, L. S. Nair, C. T. Laurencin and H. R. Allcock, *Macromolecules*, 2008, **41**, 8467.
73. H. R. Allcock and N. L. Morozowich, *Polym. Chem.*, 2012, **3**, 578.
74. H. R. Allcock, T. J. Fuller, D. P. Mack, K. Matsumura and K. M. Smeltz, *Macromolecules*, 1977, **10**, 824.
75. H. R. Allcock, S. R. Pucher and A. G. Scopelianos, *Biomaterials*, 1994, **15**, 563.
76. H. R. Allcock, S. R. Pucher and A. G. Scopelianos, *Macromolecules*, 1994, **27**, 1071.
77. J. H. Goedemoed, E. G. H. Mense, K. De Groot, A. M. E. Claessen and R. J. Scheper, *J. Controlled Release*, 1991, **17**, 245.
78. J. Vandorpe, E. Schacht, S. Stolnik, M. C. Garnett, M. C. Davies, L. Illum and S. S. Davis, *Biotechnol. Bioeng.*, 1996, **52**, 89.
79. F. M. Veronese, F. Marsilio, P. Caliceti, P. De Filippis, P. Giunchedi and P. Lora, *J. Controlled Release*, 1998, **52**, 227.
80. N. L. Morozowich, R. J. Mondschein and H. R. Allcock, *Polymer Chemistry*, 2012, **3**(3), 778.
81. J. L. Nichol, N. L. Morozowich and H. R. Allcock, *Polym. Chem*, 2013, **4**, 600.
82. H. R. Allcock, *Angew. Chem. Int. Ed. Eng.*, 1977, **16**, 147.
83. M. Deng, L. S. Nair, S. Nukavarapu, S. G. Kumbar, N. R. Krogman, H. R. Allcock and C. T. Laurencin, *Biomaterials*, in press.
84. M. D. Hindenlang, A. A. Soudakov, G. H. Imler, C. T. Laurencin, L. S. Nair and H. R. Allcock, *Polym. Chem.*, 2010, **1**, 1467.
85. A. L. Weikel, N. R. Krogman, N. Q. Nguyen, C. T. Laurencin and H. R. Allcock, *Macromolecules*, 2009, **42**, 636.
86. N. R. Krogman, N. Q. Nguyen, A. L. Weikel, L. S. Nair, C. T. Laurencin and H. R. Allcock, *Macromolecules*, 2008, **41**, 7824.
87. S. Sethuraman, L. S. Nair, S. El-Amin, M. N. Nguyen, Y. E. Greish, J. D. Bender, P. W. Brown, H. R. Allcock and C. T. Laurencin, *J. Biomed. Mater. Res., A*, in press.
88. S. F. El-Amin, M. S. Kwon, T. Starnes, H. R. Allcock and C. T. Laurencin, *J. Inorg. Organomet. Polym. Mater.*, 2006, **16**, 387.
89. A. Singh, N. Krogman, S. Sethuraman, L. S. Nair, J. L. Sturgeon, P. W. Brown, C. T. Laurencin and H. R. Allcock, *Polym. Prepr. (Am. Chem. Soc., Div. Polym. Chem.)*, 2005, **46**, 713.

90. R. J. Davidson, E. W. Ainscough, A. M. Brodie, G. B. Jameson, M. R. Waterland, H. R. Allcock, M. D. Hindenlang, B. Moubaraki, K. Murray, K. Gordon, R. Horvath and G. L N. Jameson, *Inorg. Chem.*, 2012, **51**, 8307.

91. E. W. Ainscough, H. R. Allcock, A. M. Brodie, K. C. Gordon, M. Hindenlang, M. D. R. Horvath and C. A. Otter, *Eur. J. Inorg. Chem.*, 2011, **25**, 3691.

92. U. Diefenbach, A. M. Cannon, B. E. Stromberg, B. E. Olmeijer and H. R. Allcock, *J. Appl. Polym. Sci.*, 2000, **78**, 650.

93. H. R. Allcock and K. D. Lavin, and G. H. Riding, *Macromolecules*, 1985, **18**, 1340.

94. H. R. Allcock, T. X. Neenan and B. Boso, *Inorg. Chem.*, 1985, **24**, 2656.

95. H. R. Allcock, G. H. Riding and K. D. Lavin, *Macromolecules*, 1987, **20**, 6.

96. H. R. Allcock, M. F. Welker and M. Parvez, *Chem. Mater.*, 1992, **4**, 296.

97. H. R. Allcock, D. J. Brennan, J. M. Graaskamp and M. Parvez, *Organometallics*, 1986, **5**, 2434.

98. H. R. Allcock, D. J. Brennan, B. S. Dunn and M. Parvez, *Inorg. Chem.*, 1988, **27**, 3226.

99. H. R. Allcock, W. D. Coggio, M. Parvez and M. L. Turner, *Organometallics*, 1991, **10**, 677.

100. H. R. Allcock and D. J. Brennan, *J. Organomet. Chem.*, 1988, **341**, 231.

101. H. R. Allcock and S. E. Kuharcik, *J. Inorg. Organomet. Polym*, 1996, **6**, 1.

102. H. R. Allcock, S. E. Kuharcik and C. J. Nelson, *Macromolecules*, 1996, **29**, 3686.

103. H. R. Allcock, S. D. Reeves, J. M. Nelson, C. A. Crane and I. Manners, *Macromolecules*, 1997, **30**, 2213.

104. J. M. Nelson, A. P. Primrose, T. J. Hartle, H. R. Allcock and I. Manners, *Macromolecules*, 1998, **31**, 947.

105. H. R. Allcock, J. M. Nelson, R. Prange, C. A. Crane and C. R. de Denus, *Macromolecules*, 1999, **32**, 5736.

106. R. Prange and H. R. Allcock, *Macromolecules*, 1999, **32**, 6390.

107. H. R. Allcock, C. R. de Denus, R. Prange and C. J. Nelson, *Macromolecules*, 1999, **32**, 7999.

108. H. R. Allcock, S. D. Reeves, J. M. Nelson and I. Manners, *Macromolecules*, 2000, **33**, 3999.

109. Y. Chang, S. C. Lee, C. Kim, S. D. Reeves and H. R. Allcock, *Macromolecules*, 2001, **34**, 269.

110. R. Prange, S. D. Reeves and H. R. Allcock, *Macromolecules*, 2000, **33**, 5763.

111. H. R. Allcock, R. Prange and T. J. Hartle, *Macromolecules*, 2001, **34**, 5463.

112. Y. Chang, J. D. Bender, M. V. B. Phelps and H. R. Allcock, *Macromolecules*, 2002, **3**, 1364.

113. H. R. Allcock, S. Y. Cho and L. B. Steely, *Macromolecules*, 2006, **39**, 8334.

114. Z. Tian, X. Liu, C. Chen and H. R. Allcock, *Macromolecules*, 2012, **45**, 2502.
115. X. Liu, H. Zhang, Z. Tian, A. Sen and H. R. Allcock, *Polym. Chem.*, 2012, **3**, 2082.
116. X. Liu, Z. Tian, C. Chen and H. R. Allcock, *Polym. Chem.*, 2013, **4**, 1115.
117. X. Liu, Z. Tian, C. Chen and H. R. Allcock, *Macromolecules*, 2012, **45**, 1417.
118. J. M. Nelson and H. R. Allcock, *Macromolecules*, 1997, **30**, 1854.
119. S. Y. Cho and H. R. Allcock, *Macromolecules*, 2007, **40**, 3115.
120. H. R. Allcock, D. T. Welna and D. A. Stone, *Macromolecules*, 2005, **38**, 10406.
121. C. W. Allen, J. C. Shaw and D. E. Brown, *Macromolecules*, 1988, **21**, 2053.
122. I. Manners, G. Renner, O. Nuyken and H. R. Allcock, *J. Am. Chem. Soc.*, 1989, **111**, 5478.
123. H. R. Allcock, S. M. Coley, I. Manners, G. Renner and O. Nuyken, *Macromolecules*, 1991, **24**, 2024.
124. J. A. Dodge, I. Manners, H. R. Allcock, G. Renner and O. Nuyken, *J. Am. Chem. Soc.*, 1990, **112**, 1268.
125. H. R. Allcock, J. A. Dodge and I. Manners, *Macromolecules*, 1993, **26**, 11.
126. M. Liang and I. Manners, *J. Am. Chem. Soc.*, 1991, **113**, 4044.
127. M. Liang and I. Manners, *Macromolecules*, 1996, **29**, 3401.
128. S. H. Rose, *J. Polym. Sci., Part B: Polym. Lett.*, 1968, **6**, 837.
129. K. A. Reynard and S. H. Rose, *U.S. Pat.*, 3 700 629, 1972.
130. D. P. Tate, *J. Polym. Sci., Polym. Symp.*, 1974, **48**, 33.
131. R. E. Singler, N. S. Schneider and G. L. Hagnauer, *Polym. Eng. Sci.*, 1975, **15**, 321.
132. C. H. Kolich, W. D. Klobular and J. T. Books, *U.S. Pat.*, 4 945 193, 1990.
133. H. R. Allcock, A. E. Maher and C. M. Ambler, *Macromolecules*, 2003, **36**, 5566.
134. A. E. Maher, C. M. Ambler, E. S. Powell and H. R. Allcock, *J. Appl. Polym. Sci.*, 2004, **92**, 2569.
135. A. E. Maher and H. R. Allcock, *Macromolecules*, 2005, **38**, 641.
136. L. Gettleman, J. M. Vargo, P. H. Gebert, R. J. Leboef and H. R. Rawls, *Polym. Sci. Technol.*, 1987, **35**, 55.
137. P. M. Blonsky, D. F. Shriver, P. E. Austin and H. R. Allcock, *J. Am. Chem. Soc.*, 1984, **106**, 6854.
138. H. R. Allcock, P. E. Austin, T. X. Neenan, J. T. Sisko, P. M. Blonsky and D. F. Shriver, *Macromolecules*, 1986, **19**, 1508.
139. P. M. Blonsky, D. F. Shriver, P. E. Austin and H. R. Allcock, *Polym. Mater. Sci. Eng.*, 1985, **53**, 118.
140. J. L. Bennett, A. A. Dembek, H. R. Allcock, B. J. Heyen and D. F. Shriver, *Chem. Mater.*, 1989, **1**, 4.
141. M. Lerner, A. Tipton, D. F. Shriver, A. A. Dembek and H. R. Allcock, *Chem. Mater.*, 1991, **3**, 1117.
142. C. J. Nelson, W. D. Coggio and H. R. Allcock, *Chem. Mater.*, 1991, **3**, 786.
143. H. R. Allcock, M. E. Napierala, C. G. Cameron and S. J. M. O'Connor, *Macromolecules*, 1996, **29**, 1951.

144. H. R. Allcock, S. J. M. O'Connor, D. L. Olmeijer, M. E. Napierala and C. G. Cameron, *Macromolecules*, 1996, **23**, 7544.

145. H. R. Allcock, R. Ravikiran and S. J. M. O'Connor, *Macromolecules*, 1997, **30**, 3184.

146. H. R. Allcock, W. R. Laredo and R. V. Morford, *Solid State Ionics*, 2001, **139**, 27.

147. S.-T. Fei and H. R. Allcock, *Mater. Res. Soc. Symp. Proc.*, 2009, 1127-T01-05.

148. H. R. Allcock and S. Kwon, G. H. Riding, R. J. Fitzpatrick and J. L. Bennett, *Biomaterials*, 1988, **19**, 509.

149. H. R. Allcock, S. R. Pucher, M. L. Turner and R. J. Fitzpatrick, *Macromolecules*, 1992, **25**, 5573.

150. H. R. Allcock, S. R. Pucher and K. B. Visscher, *Biomaterials*, 1994, **15**, 502.

151. A. Andrianov, S. Cohen, R. S. Langer, K. B. Visscher and H. R. Allcock, *J. Controlled Release*, 1993, **27**, 69.

152. S. Cohen, H. R. Allcock and R. S. Langer, *Minutes (Editions de Sante, France)*, 1993, 36.

153. A. K. Andrianov, L. G. Payne, K. B. Visscher, H. R. Allcock and R. S. Langer, *J. Appl. Polym. Sci.*, 1994, **53**, 1573.

154. H. R. Allcock and A. M. A. Ambrosio, *Biomaterials*, 1996, **17**, 2295.

155. H. R. Allcock and A. M. A. Ambrosio, *ACS Symp. Ser.*, 2002, **833**, 82.

156. A. A. Andrianov (Ed.). *Polyphosphazenes for Biomedical Applications*, Wiley, New York, 2009.

157. C. A. Reed, J. P. Taylor, K. S. Guigley, K. S. Kully, K. A. Bernheim, M. M. Coleman and H. R. Allcock, *J. Polym. Sci. Eng.*, 2000, **40**, 465.

158. R. V. Morford, E. C. Kellam, M. A. Hofmann, R. Baldwin and H. R. Allcock, *Solid State Ionics*, 2000, **133**, 171.

159. C. Chen, X. Liu, Z. Tian and H. R. Allcock, *Macromolecules*, 2012, **45**, 9085.

160. D. A. Stone, D. T. Welna and H. R. Allcock, *Polym Prepr. (Am. Chem. Soc., Div. Polym. Chem.)*, 2004, **45**, 696.

161. X. Zhou, J. Weston, E. Chalkova, S. N. Lvov, M. A. Hofmann, C. M. Ambler and H. R. Allcock, *Electrochim. Acta*, 2003, **48**, 2173.

162. M. V. Fedkin, X. Y. Zhou, M. A. Hofmann, E. Chalkova, J. A. Weston, H. R. Allcock and S. N. Lvov, *Mater. Lett.*, 2002, **52**, 192.

163. H. R. Allcock, M. A. Hofmann, C. M. Ambler, S. N. Lvov, X. Y. Zhou, E. Chalkova and J. Weston, *J. Membr. Sci.*, 2002, **202**, 47.

164. M. A. Hofmann, C. M. Ambler, A. E. Maher, E. Chalkova, X. Y. Zhou, S. N. Lvov and H. R. Allcock, *Macromolecules*, 2002, **35**, 6490.

165. H. R. Allcock, M. A. Hofmann, C. M. Ambler and R. V. Morford, *Macromolecules*, 2002, **35**, 3483.

166. X. Zhou, J. Weston, E. Chalkova, S. N. Lvov, M. A. Hofmann, C. M. Ambler and H. R. Allcock, *Electrochim. Acta*, 2003, **48**, 2173.

167. C. M. Ambler, A. M. Maher, R. M. Wood, H. R. Allcock, E. Chalkova and S. N. Lvov, *ACS PMSE Div. Prepr.*, 2003, **89**, 595.

CHAPTER 8

Phosphorus-Based Monomers Used for Dental Application

NORBERT MOSZNER* AND YOHANN CATEL

Ivoclar Vivadent AG, Bendererstrasse 2, FL-9494 Schaan, Liechtenstein
*Email: norbert.moszner@ivoclarvivadent.com

8.1 Introduction to Dental Materials

Basically, dental materials can be divided into clinical and technical materials. Clinical materials are mainly used by the dentist in dental surgery, while technical materials are mostly employed by the dental technician in order to fabricate, for example, dentures, crowns, or bridges. Clinical dental materials, such as restorative composites, cements, or adhesives, are interesting fields of application for new components, which enable important properties of these materials to be improved. The currently used restorative composites, composite cements, and/or adhesives are based on a mixture of different monomers and fillers, an initiator system, and further additives, such as stabilizers, pigments, *etc.* The monomers used in these materials have to meet substantial requirements, which include various physico-chemical, processing, clinical, and toxicological properties or demands. Thus, suitable dental monomers should be liquid or low-melting solids, colorless compounds, which show, for example, a high rate in free-radical homopolymerization or copolymerization with other monomers, long-term stability against premature polymerization during storage at room temperature or in the refrigerator, excellent resistance to oral environment, high light and discoloration stability of the polymers thereof formed, a low oral toxicity and cytotoxicity, and no mutagenic or carcinogenic potential.

RSC Polymer Chemistry Series No. 11
Phosphorus-Based Polymers: From Synthesis to Applications
Edited by Sophie Monge and Ghislain David
© The Royal Society of Chemistry 2014
Published by the Royal Society of Chemistry, www.rsc.org

Furthermore, monomers for restorative composites or composite cements should exhibit a low volume contraction during polymerization, excellent mechanical properties after polymerization, and low water solubility. The water sorption of the formed polymer should also be low. The currently used direct restorative composites, composite cements, and enamel/dentin adhesives are largely based on methacrylate chemistry using mixtures of different methacrylates, such as crosslinking and functionalized methacrylates, which can be cured by free-radical polymerization.[1,2]

8.2   Enamel/Dentin Adhesives

8.2.1   Dental Adhesive Systems

Nowadays, durable esthetic tooth-colored restorative materials, particularly direct resin-based filling composites in combination with efficient enamel/dentin adhesives, play an important role in modern dentistry. In this context, the improvement of the dental adhesive technology has extensively influenced the clinical performance of modern dental restoratives.[3] The challenge of these adhesives is to achieve a strong bond between a highly hydrophobic restorative material and the hydrophilic dental hard tissues. Indeed, dentin is made up of hydroxyapatite (HAP) (47%), organic matter (30%, mainly collagen), and water (23%) and enamel is composed of HAP (92%), organic matter (2%), and water (6%). Nearly 60 years ago, Buonocore was the first who demonstrated that acid etching of enamel with phosphoric acid led to improved resin-enamel bonds.[4] Today, bonding to enamel has been proven to be durable and bonding to dentin has been significantly improved. The currently used enamel/dentin adhesives for restorative composites can be classified according to the clinical approach into etch-and-rinse adhesives (E&RAs) and self-etch adhesives (SEAs).[5,6]

The adhesion strategy of E&RAs involves three or at least two steps. The use of a three-step E&RA first of all requires the application of an acid etchant on the tooth substrate, commonly a 32–37% phosphoric acid gel (pH 0.1–0.4), followed by a rinsing step. A hydrophilic primer is then applied, followed by a separate hydrophobic adhesive resin. After a subsequent photopolymerization, the restorative material is finally integrated. In the simplified two-step systems, the second and third steps are combined. The E&RA technique is the most effective approach to achieve an efficient, strong, and stable bond to enamel. The acid etching results in a selective dissolution of hydroxyapatite crystals. The adhesive resin fills the formed retentive etch pattern and polymerizes *in situ*.[7] Regarding the adhesion to dentin, the application of the primer and adhesive resin results in the infiltration of monomers into the collagen network, which replaces the water between the collagen fibrils, and a subsequent *in situ* polymerization leads to the formation of the so-called hybrid layer and resin tags, which seal the unplugged dentin tubules.[8] Primers are usually based on water and 2-hydroxyethyl methacrylate (HEMA) containing solutions that ensure complete expansion of the collagen network and wet the collagen fibrils with hydrophilic monomers.

In contrast to E&RAs, SEAs do not require a separate etching step.[9] They are based on acidic monomers which are able to simultaneously condition and prime the dental hard tissues and therefore demineralize and infiltrate enamel and dentin to the same depth simultaneously. Consequently, the application of SEAs does not require the technique-sensitive rinsing step (wet bonding) inherent to the E&RAs. SEAs are used as two-step or one-step adhesives, depending on whether a self-etching primer and adhesive resin are separately provided or are combined into one single mixture. The nature of the dentin-adhesive interface formed by SEAs depends to a large extent on the structure of the functional monomers, which interact with the dentin. Thus, when decreasing the pH of the self-etch solutions, the hybrid layer depth increases from a few hundreds of nanometers in the case of ultra-mild SEAs (pH > 2.5), to around 1 µm for mild SEAs (pH ≈ 2), and several micrometers in the case of strong SEAs (pH ≤ 1).

8.2.2 Composition of Self-Etch Adhesives

In addition to the acidic monomer, contemporary SEAs typically contain a mixture of different co-monomers, polymerization initiators (mainly camphorquinone combined with a tertiary amine), inhibitors, fillers, and further additives.[10] These adhesives also contain solvents which influence their viscosity, wetting, and flowing behavior. Water, ethanol, isopropanol, and acetone are the most commonly used solvents. It should be mentioned that water is an indispensable component of SEAs, in order to ionize the strongly acidic monomers. The monomers are the main component of SEAs. The adhesive monomers should present a good solubility in solutions of ethanol, acetone, or other nontoxic polar solvents or their mixtures with water, sufficient hydrolytic stability in water-based adhesives, as well as optimal wetting and film-forming properties.[11]

As mentioned before, most of the currently used monomers are methacrylates.[12] According to their functionality, the used methacrylates can be divided into non-acidic and acidic functionalized monomethacrylates, crosslinking dimethacrylates or multifunctional methacrylates. The most frequently used non-acidic functionalized monomethacrylate is HEMA. HEMA is a water soluble, low-viscosity monomer that improves the miscibility and solubility of the polar and nonpolar adhesive components and the wetting behavior of the liquid adhesive on the dental hard tissues. Moreover, HEMA may stabilize the collagen fibril network and improve the dentinal permeability and monomer diffusion. Crosslinking dimethacrylates are used in enamel/dentin adhesives to form a polymer network, which leads to a number of favorable effects. Firstly, the polymerization rate is significantly increased because of the gel effect. Secondly, the mechanical properties of the polymer network are improved compared to linear polymers. Thirdly, the formed crosslinked layer is not water soluble and the degree of swelling decreases with the polymer network density. The most popular crosslinking dimethacrylates used in enamel/dentin adhesives are 2,2-bis[4-(2-hydroxy-3-methacryloyloxypropoxy)phenyl]propane (BisGMA), 1,6-bis[(2-methacryloyl-oxyethoxy)carbonylamino]-2,4,4-trimethylhexane (UDMA), triethylene glycol dimethacrylate (TEGDMA), and glycerol dimethacrylate (GDMA). Unfortunately, all these dimethacrylates are not hydrolytically stable in aqueous acid solutions and degrade under formation of the corresponding alcohols and methacrylic acid. In this context, we synthesized new bisacrylamides which showed an adequate reactivity in free-radical polymerization as well as improved hydrolytic stability under aqueous acidic conditions compared to the dimethacrylate crosslinkers.[13,14] Thus, bisacrylamides such as *N,N'*-diethyl-1,3-bis(acrylamido)propane (DEBAAP) were introduced as hydrolytically stable crosslinkers in current enamel/dentin adhesives.[15]

Acidic monomers are a key component in SEAs. They are able to etch the dental hard tissues and to promote the infiltration of the adhesive into the demineralized surfaces. Moreover, the ability of some acidic monomers to chemically adhere to the dental tissues (HAP) was clearly demonstrated. The "adhesion-decalcification" concept, described by Yoshioka *et al.*, states that the acidic monomer could either demineralize or bond to the HAP.[16] The first step consists in an ionic interaction between the acidic group and the calcium of HAP. If the resulting calcium-monomer complex is unstable, the tooth substrate will be demineralized. On the other hand, a good stability of this complex will lead to a chemical adhesion of the monomer to the HAP. These interactions may extend the bonding longevity by enhancing the sealing of the material for the prevention of nano-leakage. The prevalence of chemical adhesion is obviously directly related to the structure of the acidic monomer.[17,18] It has been reported that the functional monomers with corresponding calcium salts exhibit a low dissolution rate in water and are able to strongly adhere to the tooth surface.

Acidic monomers could be phosphates as well as carboxylic, sulfonic, or phosphonic acids. Some examples of carboxylic acid monomethacrylates are 4-(2-methacryloyloxyethyl)trimellitic acid (4-MET) and 11-methacryloyloxy-1,1-undecanedicarboxylic acid (MAC-10). Among the functionalized monomers, free-radically polymerizable phosphonic acids (PAs) and dihydrogen phosphates (DHPs) have found wide and intensive applications as adhesive components in enamel/dentin adhesives.[2,10] In this chapter, a review of the various PAs and DHPs prepared for application in dental adhesives is provided.

8.3 Polymerizable Phosphonic Acids

The first PA monomers to be included in dental materials were vinylphosphonic acid (VPA) and 4-vinylbenzylphosphonic acid (VBPA).[19] Anbar *et al.* demonstrated that the incorporation of VBPA in a dental composite resulted in an increase of the bonding strength of the material to enamel.[19] Moreover, the precoating of enamel by VPA and VBPA led to enhanced adhesion.

These trends were attributed to a strong interaction of the PA groups with HAP. Although such materials exhibited weak adhesion properties, they clearly showed the great potential of PAs in dental materials. Over the past decades, the development of new PA monomers for use in dental materials has attracted much attention. An overview of the PA methacrylates and α-substituted acrylates developed for dental applications will first be provided. Then, we will focus our attention on the preparation of hydrolytically stable monomers.

8.3.1 Methacrylates and α-Substituted Acrylates

A wide range of PA methacrylates and α-substituted acrylates have been synthesized and evaluated in dental materials. Avci's group have synthesized the aromatic PA monomers **PA-1** and **PA-2** (Scheme 8.1).[20,21]

PA-1 was shown to be able to interact with HAP. The photopolymerization behavior of **PA-1** was also investigated. Unfortunately, it was shown that **PA-1** was not able to homopolymerize. Moreover, the addition of 10 wt% of **PA-1** to Bis-GMA resulted in a drop of both the conversion and the rate of polymerization. The same drawback was observed when **PA-2** was copolymerized with HEMA, in the presence of water.

The synthesis of (meth)acryloxyalkyl 3-phosphonopropionates was described by Ikemura *et al.* (Scheme 8.2).[22,23] Those monomers were prepared *via* the esterification of 2-carboxyethylphosphonic acid with various hydroxyalkyl (meth)acrylates. Adhesive resins based on those monomers were able to provide a strong adhesion to unetched ground enamel and sandblasted Ni/Cr alloy. Some efficient 1-SEAs containing such (meth)acryloxyalkyl 3-phosphonopropionates were also developed.

In order to evaluate the influence of the nature of the acidic group on the reactivity and adhesive properties of the acidic monomer, methacrylates **PA-3**, **PA-4**, and **PA-5** were developed by Catel *et al.* (Scheme 8.3).[24] The bisphosphonic acid **PA-4** was shown to be significantly more reactive in copolymerization with DEBAAP than both **PA-3** and **PA-5**. These results were

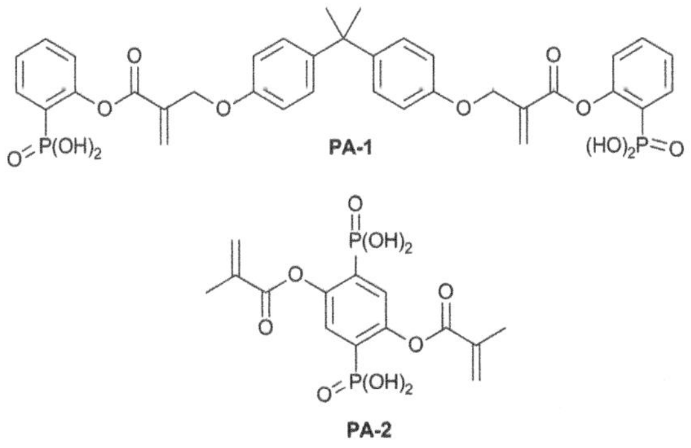

Scheme 8.1 Structure of the aromatic methacrylates **PA-1** and **PA-2**.

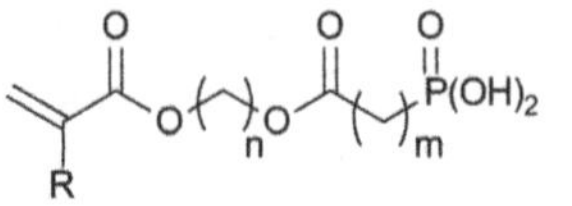

R = H or Me; 5 ≤ n ≤ 10; 1 ≤ m ≤ 2

Scheme 8.2   (Meth)acryloxyalkyl 3-phosphonopropionates.

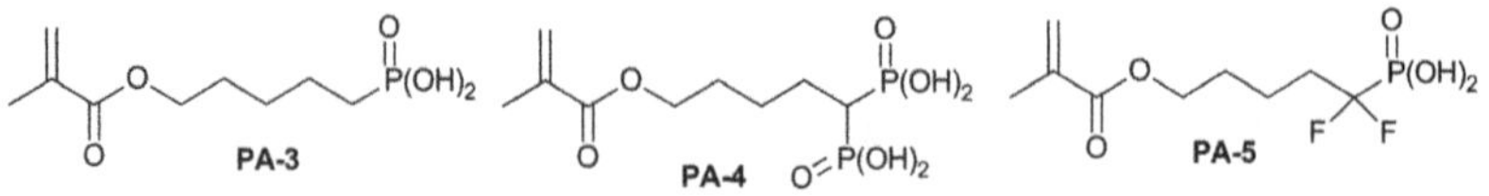

Scheme 8.3   Structures of **PA-3**, **PA-4**, and **PA-5**.

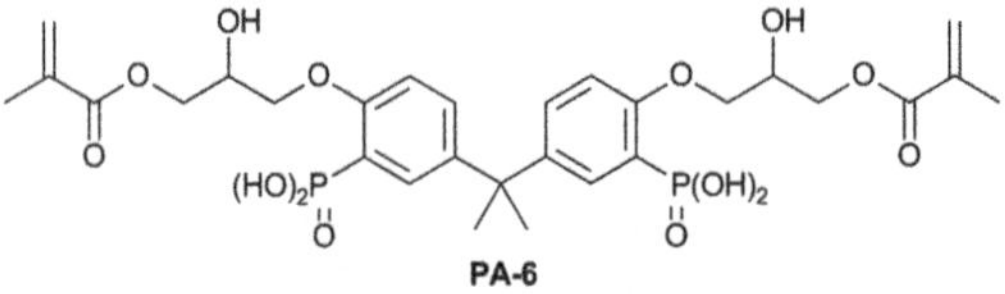

Scheme 8.4   Bis-GMA analog **PA-6**.

mainly attributed to the formation of hydrogen bonds affecting the system mobility and organization during polymerization. SEAs based on **PA-4** and **PA-5** led to better dentin shear bond strengths (SBSs) than the formulation containing **PA-3** and a commercial adhesive. Therefore, this work highlighted the important role of the nature of the acidic group on dentin adhesion.

Bis-GMA is a well-known crosslinking monomer, which is incorporated in many dental composites and adhesives. It is added in various formulations in order to significantly increase the mechanical properties. In this context, Mou *et al.* undertook the synthesis of **PA-6**, an analog of Bis-GMA bearing two PA groups on the aromatic moieties (Scheme 8.4).[25]

Photopolymerization studies showed that the homopolymerization of **PA-6** resulted in a significantly lower DC (degree of conversion) than during the polymerization of Bis-GMA.[26] This finding was attributed to an inhibition of the polymerization, induced by a higher stability of the formed free radicals. It might be also due to a partial protonation of the amine co-initiator, leading to a decrease of the amount of radicals formed during the initiation step. Unfortunately, **PA-6** was not evaluated in dental materials.

Recently, Garska *et al.* took an interest in the preparation of methacrylated calix[4]arene PAs (Scheme 8.5).[27] Moszner *et al.* had previously demonstrated that the incorporation of modified calix[4]arenes into dental materials results in a significant decrease of the polymerization shrinkage.[28] As an extension of this work, the phosphonic acid **PA-7** and the diphosphonic acid **PA-8** were prepared in three steps, starting from *p-tert*-butylcalix[4]arene. To evaluate their adhesive properties, those two monomers were

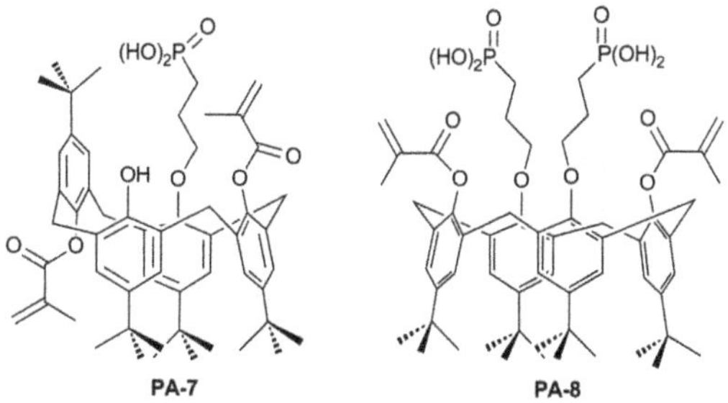

Scheme 8.5   Methacrylated calix[4]arenes bearing PA groups.

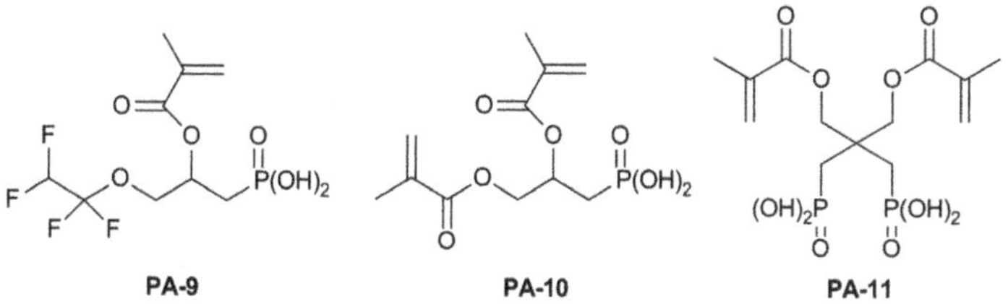

Scheme 8.6   Structures of **PA-9**, **PA-10**, and **PA-11**.

incorporated in an adhesive formulation similar to the commercially available Excite (Ivoclar Vivadent AG). The new adhesives were tested and directly compared with the original Excite. Unfortunately, significantly lower SBSs to dentin were measured when using the formulations containing **PA-7** and **PA-8**.

We contributed to the development of acidic methacrylates by synthesizing a new family of monofunctional and crosslinking methacrylates.[29] Photopolymeization experiments showed that each synthesized monomer was significantly more reactive than the commonly used HEMA. Among the isolated monomers, **PA-9**, **PA-10** and **PA-11** exhibited promising properties (Scheme 8.6). Indeed, adhesives based on these monomers were formulated and provided similar SBS to dentin as some commercial formulations.

Although some of the presented acidic methacrylates are promising candidates to enter dental materials, they also suffer a major drawback. Indeed, the methacrylate group is not stable in acidic aqueous conditions and tends to hydrolyze.[30,31] As a consequence, acidic methacrylates are not stable when stored in water. They should therefore preferably be used in materials which do not contain water, such as self-etch cements. SEAs comprising such monomers need to be stored in the refrigerator and have a limited shelf life. In order to circumvent this problem, new monomers exhibiting improved stability towards hydrolysis were synthesized.

8.3.2   Hydrolytically Stable Monomers

The first method to improve the hydrolytic stability of the acidic monomers consisted in the preparation of compounds which contained hydrolytically stable bonds between the acidic and polymerizable groups. In this context, we first synthesized the ethyl 2-[4-(dihydroxyphosphoryl)-2-oxybutyl]acrylate **PA-12**, in which the acrylate group is connected to a phosphonic acid group *via* a hydrolytically stable ether bond (Scheme 8.7).[32,33]

This monomer showed enhanced hydrolytic stability as well as great adhesive properties. Nowadays, **PA-12** is incorporated in some commercially available adhesives from Ivoclar Vivadent AG (AdhESE, Excite). Although it does not impair its adhesive potential, **PA-12** tends to slowly hydrolyze in water to form the compound **PA-13** (Scheme 8.7). To overcome this problem, we synthesized the monomers **PA-14**, **PA-15**, and **PA-16**, which were totally stable in aqueous solutions (Scheme 8.7).[33,34] Unfortunately, **PA-14** and **PA-15** showed a significant lower reactivity in radical homopolymerization than **PA-12**. Adhesives based on **PA-14** and **PA-15** were formulated: **PA-15** provided a similar SBS to dentin as **PA-12**, whereas **PA-14** exhibited very poor adhesive properties. The higher stability of **PA-16** in comparison with **PA-12** is due to the steric hindrance of the mesityl group. The presence of this mesityl group also induces a decrease of reactivity. Indeed, **PA-16** has been shown to be less reactive in free-radical polymerization than MMA or **PA-12**.[35]

Using a similar approach, Sahin *et al.* prepared the acidic monomers **PA-17**, **PA-18**, and **PA-19** (Scheme 8.8).[36] The interaction of these monomers with HAP was investigated by both ^{13}C NMR and FTIR spectroscopies. It was shown that **PA-18** was able to form a strong chemical bond with HAP. **PA-17**, **PA-18**, and **PA-19** were copolymerized with GDMA. In this photopolymerization study, an increase of reactivity was observed when **PA-17** was added to GDMA. This monomer could be a good candidate for dental applications. Unfortunately, no dental materials based on these PAs were developed.

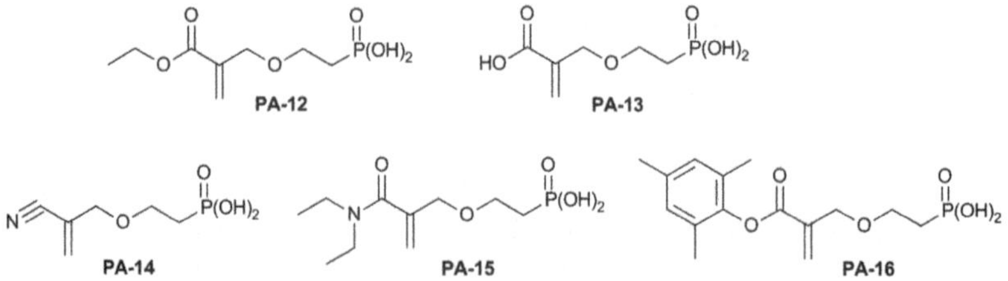

Scheme 8.7   2-[4-(Dihydroxyphosphoryl)-2-oxybutyl]acrylic acid derivatives.

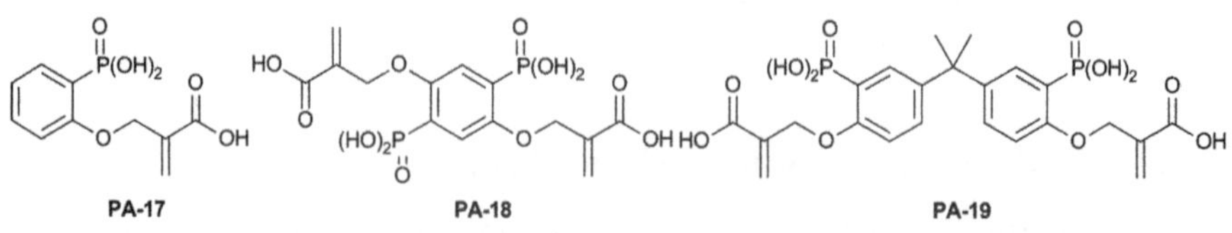

Scheme 8.8   PAs containing aromatic α-methyl-substituted acrylic acids.

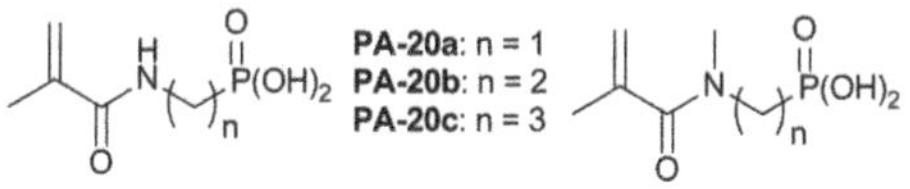

Scheme 8.9 Structure of the hydrolytically stable **PA-20a–c** and **PA-21a–c**.

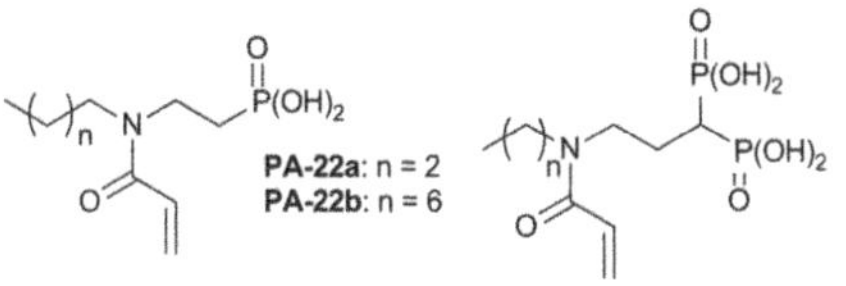

Scheme 8.10 Structure of the hydrolytically stable **PA-22a**, **PA-22b**, and **PA-23a–c**.

Another approach to improve the hydrolytic stability of the monomers consists in using (meth)acrylamides. Indeed, it is well known that amides show improved stability towards hydrolysis in acidic conditions compared with esters.[37] In this context, Nishiyama *et al.* formulated primers based on the methacrylamides **PA-20a–c** (Scheme 8.9).[38]

The influence of the alkyl chain length on both dentin and enamel adhesion was studied. Although no significant difference was observed regarding dentin adhesion, the use of **PA-20c** led to a higher enamel SBS. Similar to this work, Catel *et al.* developed the new *N*-methylacrylamides **PA-21a–c** and evaluated them in SEAs (Scheme 8.9).[39] Dentin SBS measurements showed that the longer the spacer group, the higher the SBS. Indeed, **PA-21c** provided a significantly higher dentin SBS than the other monomers. It might be explained on the basis of a higher chemical adhesion to HAP. Indeed, since **PA-21c** is more hydrophobic than **PA-21a** or **PA-21b**, it is expected that its corresponding calcium salt should present a lower solubility in aqueous environment. Recently, Klee *et al.* described the preparation of various *N*-alkyl-*N*-(phosphonoethyl)-substituted (meth)acrylamides.[40] Those monomers were synthesized in three steps, starting from diethyl ethylphosphonate. Among the isolated compounds, **PA-22a** and **PA-22b** exhibited the better adhesion to enamel and dentin (Scheme 8.10).

Finally, the synthesis of the hydrolytically stable bisphosphonic acids **PA-23a–c** was reported by Catel *et al.* (Scheme 8.10).[41] SEAs based on these monomers were able to provide a strong bond between dentin and a dental composite. The strong chelating properties of bisphosphonic acids might bring advantages such as a better sealing of the adhesive.

8.4   Polymerizable Dihydrogen Phosphates

8.4.1   (Meth)acrylates

Among the phosphorus-containing monomers, polymerizable DHPs are mainly used in current commercial enamel/dentin adhesives. One of the

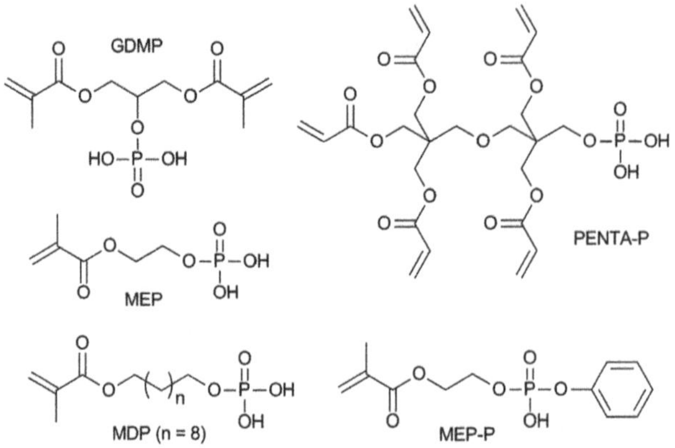

Scheme 8.11   Currently used methacrylate DHPs in dental adhesives.

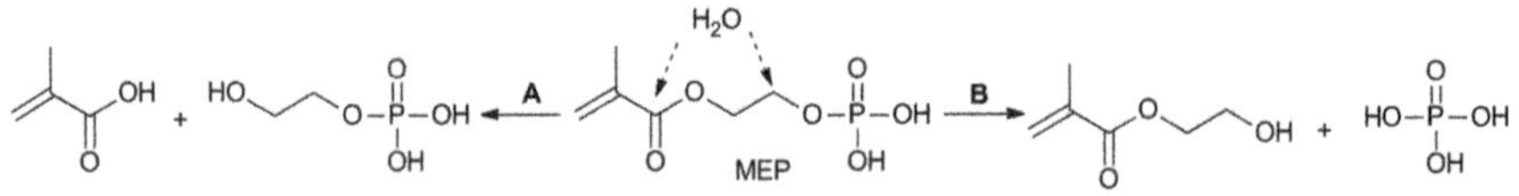

Scheme 8.12   Hydrolysis of MEP.

first chemical compounds proposed to improve the bonding to human dentin was the glycerol dimethacrylate ester of phosphoric acid (GDMP) (Scheme 8.11).[42,43] Further examples of (meth)acrylate phosphates which were used to improve the bond to dentin are the reaction product of phosphorus oxychloride with Bis-GMA,[44] methacryloyloxyethyl phenyl hydrogen phosphate (MEP-P), MDP, methacryloyloxyethyl dihydrogen phosphate (MEP), and dipentaerythrolpentaacryloyl dihydrogen phosphate (PENTA-P), which are summarized in Scheme 8.11. The polymerizable acidic phosphates shown in Scheme 8.11 have been used in commercial dental adhesives. The synthesis and dental adhesive application of MEP-P and its *p*-methoxy derivative was first described by Nakabayashi *et al.*[45,46]

The main drawback of these methacrylate DHPs is their insufficient hydrolytic stability. For MEP, we found that both the hydrolysis of the methacrylate and phosphate ester bonds took place, leading to the formation of both methacrylic acid (MAA) (*via* route A) and HEMA (*via* route B) (Scheme 8.12).[47,48] Furthermore, it was demonstrated that in the case of monomers having long alkylene spacers, like MDP, the phosphate ester bond was significantly more stable. However, aqueous solutions of MDP are also not hydrolytically stable during storage at room temperature over weeks and have to be stored in a refrigerator.[31,49]

Adhesives based on acidic monomers with long spacers between the strongly acidic and the polymerizable groups resulted in better performance. In case of MDP, the good bond strengths can be explained by a self-assembled nano-layering due to the length of the decamethylene spacer and by

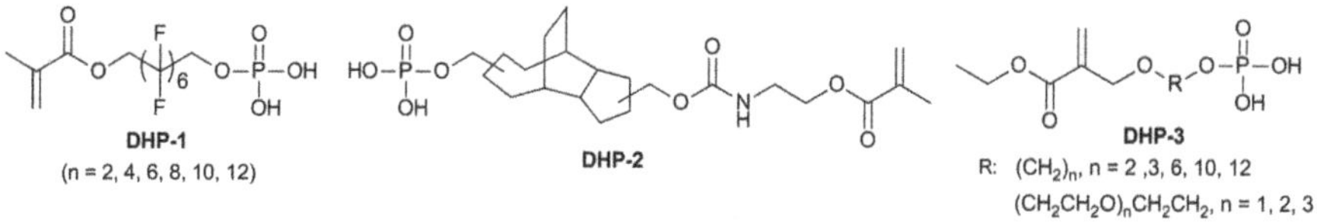

Scheme 8.13 Methacrylate DHPs claimed in dental patents.

the low solubility of the corresponding calcium salt.[50–52] Indeed, this monomer should be able to form strong chemical interactions with hydroxyapatite crystals, leading to an increase of adhesive performance.

Over the last few years, some new methacrylate DHPs were described in the patent literature. For example, the durability and coloring resistance of adhesive compositions could be improved by using DHPs such as **DHP-1** (Scheme 8.13), which contains fluorocarbon spacer groups between the polymerizable methacrylate group and the strongly acidic DHP group.[53] However, fluorine-containing monomers are very expensive and show only a limited solubility in aqueous solvent mixtures. Recently, the use of polymerizable DHPs which contain polyacyclic structural units was proposed. As an example, **DHP-2** was easily synthesized by the reaction of 2-isocyanatoethyl methacrylate with 3(4),8(9)-bis(hydroxymethyl)tricyclo[5.2.1.0^{2,6}]decane followed by phosphorylation with POCl$_3$.[54]

8.4.2 Hydrolytically Stable Monomers

As mentioned before, water is an essential component of SEAs. Consequently, especially in the case of single-bottle adhesives, the (meth)acrylates undergo hydrolysis, changing the chemical composition of the adhesive and impairing its performance. Therefore, several monomers presenting a higher stability towards hydrolysis have been prepared. Novel 2-(ω-phosphonooxy-2-oxaalkyl)acrylate monomers (**DHP-3**; Scheme 8.13) with improved hydrolytic stability were synthesized by Klee *et al.*[55] These DHPs were prepared by a four-step synthesis involving a Baylis-Hillman reaction between ethyl acrylate and formaldehyde, followed by halogenation, subsequent etherification with various diols, and phosphorylation. All-in-one SEAs based on these new DHPs were developed and evaluated for dentin and enamel adhesion and provided moderate to good SBSs. In this context, Moszner's group synthesized several (meth)acrylamides bearing DHP groups. These monomers enabled the preparation of all-in-one SEAs with high storage stability. The crosslinking **DHP-4** [1,3-bis(methacrylamido)-propane-2-yl dihydrogen phosphate] was prepared in two steps (Scheme 8.14): 1,3-diaminopropan-2-ol was first acylated with methacrylic anhydride, leading to the formation of the solid 1,3-bis(methacrylamido)-2-hydroxypropane (BMAHP).[56] In the second step, BMAHP was phosphorylated with POCl$_3$ in tetrahydrofuran in the presence of triethylamine.

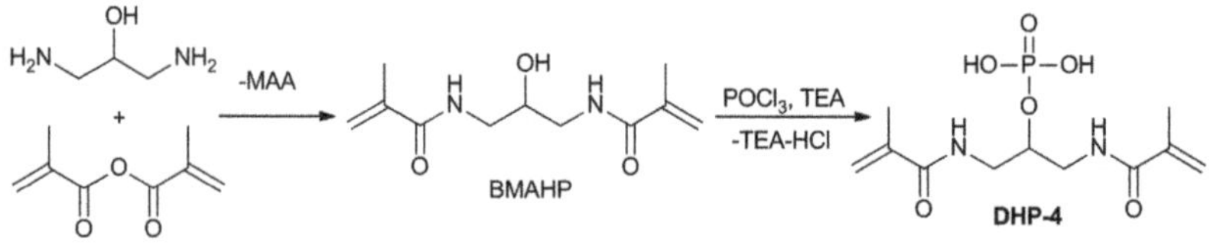

Scheme 8.14   Synthesis of a bis(methacrylamido) DHP currently used in dental adhesives.

DHP-4 is a crystalline white solid, which is very soluble in water and ethanol. This is of great importance for a dental adhesive application. As expected, **DHP-4** is significantly more stable in aqueous ethanol at 37 °C than MDP. **DHP-4** is a crosslinking monomer which showed a similar reactivity compared to GDMA in the free-radical polymerization in ethanol solution using 2,2′-azobisisobutyronitrile as initiator. The bis(acrylamide) **DHP-4** is also a strongly acidic monomer. Indeed, an aqueous solution of **DHP-4** (20 wt%) showed a pH value of 0.78. In comparison, a 20 wt% aqueous solution of H_3PO_4 has a pH of 0.21. SEAs containing **DHP-4** provided excellent SBSs to enamel and dentin. Furthermore, **DHP-4** did not show any cytotoxic effects. Based on these results, **DHP-4** was incorporated in our current available SEA AdheSE One F (Ivoclar Vivadent AG).[15] Because of its crosslinking properties, **DHP-4** is highly reactive in free-radical polymerization. Therefore, it has to be carefully handled during synthesis, purification, storage, and applications. Obviously, monomers bearing only one polymerizable group are much easier to handle. Therefore, we synthesized a number of new monofunctional (meth)acrylamido DHPs.[57] 5-(Methacrylamido)pentyl, 10-(*N*-methylacrylamido)decyl, and 11-(*N*-methylacrylamido)undecyl dihydrogen phosphates (**DHP-5**) (Scheme 8.15) were synthesized by acylation of the corresponding ω-aminoalkan-1-ols with methacrylic anhydride or acryloyl chloride, followed by phosphorylation. The hydrolytic stability of the synthesized (meth)acrylamides **DPH-5**, as well as of the methacrylates MDP and MEP, was investigated. Each monomer was dissolved at 20 wt% in a water/ethanol (1 : 1 v/v) mixture and the solution was heated at 37 °C. NMR investigations clearly demonstrated that the hydrolytic stability of the studied monomers decreased in the following order: **DPH-5** > MDP ≫ MEP. Compared to **DHP-4**, the monofunctional (meth)acrylamido dihydrogen phosphates **DHP-5** showed both similar enamel and dentin etching properties. Model SEAs based on the monomers **DHP-5** produced high SBSs to dentin and enamel.

Furthermore, we synthesized **DHP-6**, a crosslinking monomer containing an undecamethylene spacer. Unfortunately, this monomer did not show any improved performance compared to **DHP-4** or **DHP-5**. We also took an interest in the synthesis of *O*-alkylated acrylic acid hydroxamides, such as the 10-(*N*-acryloyl-*N*-methoxyamino)decyl dihydrogen phosphate **DHP-7**.[58] However, these monomers were less hydrolytically stable compared to the corresponding (meth)acrylamides. Over the last few years, several

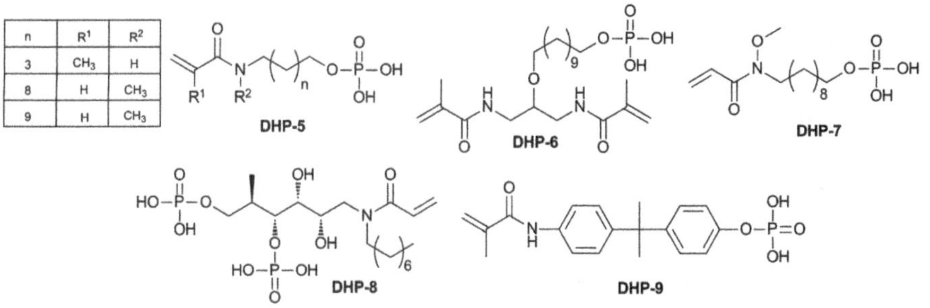

Scheme 8.15 (Meth)acrylamide DHPs.

(meth)acrylamido DHPs were also described in the patent literature. Examples are the very hydrophilic **DHP-8** (Scheme 8.15),[59] which contains two hydroxyl and two DHP groups, and the aromatic phosphate **DHP-9**,[60] which is poorly soluble in water-based adhesive formulations.

8.5 Conclusions

The use of acidic monomers to etch and prime dental tissues has revolutionized restorative filling therapy. Since the beginning of SEAs technology, significant progress has been made. The development of new acidic monomers, and more particularly of PAs and DHPs, has led to significant improvements. Over the last few years, a wide range of PA and DHP monomers were synthesized for application in dental adhesives. SEAs based on some of these monomers were formulated and evaluated *via* SBS tests to dentin and enamel. From those studies, it can be asserted that the aliphatic monomers having a long spacer group (typically C_{10}) exhibit the best adhesive properties. This observation might be explained on the basis of the strong chemical adhesion between the acidic monomer and HAP. Moreover, the replacement of (meth)acrylates by (meth)acrylamides has been a crucial step in the development of dental adhesives. Indeed, it allows the preparation of formulations which are storage stable at room temperature. Therefore, monomers such as the phosphonic acid **PA-21c** and the phosphates **DHP-5** are excellent candidates to enter future SEAs.

References

1. N. Moszner and U. Salz, *Macromol. Mater. Eng.*, 2007, **292**, 245.
2. N. Moszner and T. Hirt, *J. Polym. Sci., Part A: Polym. Chem.*, 2012, **50**, 4369.
3. M. V. Cardoso, Y. Yoshida and B. Van Meerbeek, in *Statements: Diagnostics and Therapy in Dental Medicine Today and in Future*, ed. J.-F. Roulet and H. F. Kappert, Quintessenz, Berlin, 2009, pp. 25–43.
4. M. G. Buonocore, *J. Dent. Res.*, 1955, **34**, 849.

5. B. Van Meerbeek, M. Vargas, S. Inoue, Y. Yoshida, M. Peumans, P. Lambrechts and G. Vanherle, *Oper. Dent. Suppl.*, 2001, **6**, 119.
6. B. Van Meerbeek, J. De Munck, Y. Yoshida, S. Inoue, M. Vargas, P. Vijay, K. Van Landuyt, P. Lambrechts and G. Vanherle, *Oper. Dent.*, 2003, **28**, 215.
7. A. Kakaboura and L. Papagiannoulis, in *Dental Hard Tissues and Bonding*, ed. G. Eliades, D. C. Watts and T. Eliades, Springer, Berlin, 2005, pp. 35–51.
8. N. Nakabayashi, *Polym. Int.*, 2008, **57**, 159.
9. B. Van Meerbeek, K. Yoshihara, Y. Yoshida, A. Mine, J. De Munck and K. L. Van Landuyt, *Dent. Mater.*, 2011, **27**, 17.
10. N. Moszner, U. Salz and J. Zimmermann, *Dent. Mater.*, 2005, **21**, 895.
11. N. Moszner and U. Salz, *Prog. Polym. Sci.*, 2000, **26**, 535.
12. K. L. Van Landuyt, J. Snauwaert, J. De Munck, M. Peumans, Y. Yoshida, A. Poitevin, E. Coutinho, K. Suzuki, P. Lambrechts and B. Van Meerbeek, *Biomaterials*, 2007, **8**, 3757.
13. N. Moszner, F. Zeuner, J. Angermann, U. K. Fischer and V. Rheinberger, *Macromol. Mater. Eng.*, 2003, **288**, 621.
14. N. Moszner, U. K. Fischer, J. Angermann and V. Rheinberger, *Dent. Mater.*, 2006, **22**, 1157.
15. U. Salz and T. Bock, *J. Adhes. Dent.*, 2010, **12**, 7.
16. M. Yoshioka, Y. Yoshida, S. Inoue, P. Lambrechts, G. Vanherle, Y. Nomura, M. Okazaki, H. Shintani and B. Van Meerbeek, *J. Biomed. Mater. Res.*, 2002, **59**, 56.
17. Y. Yoshida, K. Nagakane, R. Fukuda, Y. Nakayama, M. Okazaki, H. Shintani, S. Inoue, Y. Tagawa, K. Suzuki, J. De Munck and B. Van Meerbeek, *J. Dent. Res.*, 2004, **83**, 454.
18. K. L. Van Landuyt, Y. Yoshida, I. Hirata, J. Snauwaert, J. De Munck, M. Okazaki, K. Suzuki, P. Lambrechts and B. Van Meerbeek, *J. Dent. Res.*, 2008, **87**, 757.
19. M. Anbar and E. P. Farley, *J. Dent. Res.*, 1974, **53**, 879.
20. G. Sahin, D. Avci, O. Karahan and N. Moszner, *J. Appl. Polym. Sci.*, 2009, **114**, 97.
21. S. Edizer and D. Avci, *Des. Monomers Polym.*, 2010, **13**, 337.
22. K. Ikemura, Y. Kadoma and T. Endo, *Dent. Mater. J.*, 2011, **30**, 769.
23. K. Ikemura, F. R. Tay, N. Nishiyama, D. H. Pashley and T. Endo, *Dent. Mater. J.*, 2006, **25**, 566.
24. Y. Catel, V. Besse, A. Zulauf, D. Marchat, E. Pfund, T.-N. Pham, D. Bernache-Assolant, M. Degrange, T. Lequeux, P.-J. Madec and L. Le Pluart, *Eur. Polym. J.*, 2012, **48**, 318.
25. L. Mou, G. Singh and J. W. Nicholson, *Chem. Commun.*, 2000, 345.
26. D. Adusei, S. Deb, J. W. Nicholson, L. Mou and G. Singh, *J. Appl. Polym. Sci.*, 2003, **88**, 565.
27. B. Garska, M. Tabatabai, U. Fischer, N. Moszner, A. Utterodt and H. Ritter, *Polym. Int.*, 2012, **61**, 1061.
28. N. Moszner, J. Angermann, U. Fischer, A. Gianasmidis, M. Tabatabai, H. Ritter and A. Utterodt, *Macromol. Mater. Eng.*, 2011, **296**, 937.

29. Y. Catel, U. Fischer and N. Moszner, *Macromol. Mater. Eng.*, 2013, **298**, 740–756.
30. N. Nishiyama, F. R. Tay, K. Fujita, D. H. Pashley, K. Ikemura, N. Hiraishi and N. M. King, *J. Dent. Res.*, 2006, **85**, 422.
31. U. Salz, J. Zimmermann, F. Zeuner and N. Moszner, *J. Adhes. Dent.*, 2005, **7**, 107.
32. N. Moszner, F. Zeuner, U. Fischer and V. Rheinberger, *Macromol. Chem. Phys.*, 1999, **200**, 1062.
33. N. Moszner, F. Zeuner, S. Pfeiffer, I. Schurte, V. Rheinberger and M. Drache, *Macromol. Mater. Eng.*, 2001, **286**, 225.
34. F. Zeuner, S. Quint, F. Geipel and M. Moszner, *Synth. Commun.*, 2004, **34**, 767.
35. J. Pavlinec, F. Zeuner, J. Angermann and N. Moszner, *Macromol. Chem. Phys.*, 2005, **206**, 1878.
36. G. Sahin, A. Z. Albayrak, Z. S. Bilgici and D. Avci, *J. Polym. Sci., Part A: Polym. Chem.*, 2009, **47**, 1953.
37. N. Nishiyama, K. Suzuki, H. Yoshida, H. Teshima and K. Nemoto, *Biomaterials*, 2004, **25**, 965.
38. N. Nishiyama, M. Aida, K. Fujita, K. Suzuki, F. R. Tay, D. H. Pashley and K. Nemoto, *Dent. Mater. J.*, 2007, **26**, 382.
39. Y. Catel, M. Degrange, L. Le Pluart, P.-J. Madec, T.-N. Pham and L. Picton, *J. Polym. Sci., Part A: Polym. Chem.*, 2008, **46**, 7074.
40. J. E. Klee and U. Lehmann, *Beilstein J. Org. Chem.*, 2010, **6**, 766.
41. Y. Catel, M. Degrange, L. Le Pluart, P.-J. Madec, T.-N. Pham, F. Chen and W. D. Cook, *J. Polym. Sci., Part A: Polym. Chem.*, 2009, **47**, 5258.
42. G. Buonocore, W. Wileman and F. Brudevold, *J. Dent. Res.*, 1956, **35**, 846.
43. O. Hagger (De Trey), *Br. Pat.*, 687 299, 1953.
44. N. D. Ruse and D. C. Smith, *J. Dent. Res.*, 1991, **70**, 1002.
45. J. Yamauchi, N. Nakabayashi and E. Masuhara, *Polym. Prepr. (Am. Chem. Soc., Div. Polym. Chem.)*, 1979, **20**, 594.
46. T. Nikaido and N. Nakabayashi, *Jpn. J. Dent. Mater.*, 1987, **6**, 690.
47. U. Salz, P. Burtscher, K. Vogel, N. Moszner and V. Rheinberger, *Polym. Prepr. (Am. Chem. Soc., Div. Polym. Chem.)*, 1997, **38**, 143.
48. N. Moszner, F. Zeuner, A. Rumphorst, U. Salz and V. Rheinberger, *Dental-praxis*, 2001, **18**, 105.
49. I. Teshima, *J. Dent. Res.*, 2010, **89**, 1281.
50. K. Yoshihara, Y. Yoshida, N. Nagaoka, D. Fukegawa, S. Hayakawa, A. Mine, M. Nakamura, S. Minagi, A. Osaka, K. Suzuki and B. Van Meerbeek, *Acta Biomater.*, 2010, **6**, 3573.
51. K. Yoshihara, Y. Yoshida, S. Hayakawa, N. Nagaoka, M. Irie, T. Ogawa, K. L. Van Landuyt, A. Osaka, K. Suzuki, S. Minagi and B. Van Meerbeek, *Acta Biomater.*, 2011, **7**, 3187.
52. K. Fujita, S. Ma, M. Aida, T. Maeda, T. Ikemi, M. Hirata and N. Nishiyama, *J. Dent. Res.*, 2011, **90**, 607.
53. K. Nakatsuka (Kuraray Medical), *Eur. Pat.*, 1 873 219 A1, 2006.
54. M. Stepputtis and T. Blömker (Voco), *Eur. Pat.*, 2 450 025 B1, 2011.

55. J. E. Klee and U. Lehmann, *Beilstein J. Org. Chem.*, 2010, **6**, 766.
56. N. Moszner, J. Pavlinec, I. Lamparth, F. Zeuner and J. Angermann, *Macromol. Rapid Commun.*, 2006, **27**, 1115.
57. N. Moszner, J. Angermann, U. Fischer and T. Bock, *Macromol. Mater. Eng.*, 2013, **298**, 454–461.
58. N. Moszner, I. Lamparth, U. Fischer, F. Zeuner, A. de Meijere and V. Rheinberger (Ivoclar Vivadent), *Eur. Pat.*, 1 974 712 B1, 2010.
59. J. Yang, S. B. Mitra, Y. He, B. A. Shukla, N. Karim, A. Falsafi, R. B. Ross, P. R. Klaiber and G. W. Griesgraber (3M), *World Pat.*, 2012/036838, 2011.
60. M. Takagi, A. Otsuji, A. Nagatomo and K. Suesugi (Mitsui Chemicals), *U.S. Pat.*, 2010/0024683 A1, 2007.

CHAPTER 9

Biomedical Applications of Phosphorus-Containing Polymers

EDELINE WENTRUP-BYRNE,*[a] SHUKO SUZUKI[b] AND LISBETH GRØNDAHL[c]

[a] Institute of Health and Biomedical Innovation, Queensland University of Technology, Kelvin Grove, Q 4059, Australia; [b] Queensland Eye Institute, Melbourne Street, South Brisbane, Q 4101, Australia; [c] School of Chemistry and Molecular Biosciences, The University of Queensland, St Lucia, Q 4072, Australia
*Email: edelinebyrne@gmail.com

9.1 Introduction

Biomaterials play an essential role in a vast range of biomedical applications, ranging from well-established orthopedic hip replacements and contact lenses to the rapidly developing field of tissue engineering (TE) (see Figure 9.1). The field of regenerative medicine encompasses cell therapies, gene therapies, as well as tissue engineering, and success in this field is closely linked to the development of new biocompatible materials and the engineering of a broad range of devices and scaffolds. However, as eloquently debated in his 2009 paper "On the Nature of Biomaterials",[1] Williams emphasizes that since a consensus was first reached in 1987 on the definition of a biomaterial as being "a nonviable material used in a medical device, intended to interact with biological systems", much has

RSC Polymer Chemistry Series No. 11
Phosphorus-Based Polymers: From Synthesis to Applications
Edited by Sophie Monge and Ghislain David
© The Royal Society of Chemistry 2014
Published by the Royal Society of Chemistry, www.rsc.org

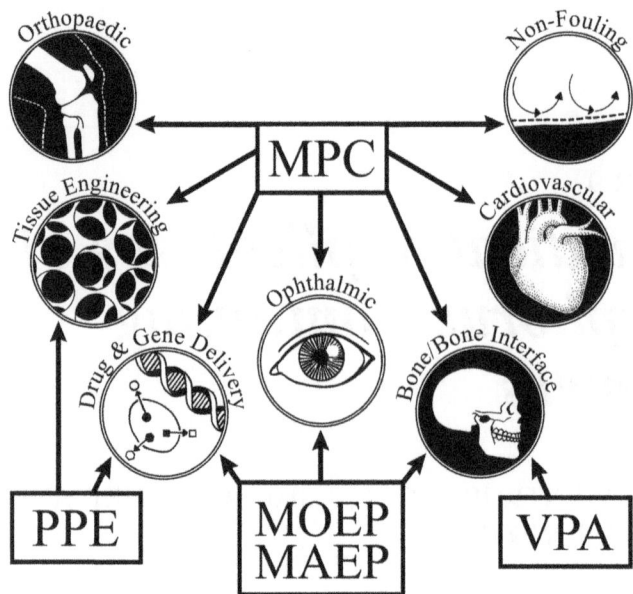

Figure 9.1   General overview of the biomedical applications of phosphorus-containing polymers. See Schemes 9.1 and 9.2.

changed in the field of biomaterials science. Already in 2004, Ratner *et al.*[2] contributed to the on-going debate by pointing out that by dropping the terms *medical* and *viable* from such definitions, the range of applications encompassed by biomaterials science expands enormously. In his *Dictionary of Biomaterials*, Williams defines a biomaterial as "a substance that has been engineered to take a form which, alone or as part of a complex system, is used to direct, by control of interactions with components of living systems, the course of any therapeutic or diagnostic procedure, in human or veterinary medicine". He elaborates and says "a biomaterial does not have to be solid, visible and tangible; it does not have to be manufactured by conventional top-down industrial processes. A biomaterial does not have to be dead. It can be a suspension of nanoparticles, or a self-assembling peptide hydrogel. It could even be an engineered viral vector or engineered re-cellularized extra-cellular matrix". He may very well revisit some or all of these definitions in his latest book, which is due to be published in 2014.[3]

Clearly at this point the reader will appreciate the impossibility in this short chapter of covering all aspects of these topics, not to mention to be confident we have indeed discovered all the relevant literature. Instead, we shall focus on a selection of seminal papers and refer to the latest research trends and directions involving phosphorus-containing polymeric bio-materials in specific medical applications.

Although TE is still a relatively young field, its interdisciplinary nature and rapid growth, combined with the enormous expectations of both the multi-disciplinary research teams involved and the potential end-users, have combined to complicate our fundamental understanding of what exactly constitutes TE and hence wherein lies its future. The actual term TE can be dated back to a National Science Foundation meeting in 1987[4] and already the following year the first TE meeting was held. In the 1990s the term regenerative medicine began to be used and was generally associated with emerging stem cell technologies. What is sometimes not recognized is that much of the underlying fundamental research which forms the foundations of TE and regenerative medicine was carried out long before these definitions became synonymous with the use of the term. In the late 1990s, advances in TE were being discussed by journalists as "the greatest scientific achievements of the twentieth century",[5] as well as a career in TE as being one of the "hottest jobs of the future". Although nearly 10 years old now, Lysaght and Hazelhurst's review "Tissue Engineering: the End of the Beginning" makes interesting background reading, especially in retrospect. Another somewhat more recent review by French researchers gives a more European perspective, which is of course very relevant when we consider the different laws governing the use of stem cells and other therapies compared to, say, in Australia and the USA.[6]

More recently, the emerging need to better understand the specific role of the *engineering* in TE and tissue regeneration has given rise to both discussion and debate. For example, to summarize Griffith's succinct interpretation of TE, she said "coaxing cells to form tissue is inherently an engineering process because they require supporting structures. However, both chemical and mechanical signals are also essential in order to create the hierarchical structures that constitute native tissue". What is clear is that, except in few cases, the promise of 10 years ago to successfully translate tissue-engineered products from the laboratory to real clinical solutions has not yet been realised.[5,7] As Williams rightly pointed out already in 2006, not only the paradigms but also some of the concepts and definitions involved may need to be redefined.[8] This in fact is what is happening. As our understanding of the biological processes and interactions that are involved grows, so too must the language and definitions that we employ.[9] Just as the definition of what constitutes a biomaterial has changed, so has our understanding of biocompatibility and what exactly it involves. All these aspects of the topics under discussion are complicated by the broad range of disciplines and experts involved and the need to communicate using different words and terms, often for the same topic. It is critical that there is a consensus because of the serious legal, economic, and regulatory issues that depend on these decisions.

Williams' 1999 definition of TE is both elegant and clear. He said it is "the persuasion of the body to heal itself, through the delivery to the appropriate sites of molecular signals, cells and supporting structures".[9] More recently, Hutmacher explained the underlying concept of TE in a way that is very relevant to this chapter.[7] A scaffold or matrix is combined with living cells and/or biologically active molecules to form a tissue engineering construct (TEC) to promote the repair and/or regeneration of tissues. The scaffold is expected to perform various functions, including the support of cell coloni-zation, migration, growth, and differentiation. Scaffolds consist of a solid support structure with an interconnected pore network, or a hydrogel matrix in which cells can be encapsulated.

This chapter has developed from a review covering "Phosphorus-Containing Polymers: A Great Opportunity for the Biomedical Field" by Monge *et al.*[10] which appeared in *Biomacromolecules* in 2011. In this chapter, our focus is on the use of polymers with phosphorus in their side groups and polyphosphoesters, pre-pared from phosphorus-containing monomers. The commercially available phosphorus-containing monomers mono(2-acryloyloxyethyl) phosphate (MAEP), 2-(methacryloyloxy)ethyl phosphate (MOEP), and 2-methacryloyloxy-ethyl phosphorylcholine (MPC), as well as vinylphosphonic acid (VPA), are highlighted because of their unique and interesting properties that make them attractive for a range of biomedical applications (see Figure 9.1). The monomer structures are shown in Scheme 9.1. As correctly pointed out by Chirila in his critical review of 2007, various names and acronyms for MOEP and MAEP have been used over the years.[11,12] Often these are not only inconsistent with the recommended IUPAC nomenclature, but have caused some confusion for the different disciplines involved in biomaterials research. In Table 9.1 we sum-marize what we have found in the literature.

A word of caution is pertinent at this point. Anyone working with bio-materials destined for biomedical applications is aware of the strict approval protocols required, and of the subsequent legal ramifications when there is a failure. The fact that a commercially available phosphate monomer con-tained almost 30% of a diene impurity went virtually unacknowledged for many years must be taken seriously because of the consequences. An un-awareness of the true monomer composition will surely affect the sub-sequent composition and even the engineering design. Hence, TE and other applications could be seriously compromised.[13] We believe that the most reliable methodology is to purify the monomers in question (washing with *n*-hexane has been used by several groups) just before use and to verify the purity using ^{31}P NMR or other techniques capable of detecting the diene.

A useful review covering polyphosphoester (PPE) synthesis and biomedical applications appeared in 2009.[14] However, a word of caution: some of the

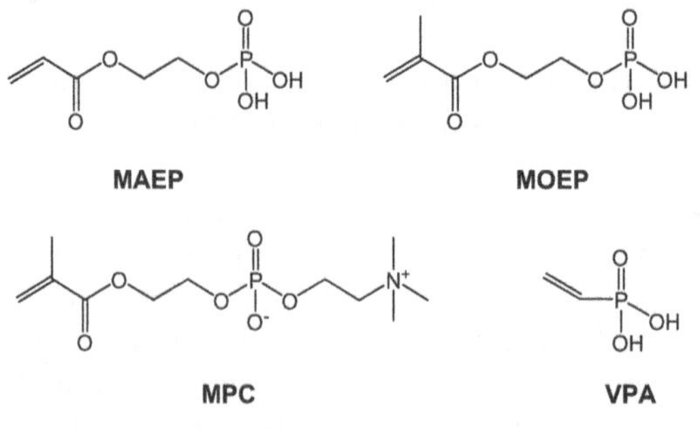

Scheme 9.1 Chemical structures of the monomers MAEP, MOEP, MPC, and VPA.

Table 9.1 Systematic, acronymic, and various literature names for monomers.

Acronym	Chemical name
Royal Society of Chemistry[a] systematic name	2-(Methacryloyloxy)ethyl phosphate[b]
MMEP[b]	(i) 2-(Methacryloyloxyethyl) phosphate[b] (ii) 2-Methacryloyloxyethyl phosphate[b]
MOEP[b]	(i) 2-(Methacryloyloxyethyl) phosphate[b] (ii) 2-(Methacryloxy) ethyl phosphate[b] (iii) 2-Methacryloxyethyl phosphate[b] (iv) (Methacryloxyethyl) phosphate[b] (v) Methacryloxyethyl phosphate[b]
MEP[b]	2-(Methacryloyloxy) ethyl phosphate[b]
EGMP[b]	Ethylene glycol methacrylate phosphate[b]
Phosmer M™ [b]	2-Methacryloyloxyethyl dihydrogen phosphate[b]
MAEP[c]	(i) Monoacryloxyethyl phosphate[c] (ii) Mono(2-acryloyloxyethyl) phosphate[c]

[a]CSID:10804856; http://www.chemspider.com/Chemical-Structure.10804856.html (accessed May 6, 2013).
[b]CAS No. 24599-21-1.
[c]CAS No. 32120-16-4.

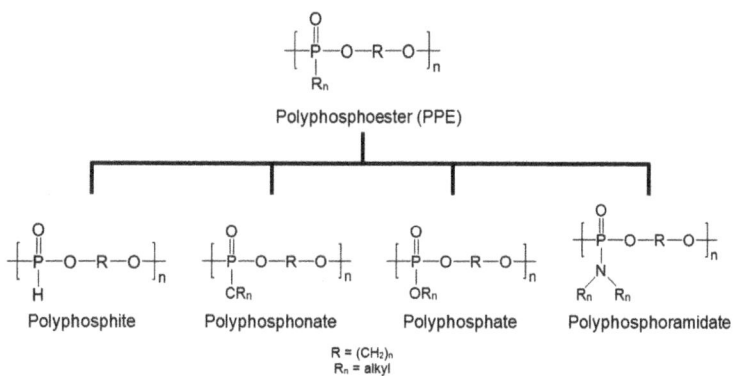

Scheme 9.2 Classification of polyphosphoesters.

references given therein, supposedly covering TE applications, appear to be inaccurate. PPEs contain phosphoester bonds in the backbone and are biodegradable through hydrolysis as well as by enzymes under physiological conditions. Their structures are shown in Scheme 9.2. Depending on the nature of their side groups, they can also be categorized as polyphosphates, polyphosphonates, polyphosphites, or polyphosphoramidates.

The biomedical applications of polyphosphazene derivatives and phosphorylated polymers will not be covered in this review, although biodegradable polyphosphazene derivatives have shown potential.[15-17] They represent a unique class of materials due to the P=N double bond in the backbone. Phosphorylation is another approach for the incorporation of the phosphorus functionality into polymeric biomaterials. This too is beyond the scope of this review, since our focus is on phosphorus-containing polymers produced from phosphorus-containing monomers.

We should point out that one of challenges in writing this chapter is that even after dividing the different applications into nice tidy sections, some of the literature papers just do not really fit the chapter sub-headings. The reason for this is that there will always be overlap between different components of the science behind a particular application. The first two applications in this chapter are a good example: non-fouling surfaces and cardiovascular applications. The many disorders of the cardiovascular system can be separated into diseases that primarily affect the blood, heart, or blood vessels. From a biomaterials perspective, devices such as vascular grafts, stents, and rotary blood pumps all contribute to the repair and regeneration of the heart and blood vessels as well as restoration of blood flow. One essential requirement of material surfaces used in cardiovascular applications is that they be non-fouling. Hence, clearly we found many literature reports which could fit into either section. In fact, the non-fouling section was like a hydra: it kept growing. In retrospect, this is not too surprising since protein fouling is seen as a first event for bacteria as well as cell adhesion. Much research has been invested into preventing adverse reactions towards biomaterials and to combat bacterial infections.

Where cardiovascular applications are concerned, polymers incorporating the phosphorylcholine (PC) head group dominate the use of phosphorus-containing polymers. An important review by Iwasaki and Ishihara on the use of PC-containing polymers for cardiovascular application has described an extensive amount of detailed *in vitro* evaluation studies.[18] A recent review on cardiac TE describes recent key advances in this field.[19]

The ageing population is a contributing factor to the growing number of joint replacements worldwide. Hence, the number of revision joint replacements is also growing. Despite many improvements in the lifetime, design, and composition of artificial joints there are still many challenges remaining. One research group dominating the field in orthopedic applications is Ishihara and his team. They have used MPC to modify material surfaces for orthopedic applications. MPC is used to impart non-frictional and lubrication properties to implants with a view to minimizing wear, and the production of PE particulates.[20,21]

In any discussion of applications involving biomaterials employed in bone and bone-interface materials the biomineralization processes will be a dominant aspect. As emphasized in a recent paper by Habibovic and coworkers, our knowledge of the fundamental biomineralization processes as well as of the best biomaterial properties for these applications has increased enormously over the past decades.[22] Phosphorus-containing polymers, in particular PMOEP, PMAEP, and PVPA, have all contributed greatly to this progress and surface graft polymerization into a range of substrates has been thoroughly investigated by several groups.[23,24] More recently, RGD peptides have been used to further functionalize the latest generation of phosphate-grafted polymers.[25]

In addition to several review chapters (not specifically for phosphorus-containing polymers) in Chirila's book "Biomaterials and Regenerative Medicine in Ophthalmology",[26] a recent chapter by Huang covering corneal TE has also appeared.[27] Because of their network structure and the fact they resemble the extracellular matrix of many tissues, hydrogels have been extensively studied for a range of medical applications (these will be discussed in more detail in later sections).[28] Where ophthalmological applications are concerned, one of the most widely researched hydrogels is poly(2-hydroxyethyl methacrylate) (PHEMA). However, its poor cell-adhesive properties have somewhat limited its use. Phosphate-containing monomers have been used for its modification and studies have shown improved attachment and growth of human corneal epithelial cells.[29]

Drug and gene delivery is one of the most rapidly growing interdisciplinary areas to be included in this chapter. A perusal of the literature reveals that the development of novel delivery systems involves a huge range of polymeric materials and covers a wide range of potential applications. Our focus is on phosphorus-containing polymers, especially PMPC[30] and PPE,[31] which as we will demonstrate have shown most encouraging results as carriers.

Although a recent rather general review on polymeric scaffolds in TE applications does not cover phosphorus-containing polymers specifically, it does have some useful tables including one of scaffold fabrication techniques and one on commercially available polymeric scaffold products.[32] In a recent overview, Hutmacher *et al.*[33] propose a different perspective on the requirements of scaffolds for bone TE. His group is focusing on developing scaffolds that remain intact as newly formed tissue matures. In their opinion, ideally the onset of degradation should only occur after the regenerated tissue has undergone remodeling at least once. Another recent review covers current strategies and challenges in osteochondral TE.[34] However, as far as we can ascertain, the most recent review specifically covering the mineralization of polymeric scaffolds for bone applications appeared in 2007.[35] Clearly it is time for an update.

9.2 Non-Fouling Applications

Biomaterials that are placed in contact with living tissue or body fluids containing living cells, or indeed any material that may come in contact with microorganisms, can malfunction due to adverse effects initiated by protein fouling. This leads to undesirable cellular responses and/or the formation of biofilm. Indeed, bacterial infection is a leading cause of revision surgery. Furthermore, a secondary implant used in revision surgery is at greater risk of infection compared to the primary implant.[36] Much research has been invested into preventing such adverse reactions towards biomaterials over recent decades and although advances have been made in certain areas (some of which are covered in this book chapter) an ideal solution is still to be identified.[36] A vast body of work involves the use of poly(ethylene glycol) for the production of non-fouling surfaces. However, these are prone to oxidation, which renders them less than ideal.[37] Inspired by the chemical composition of non-fouling lipid cell membranes, polymers incorporating the zwitterionic phosphorylcholine (PC) head-group have been studied extensively for the creation of biomaterial coatings with a view to preventing the processes initiated by protein fouling. Of particular note is the extensive research by Ishihara involving polymers and copolymers based on MPC which was developed from 1974 to 1976 in Japan.[38–41] He covers the synthetic aspects of MPC and its polymers in Chapter 5 of this book.[42] These PC-based polymers have found widespread use in a number of areas, including blood-contacting devices (Section 9.3), ophthalmic devices (Section 9.6), the lubrication of orthopedic devices (Section 9.4), and for the combat of biomedical device-related infections (discussed below). It is beyond the scope of this chapter to comprehensively cover all the research in this area, which spans more than 30 years. Instead, the focus is on recent research (from 2009) nor does it include materials intended for use as biosensors. Although similar requirements exist for biosensors, the substrates are typically different to those used in medical devices. The most recent work in the areas of cardiovascular, ophthalmic, and orthopedic applications is summarized in Table 9.2. In addition to these areas, a vast amount of work has involved the design of non-fouling drug and gene delivery devices, which is covered in Section 9.7. It can be seen from Table 9.2 that many different substrates and methods of attachment have been explored for the fabrication of non-fouling surfaces for medical devices. These methods can be categorized into (i) direct attachment (*via* adsorption or *via* reaction of catechol or silane groups in the co-polymer), (ii) graft copolymerization directly from a material or by grafting a copolymer with complementary functional groups onto a material, (iii) copolymerization, and finally (iv) the fabrication of polymer blends. One approach of direct attachment uses polymers containing biomimetic catechol groups which have a high affinity for substrates such as titania.[43,44] A recent study evaluated the use of such catechol-based copolymers for a series of substrates including poly(tetrafluoroethylene) (PTFE) and good adhesion was found in all cases.[45]

The use of MPC-based polymers for the prevention of bio-fouling was nicely reviewed in 2000 by Lewis[60] and has indeed been covered in several other reviews on antifouling surfaces.[37,61] The 2008 review by Chen *et al.*[61] gave a comprehensive overview of the field at the time. While different mechanisms have been proposed to account for the non-fouling properties of this class of polymers, it is generally accepted that the presence of the PC head-group is the most important design feature and the mechanism of its non-fouling properties has been attributed to its highly hydrated nature that

Table 9.2 Non-fouling applications of MPC-based polymers (2009–April 2013).

Application	Substrate and method of attachment	Ref.
Cardiovascular: rotary blood pump	Ti substate 1) hydroxylation; 2) silanation; 3) argon-plasma treatment and exposure to air; 4) UV radiation-induced grafting of MPC	46
Cardiovascular: rotary blood pump	Ti substrate 1) magneton sputtering to generate crystalline TiO_2 film; 2) HUPA[a] linking layer modified to produce ATRP initiator sites; 3) ATRP polymerization of MPC	47
Cardiovascular: antithrombus, improved hemocompatibility	Ti substrate MPC copolymer with monomer carrying catechol groups; direct attachment to TiO_2 by dipping process	43,44
Cardiovascular: coronary artery disease targeting late stent thrombosis	PLLA[b] stent Polymer blend with $P(MPC_{30}$-*co*-$BMA_{70})$[c]	48,49
Cardiovascular: TE small diameter vascular graft	PEUU[d] Methacrylic acid copolymer attached to ammonia gas plasma-treated surface	50
Cardiovascular: TE small diameter vascular graft	PEUU Polymer blend with PMBU[e]	51
Ophthalmologic: soft contact lenses	Silicone hydrogel Plasma-induced graft polymerization of MPC	52
Ophthalmologic: soft contact lenses	Silicone Method 1: copolymerization of P(SiMA-*co*-MPC)[f] Method 2: interpenetration network [P(SiMA-*ipn*-MPC)]	53
Ophthalmologic: soft contact lenses	Silicone Copolymerization (PMMT)[g]	54,55
Ophthalmologic: intraocular lenses (IOLs)	Acrylic IOLs Dip coating in PMPC ethanol solution	56
Orthopedic: non-frictional/lubrication	PE Benzophenone-initiated graft copolymerization of MPC	21,57,58
Orthopedic: non-frictional/lubrication	Co-Cr-Mo Method 1: adsorption of P(MPC-*co*-BMA) Method 2: immobilization P(MPC-*co*-MPSi)[h] Method 3: MPSi attachment *via* silanation followed by UV-induced grafting of MPC	59

[a]11-Hydroxyundecylphosphonic acid.
[b]Poly(L-lactic acid).
[c]BMA = *n*-butyl methacrylate.
[d]Poly(ester urethane)urea.
[e]Poly(2-methacryloyloxyethyl phosphorylcholine-*co*-methacryloyloxyethyl butylurethane).
[f]SiMA = bis(trimethylsilyloxy)methylsilylpropyl glycerol methacrylate.
[g]Poly[methyl methacrylate-*co*-2-methacryloyloxyethyl phosphorylcholine-*co*-tris(trimethylsiloxy)-3-methacryloxypropylsilane].
[h]MPSi = (3-methacryloxypropyl)trimethoxysilane.

does not allow proteins to adhere. This mechanism is based on a number of studies, including a paper published by Tegoulia *et al.*[62] who studied self-assembled monolayers (SAMs) containing PC head-groups. Interestingly, despite a reduction in protein adsorption and enhancement in protein desorption indicating a lower strength of protein adsorption on PC-containing SAMs, bacterial adhesion *in vitro* was not significantly reduced relative to the gold surface. Studies of a series of polymers incorporating the butyl methacrylate monomer (BMA), intended to mimic the lipophilic component of cell membranes, has indicated that the lipophilic nature of this component plays an important role in the performance of the coating. Thus, optimal non-fouling properties are, according to Ishihara,[42] achieved using a P(MPC-*co*-BMA) copolymer consisting of 30% MPC and having an overall molecular mass of at least 5×10^5 g mol^{-1}. It should be noted that fundamental work is on-going in furthering the understanding of the non-fouling properties of PC-based polymers.[63-65]

Recent work in the area of fabrication of non-fouling surfaces has explored the use of layer-by-layer (LbL) assembly to tailor the surface properties.[66-68] One series of studies investigated multilayer systems with an outer co-polymer polyelectrolyte incorporating monomers containing either short ethylene oxide moieties, MPC, or an MPC-like monomer in which the methylene moieties had been exchanged for ethylene oxide moieties. Protein adsorption studies of selected proteins (albumin, lysozyme, and fibrinogen), as well as serum, showed that the MPC-like monomer yielded the most effective non-fouling surface.[66] Further evaluation of these multilayer systems showed that these surfaces resist both cell (fibroblast) and fungi adhesion. This correlates well with the non-fouling properties observed in the protein adsorption study.[67] An interesting finding of this study was that the multilayer films deposited on silicone change from a cell-resisting surface to a cell-adhesive one when the material is stretched to 1.5 its original length. No change was seen in fungal adhesion under the same application of stress. Building on this work, a recent paper communicated the fabrication of multilayer films on a silicone substrate with an outer polyelectrolyte bilayer incorporating the MPC-like monomer. In addition, this system incorporated an RGD-modified polyelectrolyte below two of the bilayers of the non-fouling polyelectrolyte.[68] It was demonstrated that stretching this film to 1.5 times its original length caused the surface to change from non-adhesive to strongly fibroblast adhesive. This cyto-mechano-responsive film clearly offers new directions in biomaterial coatings.

In recent years, focus has moved to the use of living radical polymerization techniques such as atom transfer radical polymerization (ATRP)[47,69,70] and reversible addition–fragmentation chain transfer (RAFT)[71,72] for the production of well-defined polymer coatings. Both the *grafting from* and *grafting to* approaches have been explored for the attachment of MPC-based polymers onto a variety of substrates. *Grafting from* requires the incorporation of initiator sites onto the biomaterial surface prior to the polymerization reaction. This usually involves a series of steps.[47,70,72] Alternatively, in the *grafting to* approach, well-defined block copolymers are anchored onto the material surface by various chemical means, including silane groups reacting with the hydroxyl groups of cellulose[69] and phosphate groups binding to metal substrates.[71] It has generally been observed that subsequent protein adsorption and cell adhesion depended on both the graft density and the chain length, with optimal characteristics identified in most studies.[69–72]

Some work is based on the recognition that, in order to prevent bacteria adhesion and biofilm formation, the incorporation of a bactericidal agent is required in combination with a protein resistant surface.[37] In this regard, Fuchs *et al.*[73] prepared HEMA- and MPC-based hybrid films containing antimicrobial silver nanoparticles. Silver leaching from these films in water for 4 weeks was tested and found to be below the detection limit. Further studies on the evaluation of antibacterial and antifouling properties of these coatings are yet to be communicated. Current work in the area of non-fouling surfaces is looking at further mimicking the cell membrane function by incorporating biologically active molecules on top of the non-fouling layer. This approach is described in a recent review on biomimetics-inspired novel biomaterials.[42]

It was recently pointed out by Busscher *et al.*[36] that very few *in vitro* studies are able to mimic the complex *in vivo* environment where tissue-forming cells and bacteria co-exist. In addition, the choice of bacteria strain in an *in vitro* model is important as certain strains are more prevalent in certain applications, *e.g. E. coli* is the main strain found in urinary catheter infections while staphylococci are most commonly found in joint prostheses.[36] Another important aspect of implant-related bacterial infections is that in many cases these do not manifest themselves until years after the surgical procedure, which of course makes it difficult to design appropriate *in vitro* models. It is worth noting that in order to make substantial advances in this area the *in vitro* assessment methodologies will have to be further developed.

It is clear that the use of PC-based polymers for non-fouling applications is a very active, still developing, field of research which offers great promise for improved medical devices in general, as well as for specific applications as discussed in later sections.

9.3 Cardiovascular Applications

9.3.1 Introduction

The cardiovascular system is primarily made up of the heart, blood, and blood vessels. The constant circulation of blood through the body maintains homeostasis, and allows the necessary oxygen, nutrient, and metabolic waste exchanges between the bloodstream and the interstitial fluid surrounding cells to occur. Without efficient circulation, failure of the cardiovascular system ensues.[74] There are many disorders of the cardiovascular system which can be separated into diseases that primarily affect the blood, heart, or blood vessels. From a biomaterials perspective, repair and regeneration of the heart and blood vessels as well as restoration of blood flow are the primary focus and cover devices such as vascular grafts, stents, and rotary blood pumps. While non-fouling is seen as an important property of materials used for such cardiovascular applications, thrombosis is the specific biological response that requires dampening. Initiated by protein adsorption as well as their surface-induced conformational changes, the coagulation cascade of events commences and this leads to thrombin formation. The blood platelets respond to various stimuli, including thrombin. This in turn causes platelet activation and formation of a stable thrombus (or plaque) which causes narrowing of the blood vessels.[75] Since one of the essential requirements of material surfaces used in cardiovascular applications is that they are non-fouling, the use of phosphorus-containing polymers has been concerned with polymers incorporating the PC head group. An important review by Iwasaki and Ishihara on the use of PC-containing polymers for cardiovascular application has described an extensive number of detailed *in vitro* evaluation studies of P(MPC$_{30}$-*co*-BMA$_{70}$).[18] These evaluated the interactions of various blood components, including cells, proteins, and lipids, with the P(MPC$_{30}$-*co*-BMA$_{70}$) polymer and highlighted the importance of both low protein affinity and high phospholipid affinity for the design of a non-thrombogenic surface.

9.3.2   Coronary Stents

Coronary artery disease is a medical condition which involves the build-up of plaque on the interior surface of the coronary arteries, causing the cardiac performance to decrease and hence leading to coronary ischemia. This disease is the most common cause of premature mortality in the Western world with almost a million deaths in the US annually.[76] For the treatment of coronary artery disease, two commercial products using a PC-based coating are in clinical use. The BiodivYsio™ coronary stent developed by Biocompatibles International (US) was approved by the FDA in 2000. It consists of a stainless steel wire mesh stent coated with a PC-containing polymer.[77] A more recent development, approved in 2008, is the Endeavor® zotarolimus-eluting coronary stent manufactured by Medtronic Vascular (US).[78] This is a delivery system for zotarolimus, which is an immunosuppressant. The same concept for the development of drug-eluting stents has been applied to other drugs, including the transcription factor c-myc (which has been shown to reduce restenosis)[79] and DNA fragments (used in gene therapy).[80]

A recent scientific direction for the treatment of coronary artery disease is concerned with the issue of late stent thrombosis, which is caused by a thrombotic occlusion within the implanted stents and reported to occur more than one year after implantation.[48,49] In contrast to traditional stents made from biostable metals, new developments in this area make use of the biodegradable yet strong biopolymer PLLA as the base material (see Table 9.2). In order to impart non-thrombogenic properties to these biodegradable stents, PLLA is blended with P(MPC$_{30}$-*co*-BMA$_{70}$).[48,49] Research has shown that the high mechanical performance and slow degradation rate of PLLA is retained in these blend materials.[49] *In vivo* studies using a rat model showed superior performance after six months implantation of the blend materials compared to PLLA alone.[48] These new materials thus offer a promising solution to late stent thrombosis.

9.3.3 Prosthetic Vascular Grafts

Synthetic vascular grafts are used in a number of applications, including the bypass of occluded arteries (*e.g.* coronary artery bypass graft surgery) and for the replacement of aneurysmally dilated aortas. Much success has been achieved for the development of prosthetic vascular grafts of diameter >6 mm, with commercial products routinely used in clinical surgeries, including grafts of poly(ethylene terephthalate) (Dacron), expanded polytetra-fluoroethylene (ePTFE), segmented polyurethane (SPU), and polyethylene (PE). However, there is still a challenge in the supply of blood vessel grafts with a diameter <6 mm, where the formation of blood clots is currently unavoidable.[81] Studies led by Ishihara and Yoneyama[82,83] investigated the use of SPU-based polymer blends for coating onto existing small-diameter (2 mm id) vascular grafts, including Dacron[82] and a PE fibrous tubing.[83] The copolymer used in these studies was composed of 70% 2-ethylhexyl methacrylate and 30% MPC. The coated vascular grafts were evaluated in an *in vivo* rabbit model by placing the graft in the carotid arteries. It was found that the polymer coatings improved the non-thrombogenicity over 5 days[82] or 8 weeks.[83] In another study, Korematsu *et al.*[84] investigated an alternative approach to the incorporation of PC groups. In their study the substrate SPU was surface-modified by graft copolymerization of MPC onto hydroxylated SPU using a chemical initiator. This surface-modified substrate was found to have reduced platelet adhesion when tested in an *in vitro* contact study using platelet-rich plasma.

Recent work in the area of vascular grafts has explored the fabrication and testing of tissue engineered constructs.[50,51] Studies led by Soletti and Wagner (see Table 9.2) employed biodegradable poly(ester urethane)urea (PEUU) as the base material due to its ability to match native vessel compliance. Two approaches were investigated for the incorporation of PC-containing polymers. One approach involved blending PEUU with P(MPC$_{30}$-*co*-methacryloyloxyethyl butyrate$_{70}$) at various ratios, followed by electrospinning processing to fabricate tubular scaffolds (1.3 mm id).[51] These materials displayed suitable mechanical properties, significantly reduced platelet adhesion *in vitro* with increasing MPC-based polymer content, and good potency *in vivo* (8 weeks in rat model) with endothelialization and tissue integration for blends incorporating 15% MPC-based polymer. A second approach involved the incorporation of P(MPC$_{30}$-*co*-methacrylic acid$_{70}$) by the carbodiimide-mediated attachment of the carboxylate groups to an electrospun PEUU scaffold (1.3 mm id) incorporating amine groups that had been introduced using ammonia radiofrequency glow discharge plasma.[50] Unfortunately this study does not give any details of the chemical characterization of the chemically modified substrates. However, evaluation of the mechanical and biological performance is described. It was found that the "mechanical properties were compatible with native arterial conduits", that *in vitro* platelet adhesion was reduced to 10% for treated surfaces, and *in vivo* assessment in a rat model for 24 weeks found oriented collagen and elastin deposition. This was in addition to endothelialization and tissue integration. Both these approaches to a tissue engineered small-diameter vascular graft appear very promising.

9.3.4 Other Blood-Contacting Devices

Blood-contacting devices made from titanium and its alloys encompass both implants (*e.g.* prosthetic heart valves) and devices that are external to the body (*e.g.* ventricular assist devices). Similar to other devices used for cardiovascular applications, thrombosis concerns have led to surface modification of all of these using PC-based polymers. Wagner and his group[46] investigated the covalent attachment of PMPC. This was introduced *via* a series of steps, including plasma treatment and UV irradiation-induced grafting (see details in Table 9.2). A surface coating consisting of an intermediate silane layer between the metal and the polymer avoided covalent labile bonds such as amides, which posed concerns in their earlier work.[85] Another multistep approach to surface modification described by Zhao *et al.*[47] involved the formation of a crystalline titania film with a self-assembled alkyl phosphoric acid intermediate layer. This allowed the formation of initiation sites for the subsequent ATRP polymerization of MPC from the surface (see Table 9.2). An alternative approach by Ishihara and Huang[43,44] involved the use of a MPC-based copolymer (30% MPC) containing catechol groups. These were attached to the titanium substrate in a simple dipping procedure and resulted in a coating thickness of 20 nm (see Table 9.2). All three approaches resulted in significant reduction in platelet adhesion when exposed for either $2\frac{1}{2}$ hours to heparin-containing ovine blood[85] or for 1 or 2 hours to platelet-rich plasma.[44,47] Furthermore, two of the studies investigated the coating stability over a one-month period under conditions simulating blood circulation (in water or PBS). No change in surface properties was found,[85] as evaluated by water contact angle and chemical composition by XPS, or as later reported by FTIR and XPS.[44] It thus appears that both of these more thoroughly evaluated approaches may be suitable for use in medical devices, although it should be mentioned that the catechol-containing copolymer did contain an amide linkage between the polymer backbone and the catechol group.

9.4 Orthopedics

9.4.1 Introduction

Where bone and cartilage repair and regeneration, including TE applications, are concerned, it is well accepted that the surface properties, such as the functionality of the surface groups and the hydrophilicity of the polymeric biomaterials used, play essential roles in protein and cell adhesion and function. Ultimately the desired nature of the optimized surface will of course depend on the intended application.

In addition to the growing number of joint replacements worldwide, the number of revision joint replacements is also growing. Despite many improvements in the lifetime, design, and composition of artificial joints, there are still many challenges involving the materials employed. Periprosthetic osteolysis results from the production of wear particles and there are many studies confirming that PE is the most abundant of these. Any reduction in the production of PE particles should lead to a reduction in the bone resorption processes. Hence much research has focused on modifying the surface of the both the PE and metal components of orthopedic devices. Efforts at reducing osteolysis have included investigating various combinations of load-bearing surfaces such as ceramic-on-PE and metal-on-metal. This is in addition to the changes in the PE from ultra-high molecular weight polyethylene (UHMWPE) to crosslinked polyethylene (CLPE). In addition to PE, UHMWPE, and CLPE, various metal combinations are among the substrates investigated. Not surprisingly, this is an area of research attracting much interest. In orthopedics, MPC is used to impart non-frictional and lubrication properties to implants which are associated with excessive wear, in other words with those materials used to replace joints.

9.4.2 Polyethylene Non-Frictional Applications

In research spanning many years, Ishihara and his group have used MPC to modify material surfaces for orthopedic applications. In a recent paper entitled "Biomimetic hydration lubrication with various polyelectrolyte layers on cross-linked polyethylene orthopedic bearing materials", many of his current strands of relevant research in this field have been elegantly brought together.[20] Surface modification of the PE to improve lubrication to better mimic native cartilage involved using nanometer-scale hydrophilic layers with various charges. Graft polymerization was used to produce the layers, which included both MOEP and MPC. Using techniques they have developed over many years,[21,86] they produced hydrophilic nanometer-scale polyelectrolyte layers on the CLPE surface. They recognized the limitations of their study but successfully demonstrated that the modified surface of the acetabular caps had greater durability. In an earlier study they found that grafting MPC onto the PE liner surface "dramatically decreased the wear production" compared to the effect of cross-linking the PE or changing the femoral head materials (see Table 9.2).[58] They also studied the mechanism underlying the improved lubrication afforded by the hydrophilic phosphorus-containing grafted polymer.[21] They attributed the highly porous nature of the MPC polymer membrane to its high water retention capacity, which they directly linked to the wear properties.

In a more recent pilot study they investigated the effects of irradiation times on the density and stability of the brush-like PMPC chains on a CLPE surface (see Table 9.2).[57] Their data support the finding that a high density of PMPC graft chains reduces the steady wear rate by more than 90%. However, they do emphasize that further detailed clinical studies are required.

9.4.3   Cobalt–Chromium–Molybdenum Non-Frictional Applications

A cobalt–chromium–molybdenum (Co–Cr–Mo) alloy is one of the most commonly used in artificial femoral heads. Ishihara and collaborators grafted PMPC onto the C–Cr–Mo surface in order to better mimic the natural lubricity of native cartilage.[87] They identified three important issues: (i) the need for strong bonding between the PMPC and the metal surface, (ii) good mobility of the free PMPC end groups, and (iii) optimum density of the MPC layer. In order to achieve their goals they used photoinduced graft polymerization techniques to create covalent bonds between the Co–Cr–Mo substrate and the PMPC *via* a (3-methacryloxypropyl)trimethoxysilane intermediate layer. In a later study of this same system they systematically investigated the surface properties, including the stability and lubricity, and found that the cartilage/PMPC-grafted Co–Cr–Mo interface, designed to mimic a natural joint, exhibited a friction coefficient as low as that of a natural cartilage interface (< 0.01) (see Table 9.2).[59]

Many of the expectations of 10 years ago regarding the transformation of laboratory research into actual clinical reality have clearly not been realized where orthopedic applications are concerned. However, this is most probably more the result of people's unrealistic hopes than any great fault of the research concerned. Progress is being made and will continue to be made as long as the interdisciplinary teams dedicated to such research continue the dialogue and hard work.

9.5   Bone and Bone-Interfacing Biomaterials

9.5.1   Introduction

Although biomineralization, a widespread complex process which ranges from fabrication of sea shells to production and maintenance of bone and teeth, has been thoroughly researched over many years, the cascade of chemical and biological events involved are not fully understood even today.[88,89] It is, however, generally accepted that all biomineralization processes are initiated by the deposition of an anionic, hydrophilic nucleator onto an organic material. This chapter cannot hope to cover the latest discussions regarding the ability of various calcium phosphate minerals to induce osteoconductivity when incorporated into biomaterials for medical applications. It is important, however, to advise readers that there is an on-going and vibrant debate regarding the fundamental mineralization processes associated with biomaterial performance that is very relevant. It appears that to some extent these studies are not always considered in the biomaterials context. As pointed out in a recent very elegant paper entitled "Osteoinductive biomaterials: current knowledge of properties, experimental models and biological mechanisms" by Habibovic and co-workers, drawing any general conclusions from even *in vivo* studies is challenging because there are not only many differences in the physicochemical and structural properties of the biomaterials, but also in animal models, implantation sites, and duration of the many studies. They point out that limited, if any, bone induction by biomaterials is observed in smaller animals, particularly in rodents. Hence, larger animal models are essential for proper evaluation of *in vivo* responses relevant to humans.[22] In spite of the cautious spirit of this paper, they also state that "the knowledge of material properties relevant for osteoinduction to occur has tremendously increased in the past decades".

Although dental implants (see Chapter 8 of this book) were one of the first applications studied with respect to improving the osseointegration of the metal implant/tissue interface, cranial and facial as well as orthopaedic applications also dominate the literature where bone biomaterials and bone-interfacing biomaterials are concerned. From the perspective of this chapter, mineralization, which dominates these applications, is particularly relevant. Another important aspect of bone applications including TE is whether or not the applications are load-bearing (orthopedic) or non-load-bearing (facial and cranial). Although it should be noted of course that it is not always possible to make a clear delineation, the jaw is a good example.

The functionalization of biomaterials, both natural and synthetic, with phosphate-containing groups in order to provide nucleating sites to trigger the biomineralization cascade and produce calcium phosphate phases has been well researched in line with the implication of anionic polymers in *in vivo* mineralization. Where synthetic polymers are concerned, mineralization studies have led to some apparently contradictory results. Chirila *et al.*[11,90] critically reviewed this topic as well as investigating the calcification capacity of acrylic hydrogels in 2007. In order to better demonstrate these apparent anomalies we will divide our discussion into polymeric hydrogels (Section 9.5.2) and grafted systems (Section 9.5.3).

9.5.2 Hydrogels

To the best of our knowledge, the first paper investigating the effect of phosphorus-containing functional groups on calcification with respect to bone induction appeared in 1976, by Swart *et al.*[91] Their results proved somewhat unexpected: calcification appeared after 2 weeks on the PHEMA hydrogels whereas none was observed even after 6 months on the phosphorylated hydrogels. Chappard's results for a set of hydrogels based on MOEP and its copolymers with 1-vinylpyrrolidin-2-one (VP) differed greatly.[92] According to Chirila,[11] their observation that calcium phosphate phases were deposited to varying extents on all the polymers studied, suggests that "the effect of the phosphate group is perhaps more complicated than initially thought". This is somewhat prophetic in light of our subsequent findings, especially regarding the purity of the MOEP monomer (see Section 9.1). Chappard reported purifying the MOEP monomer in a two-step process (24 h extraction with toluene, then distillation under pressure) before use in the polymerization reaction. After an *in vivo* study in the cranial bone in rats, Chappard *et al.* came to the conclusion that MOEP hydrogels were not suitable for bone replacement therapies. In spite of massive calcification, osteointegration was poor, no osteoblasts were observed in these calcified areas, and there was no observable bone formation or bone bonding.[92] A few years later, Chirila *et al.*[90] found that there was a decrease in calcification in the copolymers of HEMA with either 10% MAEP or MOEP when compared to PHEMA both *in vitro* and *in vivo*.

In a more recent paper, mineralization (and cell) studies of a copolymeric hydrogel of bis[2-(methacryoyloxy)ethyl] phosphate (BP) incorporated into oligo[poly(ethylene glycol)] fumarate (OPEGF) were reported.[93] Although a concomitant increase in calcium with phosphorus content was observed in the P(OEGF-*co*-BP) hydrogels *vs.* none in the pure OPEGF hydrogels, the Ca/P ratio decreased from 0.6 to 0.2 with increasing phosphate concentration on the hydrogel surface. A sample P(OEGF$_{70}$-*co*-BP$_{30}$) showed a ratio of about 0.23 and the inorganic crystals had a Ca/P ratio of 0.9, which is similar to that of the minerals brushite and monetite.

Isihara *et al.*[94] reported the formation of a P(MPC-*co*-MOEP) hydrogel intended for cartilage and bone applications. This P(MPC-*co*-MOEP) hybrid hydrogel was synthesized by random copolymerization and they stated that they had removed the diene impurities from the MOEP. The fact that the P(MPC-*co*-MOEP) hydrogels only mineralized after they had gone through a soaking process in CaCl$_2$ and Na$_2$HPO$_4$ cannot, in our opinion, be attributed to the presence of the phosphate groups, but rather to the presence of nucleating ions dispersed throughout the hydrogel matrix.

9.5.3   Grafted Systems

Graft polymerization introduces covalently tethered polymer chains with specific functional groups onto an existing polymer surface. The resulting modified surface can exhibit completely different properties than the original polymer substrate. The surface graft polymerization of MAEP and MOEP onto a range of substrates, such as silk fabrics, HDPE, poly(ethylene terephthalate) (PET), polyacrylonitrile, and a PTFE surface, has been well investigated. One study by Trettinikov and Ikada's group involved the surface modification of HDPE by graft polymerization of MOEP with the aim of improving the bone-bonding ability of the polymer surface.[24] In simulated physiologic solution, the precursor to what is now known as simulated body fluid (SBF) conditions, the modified polymer showed improved carbonated hydroxyapatite (HAp) growth *in vitro.* Subsequently, an *in vivo* study showed significant enhancement of bone growth at the material/bone interface.[23] In a later study, the same group investigated the growth of HAp on a PET surface grafted with MOEP.[95] They reported that the water contact angle for the MOEP-grafted surface was reduced from 93° to 47°. The aim of this study was to find a way to deposit a thick HAp layer more rapidly and more firmly on the polymer surface than had previously been possible. In their earlier study,[24] in addition to the HAp layer taking around 10 days to deposit, it was also partially substituted with Mg^{2+} ions. With a view to improving this process they investigated a two-step process which involved first depositing a thin layer on the substrate surface over a number of cycles and then using SBF conditions to deposit a thicker layer of calcium phosphate minerals over a period of 10 days. In the first step, a disodium hydrogenphosphate solution was added to a buffered aqueous calcium chloride solution, with vigorous stirring, before adding the polymer samples and continuing the stirring for 30 min. Further treatment was required before the samples were subjected to the SBF.

We have investigated grafting of MAEP and MOEP onto fluorinated polymers over the past 10 years.[13,96–102] Expanded PTFE (ePTFE) has been used for many years for tissue repair in craniofacial reconstruction. The surface modification by grafting hydrophilic phosphate-containing moieties onto a highly hydrophobic expanded ePTFE was intended to improve the biological response at the material/tissue interface by not only introducing suitable new functional groups, but also by altering the wettability of the surface. Our first SBF study showed no mineral growth on unmodified ePTFE.[97] This is in agreement with two earlier *in vivo* studies on PTFE.[103,104] The chemical composition of the graft copolymers formed using MAEP and MOEP were thoroughly investigated and our data highlighted that they could not simply be described as PMAEP and PMOEP graft copolymers. In addition, Suzuki *et al.*[105] were the first to report the RAFT-mediated polymerization of both MAEP and MOEP monomers to give a series of non-crosslinked soluble polymers. In order to explain the data obtained from these studies, we investigated in greater depth the starting MAEP and MOEP monomers in an effort to understand the phosphorus content and overall composition of the resulting graft copolymers and RAFT polymers, including the PMOEP formed in a two-phase monomer system.[12,13] Our findings clearly showed that chemical events such as ester cleavage and the influence of the unacknowledged diene can greatly affect the polymer topology (molecular structure) and composition of the target polymers.[12]

At this point in the phosphate-grafted PTFE membrane research odyssey, a serendipitous discovery made it possible for us to make a confident correlation between the nature of the mineral phase formed on the co-polymer and its topology.[13] Since these results most probably will have important implications in other phosphate polymer studies, where the presence of the diene impurity (discussed in Section 9.1) and its influence on crosslinking have not been considered, a more detailed discussion is warranted here. Therefore, in spite of the fact that synthesis is not the focus of this chapter, a short digression into the radiation-induced grafting synthesis of phosphate-grafted ePTFE membranes is relevant.

As discussed in detail in our earlier papers,[13,102] in radiation-induced grafting the choice of the solvent used for dilution of the monomer during simultaneous grafting is one of the critical experimental factors leading to a successful outcome. One of the desirable characteristics is a low chain transfer constant. Nasef *et al.*[106] found that the degree of grafting was strongly dependent on the inherent nature of the solvent as well as its interaction with the polymer substrate and the monomer. It must be acknowledged that the characterization of graft copolymers is most challenging. A suite of techniques is required in order to understand both their chemistry and topology, and while some are suitable for the characterization of the graft copolymer (*e.g.* XPS and FTIR), others require soluble polymers for optimal characterization (*e.g.* NMR). As has been pointed out in the literature, detailed chemical analysis of the homopolymer formed at the surface can also contribute to the analysis of the graft copolymer.[107] Our study led to the discovery that by causing grafting to occur in the aqueous phase of a two-solvent two-phase system (water/dichloromethane), the grafting process is no longer encumbered by the ubiquitous formation of insoluble homopolymer.[13] As explained earlier, the presence of the undetected diene had complicated the earlier grafting outcomes.[12] However, by grafting in the two-phase system, a homopolymer (and implied graft copolymer) uncontaminated by diene incorporation was achieved. As mentioned above, we have reported our findings in a series of papers: the effects of the monomer and its concentration, the solvent, and other methodology effects on the polymer chemistry.[99,100]

We can now return to our SBF *in vitro* mineralization studies. A range of MOEP-grafted ePTFE membranes were studied, as shown in Table 9.3. Three samples had been grafted in organic solvents: either pure methanol or methyl ethyl ketone (MEK, butan-2-one) or in a mixed solvent system. The fourth sample, consisting of linear grafted chains, was prepared in the aqueous phase of the two-solvent two-phase system as described above. It was on this sample alone that we found that a uniform coating of carbonated hydroxyapatite formed. As a result of this "ideal" mineralization it was possible to relate the nature of the mineral growth to the surface chemistry and topology of the grafted membranes. In our paper we discuss in more detail the difficulty of addressing the topology. In conclusion, these results will undoubtedly have important implications for other *in vitro* phosphate polymer mineralization and cell studies where the presence of the diene impurity and its influence on polymer crosslinking are not being considered.

Table 9.3 MAPE/MOEP polymerization techniques and their polymer topologies.

Technique	Topology	Ref.
RAFT-mediated polymerization taken to low conversion ($M_w < 20\,000$)	Branched	105
Grafting in aq. phase of two-phase solvent system (H_2O/CH_2Cl_2)	Linear	13
All other grafting systems (single organic solvent, two miscible organic solvents, or organic phase of a mixed solvent system)	Branched/crosslinked	96–102

In a recent paper, P(HEMA-*co*-MOEP) brushes were further functionalized using RGD peptides. The desired outcome was to obtain osteoconductive substrates suitable for bone implant applications.[25] Surface-initiated atom transfer radical copolymerization was used to produce the copolymer brushes. It does not appear that the MOEP was purified before use. MOEP was found to promote matrix mineralization with a concomitant increase with MOEP content. Mineral formation was most pronounced on substrates that presented both phosphate groups and the RGD peptide sequences.

In addition to *in vitro* mineralization to evaluate the bone bonding ability of a material, the study of cellular responses to materials intended for use at the bone interface is also important. Using ethyl ethylene phosphate-based polymer (EEP), Wang *et al.*[108] investigated cell behavior on a spin-coated P(LLA-*b*-EEP) scaffold. They found that osteoblast attachment and proliferation was significantly enhanced on the PEEP-modified PLLA surface.

In our work we evaluated both the ability of the grafted membranes to promote osteoblast growth[101] and to dampen proinflammatory response in macrophages.[98] Apatite formation after immersion in SBF for 7 days correlated with the phosphorus content of the grafted surfaces, whereas osteoblast-like SaOS-2 cell adhesion and spreading associated with the amount of protein adsorption, which was higher for the higher grafting surfaces despite the carbon/phosphorus ratios.[101]

Before undertaking protein and macrophage cell studies using a range of our surface-modified ePTFE membranes, we used XPS to analyze the functionalized membrane surfaces.[98] Results indicated that the *in vitro* proinflammatory response was not affected by the surface morphology or wettability but rather by the surface chemistry of the relevant set of membranes. When the surface chemistry is the result of charged acrylate groups containing phosphate and carboxylate (albeit at a low density), there is a low proinflammatory response *in vitro*. Based on the combined data from the cell and protein studies (from surface-MALDI-MS), there is compelling evidence to support a correlation between the macrophage response and the type of proteins adsorbed from the serum. It is clear that just like the stream of biomineralization events that occur in nature, the implantation of any biomaterial in the body results in a cascade of protein and cellular response that does not always lead to a positive end result. However, by better understanding and controlling the properties of surface-modified bio-materials we can go some way to improving the outcomes.

Applications relevant to this chapter involving poly(vinylphosphonic acid) (PVPA) and its derivatives appear less frequently in the literature. Ilia's 2010 review on the subject is both topical and useful.[109] Many of the references cited in the section "Biomaterials based on PVPA" lie outside the topic of the current chapter, and a perusal of the references show many concern po-tential dental applications. Phosphonates were found to induce mineral-ization of reconstituted collagen fibrils.[110] In another report, the authors found that HAp growth on the copolymers of VPA and 4-vinilyimidazole depended on the phosphonate content.[111] The influence of different func-tional groups as well as the pH was investigated in a study where functio-nalized polymeric nanoparticles were used as mineralization templates. The results indicated that in this case the amount of HAp crystals formed de-pended on the number of negative charges on the particle surface. PVPA was one of the polymers investigated. A suite of characterization and analytical techniques (SEM, TEM, ED, XRD and DLS) was used to fully characterize the resulting materials.[112] Fundamental studies of this kind may well help to open up an extended future for VPA applications in medicine.

9.6   Ophthalmological Applications

The history of biomaterials in ophthalmological applications spans over 150 years and has been eloquently documented by Chirila in his recent book.[26] The latter covers a range of topics from intraocular lenses to different ophthalmological TE applications. One conclusion on reading this work is that some remarkable commercial successes are evident in this particular field.

In general, one unique requirement of materials for eye applications is transparency. Today, a wide variety of nondegradable and degradable poly-meric materials is clinically used and still being extensively investigated. Some examples of these are poly(methyl methacrylate) (PMMA), PHEMA, silicone for contact lenses and intraocular lenses,[113,114] collagen[115,116] and silk fibroin[117–120] matrixes for corneal regeneration, and PEG-based oph-thalmologic adhesives.[121,122] Phosphorus-containing monomers, particu-larly MPC, have also found application in this field. This is attributed mainly to their high wettability, hemocompatibility, anti-biofouling properties, cell compatibility, and high enzymatic resistance.

MPC has been found to be ideally suited for use in soft contact lens (SCL) materials.[123] Proclear® is the PHEMA-based SCL-containing MPC unit (CooperVision®). The coating technique of using MPC on a silicone hydrogel has also been reported.[124] Such MPC-modified materials have been found to have good mechanical properties and have demonstrated high oxygen per-meability compared with earlier materials. An MPC hydrogel was designed for SCL use.[125,126] An MPC-based crosslinker was also synthesized. It can be used to adjust the equilibrium water content of the MPC hydrogel and increase the mechanical strength.[127,128] To improve the hydrophilicity, protein resistance, and ultimately the anti-biofouling of the silicone hydrogel, MPC was incorporated by surface grafting,[52] or an interpenetrating polymer network (IPN) structure,[53] or copolymerization (see Table 9.2).[54,55] MPC has also been used as a coating for SCL biosensors for the *in situ* monitoring of tear glucose. It proved useful as a non-invasive blood sugar assessment process.[129,130]

Bacterial adhesion is a serious concern for currently used intraocular lenses (IOLs). Problems lead to a high incidence of endophthalmitis, capsular opacification, and vision activity. MPC has been used in the surface grafting of silicone IOLs[131–133] and as a coating for acrylic IOLs (see Table 9.2).[56,134–137] In addition, as described earlier in this chapter, they have shown desirable anti-biofouling properties.

As discussed in Section 9.5, MAEP/MOEP was copolymerized with HEMA for the purposes of inhibiting calcium phosphate deposition in ophthalmologic applications.[90] Reduced mineral deposition was observed, both *in vitro* and *in vivo*, for the copolymers when compared to that of PHEMA. To enhance the cellular activity of PHEMA, MOEP has been successfully grafted using ATRP.[29] The resulting material significantly promoted the attachment and growth of human corneal epithelial cells. Although MPC has antifouling properties and hence reduces cell adhesion, it can also be used to control biological responses.[138] Collagen–MPC interpenetrating network (IPN) hydrogels were produced as corneal substitute implants.[139–141] MPC was found to improve structural properties and enzymatic resistance compared to collagen-only scaffolds. An *in vivo* study indicated that collagen–MPC hydrogels showed regeneration of the corneal tissue as well as the tear film and sensory nerves after 12 months in a mini-pig model.[141] The hydrogel was also found to promote cell and nerve repopulation in alkali-burned rabbit eyes.[139] In addition, enhanced resistance to neovascularization was noted.

The cytotoxicity of MPC-based eye drops containing 0.1% Lipidure-PMB (a copolymer of MPC and butyl methacrylate; NOF Corp.) was examined and it was found to have similar nontoxicity to other clinically approved artificial tear products.[142] From these literature reports it appears that the unique properties and nontoxicity of MPC are ideal for ophthalmological applications.

9.7 Applications in Drug and Gene Delivery

9.7.1 Introduction

In his elegant 2008 overview of the history and origins of "controlled" drug delivery systems (DDS), Hoffman comments on the future of "controlled" drug delivery: "as our knowledge of cell biology and DNA increases, so will our ability to design nano-scale DDS that are serum stable and efficiently taken up by specific sites and pathways within the cells".[143] Recent reviews highlight specific DDS areas, including gene delivery, intelligent nanomaterials, remotely triggered delivery, and other delivery routes.[144–146] The design of complex architectures and nanostructured polymeric materials became possible as a result of the development of sophisticated synthetic methodologies, which included controlled radical polymerization, especially ATRP and RAFT. Indeed, the resulting polymers have been extensively studied for drug and gene delivery systems.[147,148] Owing to space constraints and the large number of papers in this field, this section will focus mainly on the more recent papers on PMOEP, PMPC, and PPE (2009–April 2013). To facilitate the reader we have divided these into sections based on the polymer rather than the specific application. See also Figure 9.1.

9.7.2   Poly[2-(methacryloyloxy)ethyl phosphate] (PMOEP)

Negatively charged phosphate groups are capable of interacting ionically with positively charged molecules such as basic growth factors. Giglio *et al.*[149] investigated the mineralization and rhVEGF loading capability of HEMA and MOEP copolymer hydrogel coatings on titanium substrates. The incorporation of MOEP was found to enhance mineralization in SBF compared to the HEMA-only hydrogel coatings. Results also showed that the incorporation of MOEP appears to facilitate the entrapment and release of rhVEGF *in vitro*. Jeon *et al.*[150] investigated the loading of BMP-2 into PMOEP-grafted nano-hydroxyapatite which was immobilized into highly porous amine-functionalised PLGA microspheres for bone tissue regeneration applications. BMP-2 was efficiently bound onto the anionic phosphate groups on the surfaces and released gradually over a period of approximately one month in PBS (pH 7.4). It should be noted that this was accompanied by continuous shaking. pH-sensitive calcium phosphate nanoparticles coated with a P(EG-*b*-MOEP) block copolymer were prepared for creating gene delivery carrier systems.[151] Although synthesis is not the focus of this chapter, it may be useful to summarize their synthesis as an example. The MOEP precursor monomer, 2-(methacryloyloxy)ethyl dimethyl phosphate, was synthesized from HEMA and dimethyl chlorophosphate, and after block copolymerization by ATRP using PEG as the macroinitiator the phosphoryl dimethyl ester of the copolymer was demethylated by bromotrimethylsilane in order to obtain the phosphate groups. P(EG-*b*-MOEP) with a narrow molecular weight distribution was obtained. Calcium phosphate nanoparticles were formed in the presence of phosphate block copolymers, which resulted in an average size below 200 nm. pDNA was incorporated by simple mixing. Although the nanoparticles were stable at pH 7.4, entrapped pDNA was found to be released at pH 5.0 by a pH-dependent protonation of the phosphate moieties. The use of phosphoryl dimethyl ester monomer instead of MOEP made it possible to obtain narrow molecular weight polymers using the ATRP technique, which otherwise is often hampered by the presence of acid groups. However, the issue of whether or not significant hydrolysis of the phosphate or carboxyl groups occurred during demethylation is not addressed in this paper. Since retention of the phosphate groups cannot be investigated by ^{1}H NMR, other techniques such as phosphorus analysis and ^{31}P NMR should probably have been performed.[12]

9.7.3 Poly[2-(methacryloyloxy)ethyl phosphorylcholine] (PMPC)

There are some distinct advantages in using biocompatible PMPCs in DDS, such as serum resistance and their ability to avoid reticuloendothelial system recognition in blood circulation. Non-fouling PMPC coatings for cardio-vascular applications are discussed in Section 9.3. Currently, drug-eluting coatings for stents are extensively being developed.[152–154] PMPC and their copolymers have been investigated as drug releasing coatings.[79,80,155] At present, PMPC-coated drug eluting stents are commercially available (Endeavor®, Medtronic Inc.). However, despite the short-term clinical success for drug eluting stents, it has been suggested that the polymer surface which released almost all the drugs possibly causes thrombogenicity in the long term.[152,156]

Enomoto *et al.*[157] investigated drug release from three types of biocompatible polymers, including PMPC. These had coatings of micro-patterned diamond-like carbon (DLS), which is a biocompatible and antithrombotic carbonaceous material used in drug-eluting stent applications. An *in vitro* drug release study showed that the usual initial burst of the drug release from the polymer matrix was prevented by the micro-patterned DLS coating, and the release profile could be controlled by changing the cover area of the DLS coating on the polymers.

Polymeric nanocarriers such as micelles and polymersomes have been extensively investigated for drug and gene delivery applications because of their increased stability, site specificity, and blood circulation resistance.[158] Polymersomes are composed of amphiphilic block copolymers forming a shell that can encapsulate hydrophilic components inside the aqueous core and entrap hydrophobic components within the shell itself. Stimuli-responsive polymersomes, which change properties after applying a specific external trigger, based on the hydrophobic segment of the block copolymers, have also been developed.[159] PEG has been extensively studied for the hydrophilic segment of the block copolymers and many of these products are currently undergoing clinical trials.[160] PMPC has also been investigated for this purpose and has been recently reviewed.[30,161]

The water-soluble amphiphilic copolymers of MPC and *n*-butyl methacrylate (BMA) have been extensively studied for DDS applications. The PBMA units assemble in water as a result of the hydrophobic effect and the copolymers form aggregates that can solubilize hydrophobic bioactive agents in the core PBMA segments. Paclitaxel, a very hydrophobic anticancer drug that is poorly soluble in aqueous medium (<0.3 µg mL^{-1}), can be better dissolved at more than 2 mg mL^{-1} using P(MPC-*b*-BMA) block copolymers.[162] Recently, Ishikawa *et al.*[163] developed an emulsion lotion containing commercially available P(MPC-*co*-BMA) for the controlled release of water-insoluble diphenhydramine, which is a widely used antihistamine for allergy relief. Using RAFT polymerization, Kitagawa *et al.*[164] copolymerized P(MPC-*b*-BMA) with *p*-nitrophenylcarbonyloxyethyl methacrylate, which contains active ester groups. The resulting polymer was conjugated with the pre-S1 domain of the hepatitis B surface antigen (HBsAg) in order to generate specificity for liver cells. Paclitaxel was incorporated during micelle formation and the efficacy and safety of these micelle–drug complexes were investigated using *in vitro* cell studies and human tumor models xenografted in athymic mice. A tissue-specific paclitaxel delivery system resulted in an enhanced capacity to carry the drug to the target site with reduced adverse side effects. The same group later conjugated an anti-epidermal growth factor receptor antibody with the terpolymer, mixing it with verteporfin, which is a hydrophobic photosensitizing drug.[165] The resulting complexes were intravenously injected into tumor-bearing mice, and irradiation was performed. The complex conjugated with the antibody significantly decreased mice tumor size within 8 days compared to the complex without antibody conjugation.

Armes and co-workers reported a pH-responsive vesicle using P[MCP-*b*-(2-(diisopropylamino)ethyl methacrylate] [P(MPC-*b*-DPA)] block copolymers synthesized by ATRP. The PDPA block is pH sensitive (pK_a = 5.8–6.6 depending on the ionic strength). At physiological pH (7.4), which is above the pK_a, the PDPA segment is uncharged and hydrophobic, whereas it becomes highly ionized below the pK_a. This causes destabilization of the polymersomes and leads to release of the loaded drugs. Recently, folic acid-functionalized P(MPC-*b*-DPA) diblock copolymer micelles were synthesized using ATRP. Their ability for uptake of two poorly water-soluble anti-tumor drugs, tamoxifen and paclitaxel, and delivery to cancer cells through folate-receptor targeting was evaluated *in vitro*.[166] The result confirmed that folate-conjugated micelles led to increased drug uptake within the cancer cells; selectivity towards the tumor cells was also demonstrated. The effect of the PDPA block length on the binding affinity of the block copolymer to plasmid DNA was investigated by Lomas *et al.*[167] Large molecules such as antibodies have also been successfully delivered into live cells using P(MPC-*co*-DPA) micelles as delivery vesicles. There was no significant loss of cell viability or metabolic activity.[168]

Thermoresponsive, acid-degradable micelles with a positively charged core and a hydrophilic PMPC shell were prepared using RAFT polymerization.[169] The PMPC macro-RAFT agent was prepared and chain extended with methoxydiethylene glycol methacrylates and 2-aminoethyl methacrylamide hydrochloride as well as an acid-degradable crosslinker, 2,2-dimethacryloxy-1-ethoxypropane, in water/propan-2-ol. The resulting positively charged core of the micelles facilitated the encapsulation of oppositely charged proteins by incubating the micelles with suitable molecules at 4 °C. At low temperatures, the core became hydrophilic, which enabled the proteins to diffuse into the micelles. With increasing temperature, the hydrophobic core entrapped the proteins. The incorporated proteins were released under acidic conditions through degradation of the micellic crosslinkers.

Biodegradable polymeric nanocarriers incorporating controlled break-down for release can be prepared using an aliphatic biodegradable polyester, such as poly(lactic acid) (PLA) or poly(ε-caprolactone) (PCL), as the hydro-phobic segment.[170] Well-defined P(LA-*b*-MPC) copolymers were synthesized using ATRP and their micelles were investigated for the intracellular delivery of an anticancer drug, doxorubicin (DOX).[171] The micelle-loading efficacy of DOX was found to be around 44–67%, with efficient delivery of the drugs into the cancer cells reported. The same group prepared P(LA-*b*-MPC) poly-mersomes *via* a dialysis method (*i.e.* dissolving the copolymers in an organic solvent, followed by the addition of water and then dialyzing against water).[172] Hydrophilic DOX · HCl or hydrophobic DOX was added during the formation of the polymersomes. They were loaded into either the polymer-some interior or the shell membrane, respectively. Significantly faster drug release was found at mildly acidic pH (5.0) compared to physiological pH 7.4 *in vitro* due to the hydrolysis of the PLA segment. The drug-loaded poly-mersomes rapidly entered the cancer cells and released their drugs owing to degradation in the acidic endocytic compartment. Du *et al.*[173] found that the P(CL-*b*-MPC) instantly formed polymersomes upon adding dried copolymer powder into hot water (70 °C) without the need to use organic co-solvents, pH adjustment, or even stirring. This is highly advantageous for encapsu-lation of biomacromolecules. They stabilized these vesicles using aqueous sol–gel chemistry by using tetramethyl orthosilicate as the silica precursor in the absence of any external catalyst.

Another interesting field in DDS is that of micelles formed by using amphiphilic copolymers with various shapes such as stars and combs rather than their linear analogues. Interest is mainly because of their unique bulk and solution properties. Using ATRP, Tu *et al.*[174] synthesized six-arm star P(CL-*b*-MPC) polymers which self-assembled into micelles, and their pacli-taxel loading and cellular uptake were evaluated. The six-arm star P(CL-*b*-MPC) loaded with paclitaxel showed higher cytotoxicity against cancer cells than P(CL-*b*-EG) micelles, in response to the higher efficiency of cellular uptake. Liu *et al.*[175] investigated the delivery of hydrophobic paclitaxel loaded in the polymersomes formed from comb-shaped P(MPC-*g*-EG-*b*-LA) copolymers that were synthesized using radical polymerization.

Hydrophilic P(EO-*b*-MPC) block copolymers were synthesized and polymersome vesicles were prepared by complexing PEO segments with α-cyclodextrins. The latter are cyclic oligosaccharides containing a hydro-phobic internal cavity that acts as a host for various guest molecules.[176] The polymersomes formed in aqueous medium without organic solvents and were found to be cytocompatible. DOX · HCl has been successfully loaded into such polymersomes and efficiently delivered to cancer cells *in vitro*.

For gene delivery applications, cationic polymers that form complexes with nucleic acids through electrostatic interactions are often used as non-viral vectors that provide protection of DNA from enzyme degradation and facilitate cellular uptake. Monosaccharides or polysaccharides have been well studied for the production of low-toxicity gene delivery vectors.[177,178]

Copolymers of carbohydrates and PMPC-based polymers have been examined for efficient gene delivery.[179–183] In a recent study, Ahmed prepared "block-statistical" copolymers based on carbohydrates and PMPC using RAFT polymerization for carriers of plasmid DNA.[184] It was found that, in contrast to cationic diblock copolymers, the "block-statistical" copolymers were significantly less cytotoxic and proved to be excellent DNA carriers to cells.

Hemp *et al.*[185] successfully synthesized well-defined phosphonium-based block copolymers using RAFT polymerization for gene delivery applications. MPC or oligo(ethylene glycol)$_9$ methyl ether methacrylate were first polymerized for the stabilizing block. Subsequently, these macro-RAFT agents were chain extended with tributyl(4-vinylbenzyl)phosphonium chloride to various degrees of polymerization (25, 50, and 75 for the phosphonium-containing block). All block copolymers were found to bind DNA efficiently and formed nanoparticles.

9.7.4 Polyphosphoester (PPE)

Over the past 30 years, PPEs have been extensively explored for biomedical applications, especially for drug and gene delivery.[31] In recent years, the versatility of the PPE pendant groups allows the incorporation of interesting features that can be used to conjugate various molecules. Numerous PPE copolymers have also been widely investigated for these applications. Recent progress of PPE in DDS has been covered in several reviews.[14,186]

Hydrophobic PPEs have been investigated as biodegradable drug delivery vesicles for low molecular weight drugs, proteins, DNA, and plasmids. Copolymers of PPE and PLA have been used as drug carriers in several preclinical and clinical studies. A Phase I trial of Paclimer® microspheres (Guilford Pharmaceuticals, median diameter ~ 50 μm), which consists of P(PE-*co*-LA) microspheres that contain 10% w/w paclitaxel, has been carried out.[187] However, subsequently the clinical development of Paclimer® microspheres was suspended by the manufacturer.

Wang *et al.*[188] investigated a series of copolymers of poly(ethyl ethylene phosphate) (PEEP) for DDS applications. Poly(ε-caprolactone)-*b*-poly(ethyl ethylene phosphate) [P(CL$_{150}$-*b*-EEP$_{30}$)] was synthesized by ring-opening polymerization and found to form nano-sized vesicles in aqueous solution using a thin-film hydration method. Using the acid gradient method, DOX was loaded into the vesicles and the DOX-loaded vesicles were internalized by A549 (human lung carcinoma cell line) cells and inhibited cell proliferation. Triblock copolymers of poly(propylene oxide) and poly(ethyl ethylene phosphate) [P(EEP-*b*-PO-*b*-EEP)] were synthesized and their thermoresponsive behavior was investigated.[189] Thermo-induced aggregation and gelation of the aqueous solutions of copolymers were found to be affected by the polymer structures, *i.e.* by the molecular weights of the PEEP blocks. A thermoresponsive hydrogel was used for DOX incorporation and demonstrated sustained release of the drugs *in vitro*. In another study, the diacrylate version triblock copolymer of P(EEP$_{151}$-*b*-EG$_{2k}$-*b*-EEP$_{151}$) was synthesized and thermosensitively assembled into nanoparticles in aqueous solution at a temperature higher than its lower critical solution temperature.[190] The nanoparticules were further crosslinked by reacting the diacrylates, and the efficiency of the swelled gels at physiological temperature was tested for drug delivery of DOX using A549 cells. Cuong *et al.*[191] also investigated a series of star-shaped P(CL-*co*-PE) polymers as potential DOX carriers. The micelles were formed in aqueous solution with the cores encapsulating the DOX. The *in vitro* release of the drug was found to be faster at pH 5.4 than that at pH 7.4 *in vitro*.

Reversibly crosslinked micelles based on a triblock copolymer, P(CL-*b*-PE$_{SH}$-*b*-EG), were investigated for DOX-DDS.[192] The middle phosphoester block contains thiol side groups which enables disulfide crosslinking of the block copolymers. The crosslinking within the shell reduces their critical micellization concentration and enhances their stability against severe conditions. A sophisticated system of dual pH-sensitive PPE-based polymer–DOX conjugate nanoparticles was developed by the same group.[193] A diblock copolymer, monomethoxyl poly(ethylene glycol)-*b*-poly(allyl ethylene phosphate) [P(mEG-*b*-AEP)], was prepared and reacted with cystamine by a UV-induced thiol–ene click method. A proportion of amines on the PAEP segments were converted with thiol groups by reaction with 2-iminothiolane. The thiol-reactive derivative of DOX was reacted with copolymers and formed conjugates through an acid-labile hydrazone bond. The conjugate nanoparticles were found to respond to the tumor extracellular pH ($\sim$6.8) with reversal of the surface charges from negative to positive to enhance cellular uptake. The conjugates released DOX under acidic conditions (pH $\approx$ 5.0), which is the intracellular environment of tumor cells. With a view to delivering antibiotics to treat bacterial infections, one group created triple-layered nanogels based on the triblock copolymers of PEG, PCL, and PPE. It is worth noting that PCL is bacterial lipase-sensitive and formed an interlayer which surrounded the core, a crosslinkable polyphosphoester segment forms the core, and the poly(ethylene glycol) becomes the shell.[194] The antibiotics were protected inside the polyphosphoester core and released at the bacterial infection sites where secreted lipases degraded the PCL segments.

He *et al.*[195] synthesized α-methacryloyloxyethyl ω-acryloyl poly(ethyl ethylene phosphate) (HEMA-PEOP-Ac) as a macro-crosslinker, and polymerized it with acrylic acid using ammonium persulfate and *N,N,N′,N′*-tetramethylethylenediamine (APS/TEMED) as the initiator and accelerator, respectively. The swelling ratio and the degradation rate of the resulting hydrogels were strongly influenced by the crosslinking density and pH values of the media. Bovine serum albumin (BSA) was used as a model drug and was loaded into the hydrogel by simple soaking of the dried gel in the BSA solution. It was found that increasing the crosslinking density reduced the swelling and release abilities. More recently, they investigated *in situ* rapidly curable PPE-based hydrogels for injectable drug delivery vehicles.[196] It was found that HEMA-PEOP-Ac rapidly polymerized with 2-(dimethylamino)ethyl methacrylate (DMAEMA) under mild conditions [in deionized water (pH $\approx$ 6.5), at room temperature] in the presence of ammonium persulfate. The gelation was completed within minutes, with the drugs encapsulated into the hydrogels homogeneously during gel formation. Hence the drug loading content can be increased compared to adsorption of the drugs into the hydrogels. The drug (DOX) release rate *in vitro* was found to correlate with the crosslinking density.

For bone-specific drug delivery applications, amphiphilic PPE ionomers, which have several ionized units of phosphates in the segments, were synthesized from 2-ethoxy-2-oxo-1,3,2-dioxaphosphorane (EP) and 2-benzyloxy-2-oxo-1,3,2-dioxaphosphorane (BP) in the presence of organocatalysts. This was followed by deprotection of the benzyl groups of BP.[197] They were immobilized onto 1,2-dioleoyl-*sn*-glycero-3-phosphocholine vesicles, and showed improved stability of the vesicles. The affinity of the vesicles to calcium deposits generated by MC3T3-E1 cells was enhanced by the amphiphilic polyphosphoester ionomer modification.

Water-soluble PPEs have also been investigated, mainly for gene delivery applications. As mentioned in Section 9.7.3, the electrostatic interactions between the cationic polymer and the negative DNA molecules form complexes or nanoparticles. Liu *et al.*[198] synthesized pH- and temperature-responsive double-hydrophilic diblock copolymers, poly(ethyl ethylene phosphate)-*b*-PDMAEMA [P(EEP-*b*-DMAEMA)] in a two-step polymerization: ring-opening polymerization of EEP using 2-hydroxyethyl 2-bromoisobutyrate to form a PEEP block with a bromine-terminated end group (PEEP-Br), followed by ATRP of DMAEMA monomer with PEEP-Br as a macroinitiator. The tertiary amine groups of DMAEMA become deprotonated under basic conditions, which cause the hydrophobic DMAEMA segments to aggregate and form micelles. At pH 7.4, partially deprotonated amine groups form pockets of hydrophobic segments that form smaller and looser aggregates. They demonstrated that self-assembly of the diblock copolymers was also temperature responsive and they could effectively condense DNA. The same group synthesized pH-sensitive brush copolymers, P[(HEMA-*g*-EEP)-*b*-DMAEMA]. The self-assembled aggregates were subsequently investigated for gene delivery purposes.[199] The brush-type hydrophilic polymer on the surface of the micelle–gene complex was thought to reduce nonspecific adsorption of proteins under blood circulation conditions. The brush copolymers self-assembled into aggregates at pH 7.4, and the positively charged PDMAEMA segments effectively condensed DNA to form complexes.

As we have clearly demonstrated in this section, DDS is one of the most rapidly growing applications using phosphorus-containing polymers covered in this chapter. It is one which continues to expand as our ability to design nano-scale DDS increases. Currently, using *in vitro* cellular studies, many papers indicate the efficacy of, in particular, PMPC and PPE. However, future progress clearly requires *in vivo* studies and further evaluation. Nevertheless, these polymers currently continue to be promising materials for drug and gene delivery.

9.8 Tissue Engineering and Other Miscellaneous Applications

9.8.1 Nerve Regeneration

Bridging nerve gaps and facilitating nerve regeneration[200] using bio-degradable artificial nerve guides is an area where polyphosphoesters (PPEs) have been investigated as suitable polymeric biomaterials for the supporting scaffold/matrix. Leong's group produced rather elegant work with the claim of "A new nerve guide conduit material composed of a biodegradable polyphosphoester".[201] They also investigated and created a suitable porosity.[202] However, these reports appear in isolation and current efforts in the field do not appear to have followed the PPE pathway. Leong's group also reported the synthesis of a new poly(6-aminohexyl propylene phosphate) biodegradable, photo-crosslinkable, polyphosphoester hydrogel.[203] He concludes that such hydrogels might find application as "injectable TE scaffolds" which is of course the focus of his work on nerve guide conduit materials as just discussed.

9.8.2 Nanocomposites

Nano-HAp was surface-treated with PMOEP in order to introduce phosphate-containing functional groups which were subsequently bonded to ethylene diamine-functionalised PLLA scaffolds or imbedded into PLLA scaffolds during fabrication.[204] A *grafting from* approach made it possible to modify the surface of nano-HAp crystals. Thiol-functionalized HAp nanocrystals were used in the grafting polymerization of MOEP (unpurified). The results indicated that the PMOEP-grafted HAp nanocrystals exhibited increased colloidal stability in water.[205] These modified nano-HAp particles were subsequently used for the fabrication of nanocomposite scaffolds.[204]

An interesting fundamental study involved a biomimetic route to producing calcium-coated polymeric nanoparticles intended for several bone-related applications.[112] This study leads us nicely into our next topic.

9.8.3 Biomimetics

Biomimetics is an important strategy for the fabrication of modern biomaterials that provides solutions which closely resemble those of living systems.[89] In the area of biomimetic supramolecular chemistry, a surface-grafted PMOEP polymer has recently been found to impart mesoporous silica with switchable ion channel transport properties.[206] This was enabled by the dual protonation and Ca^{2+} chelation ability of the polymer phosphate groups. A series of studies made use of PMAEP-based copolymers in the fabrication of biomimetic adhesives which were inspired by the sandcastle worm that produces peptide polyelectrolytes rich in phosphorylated serine.[207–209] These adhesives were aimed at providing molecular solutions to the "wet environment of open surgery" and contained mixtures of poly-electrolytes containing phosphate, amine, and catechol side groups. They displayed complex phase behavior triggered by pH and temperature, as well as by divalent metal ions.[207,208] The initial polyelectrolyte systems, however, displayed low adhesive strength (300 kPa). In order to overcome this and produce an adhesive suitable for biomedical applications, a multiphase system incorporating a PEG-based network made it possible to achieve an adhesive strength of 1.2 MPa.[209]

9.8.4   Hydrogels

The earliest phosphorus-containing polymeric hydrogels contained pendent phosphate groups.[210,211] With biomedical applications in view, a study into the diffusion of water into P(HEMA-*co*-MOEP) hydrogels was published 10 years later.[212] As evidenced by a useful, but somewhat non-comprehensive, review on phosphorus-containing hydrogels that appeared in 2009, interest has not waned and the chemistry is now better understood.[12,213]

Hydrogel-based skin-wound dressings are also of interest. However, as far as we can ascertain, only one paper involving a triblock copolymeric hydrogel with a phosphorus-containing component, namely MPC, has appeared.[214]

Not surprisingly, the degree of crosslinking is an issue that appears frequently in hydrogel papers, including in a MPC hydrogel[215] and another using a cyclic phosphoester monomer, 2-(2-oxo-1,3,2-dioxaphos-pholoyloxy)ethyl methacrylate.[216] As pointed out in a paper on the "design of hydrogel delivery vehicles" for craniofacial tissue regeneration,[217] the mechanism of the crosslinking reaction is important in the design of *in situ*-forming gel systems. The degree and nature of the crosslinking are clearly of equal importance. This becomes particularly relevant when studying these hydrogels as cell culture systems. The incorporation of cells such as mesenchymal stem cells (MSC) into a 3D hydrogel system is one vital step in opening up a wide range of applications.[217] However, from a chemistry point of view, caution is required in interpreting complex cell studies. Since much research is focused on establishing the most suitable functional groups for promoting a specific function, for example, phosphate groups in biomineralization, knowledge of the exact composition of the "functionalized" hydrogel is essential for the interpretation of results. As referred to earlier, in 2008 we published our findings on the "un-detected" diene impurity present in commercial MOEP.[12] Any hydrogels that include this diene will undoubtedly have the modulus of the resulting phosphorus-containing hydrogels also changed due to the higher cross-linking density caused by the undesirable diene. It is known that cells respond to substrate modulus[218] and hence any conclusions regarding the influence of the phosphorus groups on both mineralization and control-ling MSC differentiation must take this into account and some may even need to be reconsidered.[219]

9.8.5 Pancreas

With respect to diabetes therapies, there are currently two approaches involving biomaterials. The first involves the creation of an artificial pancreas; the second a completely synthetic pancreas.[220] An artificial pancreas consists of insulin-producing cells embedded in a suitable polymeric biomaterial. The second is a closed-loop device combining a continuous glucose sensor monitor, an insulin infusion pump, and a control algorithm. Ongoing research efforts for developing one of these options have not diminished, which is not surprising given the increase in the world's population suffering from diabetes. The prevention of biofouling and inflammation are critical requirements for the implantable electrochemical sensor. Cytotoxicity is of course also an issue and has been investigated both *in vitro* and *in vivo* in a subcutaneous rat model.[221] All these challenges are ones familiar to both biomaterial scientists and tissue engineering groups.

To the best of our knowledge, the use of phosphorus-containing polymers in this field is limited to MCP. Ishihara and co-workers in 2002 were the first to report the synthesis of a biocompatible polymer alloy membrane with good mechanical properties and permeability for use in an implantable artificial pancreas.[222,223] In an earlier article, Nishida had investigated the biocompatibility of MPC in an *in vivo* rat model with the aim of future applications.[224] PMPC has been explored as a biocompatible coating or membrane for the sensor by several groups.[221,225] More recently, a pH-sensitive molecularly imprinted nanospheres/hydrogel composite coating, with the hydrogel including a MPC moiety, has been reported.[226]

According to Sambanis, the current limiting factor is "a reliable glucose monitor with sufficient longevity".[227] However, it is only one of many. As pointed out in a recent progress report,[220] in addition to biocompatibility issues and the diffusion demands of the materials involved, there still remains very real limits in "translating islet encapsulated devices from the bench top to clinical trails" because of the limitations of the mouse models: "to date it has been difficult to translate results from rodents to primates in this area". In spite of this, Sambanis optimistically points out in his chapter on artificial organs/pancreas that "with challenges arise opportunities".[227]

Continuous glucose monitoring using a glucose monitor coupled to an insulin and glucogen delivery system, which in turn uses an algorithm to control the dosing, is an essential component of any substitute pancreas. As pointed out above, the development of a reliable, long-term, biocompatible implanted electrochemical glucose sensor remains the main hurdle. Various sensor coatings have been investigated with a view to improving their biocompatibility. However, in spite of an initial burst of interest in MPC as a possible coating, apart from one paper published in 2010[226] which investigated a nanosphere/hydrogel composite P(HEMA-*co*-NVP-*co*-MPC) as a coating for implantable glucose biosensors, there have been no other papers as far as we can ascertain. The most recent review,[220] "Materials for diabetes therapeutics", includes a useful table for those with greater interest in this field.

9.9   Summary and Future Directions

As we have demonstrated in this chapter, phosphorus-containing polymers have contributed to the development of both new and modified biomaterials with potential in a wide range of medical applications and devices. From the literature reports we have considered it will be clear to the reader that sometimes their promise proved more than their actual delivery, such as in the case of artificial pancreas development. Other applications such as non-fouling and drug and gene delivery appear to have a very healthy future. However, biomaterials research (or all research for that matter), even for a specific intended application, is never without the possibility of exerting an important influence in another area. Mineralization studies are a good example. There is on-going research into surfaces that promote mineralization while concomitant research into its prevention, for example in ophthalmological applications, is also underway. As a result of research reported in this chapter, these phenomena are far better understood than was previously the case. We hope that, at least to some extent, we have fulfilled the reader's expectations with this chapter, and may even inspire some future research in this field.

Acknowledgements

Our grateful thanks are due to Stephen Monteith who designed Figure. 9.1 and Sybil Curtis who created the wonderful drawings. SS thanks Professor Traian Chirila and the Queensland Eye Institute for their generous support while writing this chapter. EWB and LG acknowledge the hard work and inspiration of their students and collaborators who contributed to this research over the years.

References

1. *Definitions in Biomaterials*, ed. D. F. Williams, Elsevier, Amsterdam, 1987.
2. B. D. Ratner, A. S. Hoffman, F. J. Schoen & J. E. Lemons, in *Biomaterials Science: An Introduction to Materials in Medicine*, ed. B. D. Ratner, A. S. Hoffman, F. J. Schoen and J. E. Lemons, Elsevier Academic Press, Amsterdam, 2004.
3. D. F. Williams, *Essential Biomaterials Science*, Cambridge University Press, Cambridge, 2014.
4. R. Nerem, *Tissue Eng.*, 2006, **12**, 1143.
5. M. Lysaght and A. Hazlehurst, *Tissue Eng.*, 2004, **10**, 309.
6. J.-F. Stoltz, D. Bensoussan, V. Decot, A. Ciree, P. Netter and P. Gillet, *Bio-Med. Mater. Eng.*, 2006, **16**, S3.
7. F. P. W. Melchels, M. A. N. Domingos, T. J. Klein, J. Malda, P. J. Bartolo and D. Hutmacher, *Prog. Polym. Sci.*, 2011, **37**, 1079.
8. D. F. Williams, *Trends Biotechnol.*, 2006, **24**, 4.

9. D. F. Williams, *The Williams Dictionary of Biomaterials*, Liverpool University Press, Liverpool, 1999.
10. S. Monge, B. Canniccioni, A. Graillot and J. Robin, *Biomacromolecules*, 2011, **12**, 1973.
11. T. Chirila and Zainuddin, *React. Funct. Polym.*, 2007, **67**, 165.
12. L. Grøndahl, S. Suzuki and E. Wentrup-Byrne, *Chem. Commun.*, 2008, 3314.
13. E. Wentrup-Byrne, S. Suzuki, J. J. Suwanasilp and L. Grøndahl, *Biomed. Mater. (Bristol, U. K.)*, 2010, **5**, 045010/1.
14. Y. Wang, Y. Yuan, J. Du, X. Yang and J. Wang, *Macromol. Biosci.*, 2009, **9**, 1154.
15. M. Deng, S. G. Kumbar, Y. Wan, U. S. Toti, H. R. Allcock and C. T. Laurencin, *Soft Matter*, 2010, **6**, 3119.
16. I. Teasdale and O. Brüggemann, *Polymers*, 2013, **5**, 161.
17. H. R. Allcock, *Soft Matter*, 2012, **8**, 7521.
18. Y. Iwasaki and K. Ishihara, *Anal. Bioanal. Chem.*, 2005, **381**, 534.
19. L. L. Chiu and M. Radisic, *Curr. Opin. Chem. Eng.*, 2013, **2**, 41.
20. M. Kyomoto, T. Moro, K. Saiga, M. Hashimoto, H. Ito, H. Kawaguchi, Y. Takatori and K. Ishihara, *Biomaterials*, 2012, **33**, 4451.
21. H. Sawano, S. Warisawa and S. Ishihara, *Wear*, 2010, **268**, 233.
22. A. Barradas, H. Yuan, C. van Blitterswijk and P. Habibovic, *Eur. Cells Mater.*, 2011, **21**, 407.
23. S. Kamei, N. Tomita, S. Tamai, K. Kato and Y. Ikada, *J. Biomed. Mater. Res.*, 1997, **37**, 384.
24. O. Tretinnikov, K. Kato and Y. Ikada, *J. Biomed. Mater. Res.*, 1994, **28**, 1365.
25. D. Paripovic, H. Hall-Bozic and H.-A. Klok, *J. Mater. Chem.*, 2012, **22**, 19570.
26. *Biomaterials and Regenerative Medicine in Ophthalmology*, ed. T. V. Chirila, CRC Press, Cambridge, 2010.
27. Y.-X. Huang, in *Biointegration of Medical Implant Materials, Science and Design*, ed. C. P. Sharma, CRC Press, Cambridge, 2010.
28. J. L. Drury and D. J. Mooney, *Biomaterials*, 2003, **24**, 4337.
29. Zainuddin, Z. Barnard, I. Keen, D. J. T. Hill, T. V. Chirila and D. G. Harkin, *J. Biomater. Appl.*, 2008, **23**, 147.
30. R. Matsuno and K. Ishihara, *Nano Today*, 2011, **6**, 61.
31. Z. Zhao, J. Wang, H.-Q. Mao and K. W. Leong, *Adv. Drug Delivery Rev.*, 2003, **55**, 483.
32. B. Dhandayuthapani, Y. Yoshida, T. Maekawa and D. S. Kumar, *Int. J. Polym. Sci.*, 2011, DOI: 10.1155/2011/290602.
33. M. A. Woodruff, C. Lange, J. Reichert, A. Berner, F. Chen, P. Fratzl, J.-T. Schantz and D. W. Hutmacher, *Mater. Today*, 2012, **15**, 430.
34. S. P. Nukavarapu and D. L. Dorcemus, *Biotechnol. Adv.*, 2013, **31**, 706.
35. J. Kretlow and A. Mikos, *Tissue Eng.*, 2007, **13**, 927.
36. H. Busscher, H. van der Mei, G. Subbiahdoss, P. Jutte, J. van den Dungen, S. Zaat, M. Schultz and D. Grainger, *Sci. Transl. Med.*, 2012, **4**, 153rv10.

37. I. Banerjee, R. Pangule and R. Kane, *Adv. Mater.*, 2011, **23**, 690.
38. T. Doiuchi, T. Nakaya and M. Imoto, *Macromol. Chem.*, 1974, **175**, 43.
39. T. Kimura, T. Nakaya and M. Imoto, *Macromol. Chem.*, 1975, **176**, 1945.
40. K. Inaishi, T. Nakaya and M. Imoto, *Macromol. Chem.*, 1975, **176**, 2473.
41. T. Kimura, T. Nakaya and M. Imoto, *Macromol. Chem.*, 1976, **177**, 1973.
42. K. Ishihara, Y. Goto, M. Takai, R. Matsuno, Y. Inoue and T. Konno, *Biochim. Biophys. Acta*, 2011, **10**, 268.
43. Y. Yao, K. Fukazawa, N. Huang and K. Ishihara, *Colloids Surf., B*, 2011, **88**, 215.
44. Y. Yao, K. Fukazawa, W. Ma, K. Ishihara and N. Huang, *Appl. Surf. Sci.*, 2012, **258**, 5418.
45. Y. Gong, L. Liu and P. Messersmith, *Macromol. Biosci.*, 2012, **12**, 979.
46. S. H. Ye, C. A. Johnson, J. R. Woolley, H.-I. Oh, L. J. Gamble, K. Ishihara and W. R. Wagner, *Colloids Surf., B*, 2009, **74**, 96.
47. Y. Zhao, Q. Tu, J. Wang, Q. Huang and N. Huang, *Appl. Surf. Sci.*, 2010, **257**, 1596.
48. H. Kim, K. Ishihara, S. Lee, J. Seo, H. Kim, D. Suh, M. Kim, T. Konno, M. Takai and J. Seo, *Biomaterials*, 2011, **32**, 2241.
49. H. I. Kim, M. Takai and K. Ishihara, *Tissue Eng., Part C*, 2009, **15**, 125.
50. L. Soletti, A. Nieponice, Y. Hong, S.-H. Ye, J. J. Stankus, W. R. Wagner and D. A. Vorp, *J. Biomed. Mater. Res., A*, 2011, **96A**, 436.
51. Y. Hong, S.-H. Ye, A. Nieponice, L. Soletti, D. A. Vorp and W. R. Wagner, *Biomaterials*, 2009, **30**, 2457.
52. F. Sun, X. Li, J. Xu, P. Cao and L. Xiao, *e-Polym.*, 2011, 042.
53. T. Shimizu, T. Goda, N. Minoura, M. Takai and K. Ishihara, *Biomaterials*, 2010, **31**, 3274.
54. L. Li, J.-H. Wang and Z. Xin, *Eur. Polym. J.*, 2011, **47**, 1795.
55. L. Li and Z. Xin, *Colloids Surf., A*, 2011, **384**, 713.
56. N. Ishikawa, T. Miyamoto, Y. Okada and S. Saika, *J. Cataract Refract. Surg.*, 2011, **37**, 1339.
57. M. Kyomoto, T. Moro, Y. Takatori, H. Kawaguchi and K. Ishihara, *Clin. Orthop. Relat. Res.*, 2011, **469**, 2327.
58. T. Moro, H. Kawaguchi, K. Ishihara, M. Kyomoto, T. Karita, H. Ito, K. Nakamura and Y. Takatori, *Biomaterials*, 2009, **30**, 2995.
59. M. Kyomoto, T. Moro, K. Saiga, F. Miyaji, H. Kawaguchi, Y. Takatori, K. Nakamura and K. Ishihara, *Biomaterials*, 2010, **31**, 658.
60. A. L. Lewis, *Colloids Surf., B*, 2000, **18**, 261.
61. H. Chen, L. Yuan, W. Song, Z. Wu and D. Li, *Prog. Polym. Sci.*, 2008, **33**, 1059.
62. V. Tegoulia, W. Rao, A. Kalambur, J. Rabolt and S. Cooper, *Langmuir*, 2001, **17**, 4396.
63. S. Chen, L. Li, C. Zhao and J. Zheng, *Polymer*, 2010, **51**, 5283.
64. M. Kobayashi, Y. Terayama, H. Yamaguchi, M. Terada, D. Murakami, K. Ishihara and A. Takahara, *Langmuir*, 2012, **28**, 7212.
65. C. Rodriguez Emmenegger, E. Brynda, T. Riedel, Z. Sedlakova, M. Houska and A. Alles, *Langmuir*, 2009, **25**, 6328.

66. A. Reisch, J.-C. Voegel, E. Gonthier, G. Decher, B. Senger, P. Schaaf and P. J. Mésini, *Langmuir*, 2009, **25**, 3610.
67. A. Reisch, J. Hemmerlé, A. Chassepot, M. Lefort, N. Benkirane-Jessel, E. Candolfi, P. Mésini, V. Letscher-Bru, J.-C. Voegel and P. Schaaf, *Soft Matter*, 2010, **6**, 1503.
68. J. Davila, A. Chassepot, J. Longo, F. Boulmedais, A. Reisch, B. Frisch, F. Meyer, J. Voegel, P. Mésini, B. Senger, M. Metz-Boutigue, J. Hemmerlé, P. Lavalle, P. Schaaf and L. Jierry, *J. Am. Chem. Soc.*, 2012, **134**, 83.
69. B. Yuan, Q. Chen, W. Ding, P. Liu, S. Wu, S. Lin, J. Shen and Y. Gai, *ACS Appl. Mater. Interfaces*, 2012, **4**, 4031.
70. Z. Jin, W. Feng, S. Zhu, H. Sheardown and J. Brash, *J. Biomater. Sci., Polym. Ed.*, 2010, **21**, 1331.
71. Y. Iwasaki, A. Matsumoto and S. Yusa, *ACS Appl. Mater. Interfaces*, 2012, **4**, 3254.
72. Q. Ma, H. Zhang, J. Zhao and Y.-K. Gong, *Appl. Surf. Sci.*, 2012, **258**, 9711.
73. A. Fuchs, C. Walter, K. Landfester and U. Ziener, *Langmuir*, 2012, **28**, 4974.
74. R. E. Klabunde, *Cardiovascular Physiology Concepts*, Lippincott Williams & Wilkins, Philadelphia, 2005.
75. P. Qi, M. F. Maitz and N. Huang, *Surf. Coat. Technol.*, 2013, **233**, 80.
76. A. S. Go, D. Mozaffarian, V. L. Roger, E. J. Benjamin, J. D. Berry, W. B. Borden, et al., *Circulation*, 2013, **127**, 6.
77. BiodivYsio™AS PC (phosphorylcholine) Coated Stent Delivery System - P000011; retrieved 23rd June, 2013, from http://www.fda. gov/MedicalDevices/ProductsandMedicalProcedures/ DeviceApprovalsandClearances/Recently-ApprovedDevices/ ucm089758.htm.
78. Endeavor® Zotarolimus-Eluting Coronary Stent on the Over-the-Wire (OTW), Rapid Exchange (RX), or Multi Exchange II (MX2) Stent Delivery Systems - P060033; retrieved 23rd June, 2013, from http://www.fda. gov/MedicalDevices/ProductsandMedicalProcedures/ DeviceApprovalsandClearances/Recently-ApprovedDevices/ ucm074326.htm.
79. K. Chan, J. Armstrong, S. Withers, N. Malik, D. Cumberland, J. Gunn and C. Holt, *Biomaterials*, 2007, **28**, 1218.
80. R. R. Palmer, A. L. Lewis, L. C. Kirkwood, S. F. Rose, A. W. Lloyd, T. A. Vick and P. W. Stratford, *Biomaterials*, 2004, **25**, 4785.
81. X. Wang, P. Lin, Q. Yao and C. Chen, *World J. Surg.*, 2007, **31**, 682.
82. T. Yoneyama, K. Ishihara, N. Nakabayashi, M. Ito and Y. Mishima, *J. Biomed. Mater. Res.*, 1998, **43**, 15.
83. T. Yoneyama, K. Sugihara, K. Ishihara, Y. Iwasaki and N. Nakabayashi, *Biomaterials*, 2002, **23**, 1455.
84. A. Korematsu, Y. Takemoto, T. Nakaya and H. Inoue, *Biomaterials*, 2002, **23**, 263.

85. S. Ye, C. J. Johnson, J. Woolley, T. Snyder, L. Gamble and W. Wagner, *J. Biomed. Mater. Res., A*, 2009, **91**, 18.

86. T. Moro, Y. Takatori, K. Ishihara, T. Konno, Y. Takigawa, T. Matsushita, U. Chung, K. Nakamura and H. Kawaguchi, *Nat. Mater.*, 2004, **3**, 829.

87. M. Kyomoto, T. Moro, F. Miyaji, T. Konno, M. Hashimoto, H. Kawaguchi, Y. Takatori, K. Nakamura and K. Ishihara, *J. Biomed. Mater. Res., B*, 2008, **84B**, 320.

88. S. Mann, D. D. Archibald, J. M. Didymus, T. Douglas, B. R. Heywood, F. C. Meldrum and N. J. Reeves, *Science*, 1993, **261**, 1286.

89. W. Murphy and D. Mooney, *J. Am. Chem. Soc.*, 2002, **124**, 1910.

90. T. V. Chirila, D. J. T. Zainuddin, A. K. Hill, Whittaker and A. Kemp, *Acta Biomater.*, 2007, **3**, 95.

91. J. G. N. Swart, A. A. Driessen and A. C. DeVisser, *ACS Symp. Ser.*, 1976, **31**, 151.

92. I. C. Stancu, R. Filmon, C. Cincu, B. Marculescu, C. Zaharia, Y. Tourmen, M. F. Basle and D. Chappard, *Biomaterials*, 2004, **25**, 205.

93. M. Dadsetan, M. Giuliani, F. Wanivenhaus, M. Brett Runge, J. E. Charlesworth and M. J. Yaszemski, *Acta Biomater.*, 2012, **8**, 1430.

94. Y. Toyomoto, R. Matsuno, T. Konno, M. Takai and K. Ishihara, *Trans. Mater. Res. Soc. Jpn.*, 2010, **35**, 123.

95. K. Kato, Y. Eika, Y. Ikada and J. Biomed, . *Mater. Res.*, 1996, **32**, 687.

96. L. Grøndahl, F. Cardona, K. Chiem and E. Wentrup-Byrne, *J. Appl. Polym. Sci.*, 2002, **86**, 2550.

97. L. Grøndahl, F. Cardona, K. Chiem, E. Wentrup-Byrne and T. Bostrom, *J. Mater. Sci.: Mater. Med.*, 2003, **14**, 503.

98. A. Chandler-Temple, P. Kingshott, E. Wentrup-Byrne, A. I. Cassady and L. Grøndahl, *J. Biomed. Mater. Res., A*, 2013, **101A**, 1047.

99. A. Chandler-Temple, E. Wentrup-Byrne, A. K. Whittaker and L. Groendahl, *J. Appl. Polym. Sci.*, 2010, **117**, 3331.

100. A. F. Chandler-Temple, E. Wentrup-Byrne, H. J. Griesser, M. Jasieniak, A. K. Whittaker and L. Grøndahl, *Langmuir*, 2010, **26**, 15409.

101. S. Suzuki, L. Grøndahl, D. Leavesley and E. Wentrup-Byrne, *Biomaterials*, 2005, **26**, 5303.

102. E. Wentrup-Byrne, L. Grøndahl and S. Suzuki, *Polym. Int.*, 2005, **54**, 1581.

103. K. Minatoya, H. Okabayashi, I. Shimada, N. Ohno, T. Nishina, T. Yokota, M. Takahashi, T. Ishihara and E. Hoover, *Ann. Thorac. Surg.*, 1996, **61**, 883.

104. F. Trueba Arguiñarena and E. Fernández del Busto, *J. Urol.*, 2004, **172**, 620.

105. S. Suzuki, M. R. Whittaker, L. Grøndahl, M. J. Monteiro and E. Wentrup-Byrne, *Biomacromolecules*, 2006, 7, 3178.

106. M. M. Nasef, *Polym. Int.*, 2001, **50**, 338.

107. T. R. Dargaville, D. Hill and S. Perera, *Aust. J. Chem.*, 2002, **55**, 439.

108. X.-Z. Yang, T.-M. Sun, S. Dou, J. Wu, Y.-C. Wang and J. Wang, *Biomacromolecules*, 2009, **10**, 2213.

109. L. Macarie and G. Ilia, *Prog. Polym. Sci.*, 2010, **35**, 1078.
110. Y. Kim, L. Gu, T. Bryan, J. Kim, L. Chen, Y. Liu, J. Yoon, L. Breschi, D. Pashley and F. Tay, *Biomaterials*, 2010, **31**, 6618.
111. Ö. Doğan and M. Öner, *Langmuir*, 2006, **22**, 9671.
112. K. Schöller, A. Ethirajan, A. Zeller and K. Landfester, *Macromol. Chem. Phys.*, 2011, **212**, 1165.
113. A. W. Lloyd, R. G. A. Faragher and S. P. Denyer, *Biomaterials*, 2001, **22**, 769.
114. J. Jacob, in *Biomaterials Science*, ed. B. D. Ratner, A. S. Hoffman, F. J. Schoen and J. E. Lemons, Academic Press, New York, NY, (3rd ed.), 2013, pp. 909–917.
115. K. Merrett, P. Fagerholm, C. R. McLaughlin, S. Dravida, N. Lagali, N. Shinozaki, M. A. Watsky, R. Munger, Y. Kato, F. Li, C. J. Marmo and M. Griffith, *Invest. Ophthalmol. Vis. Sci.*, 2008, **49**, 3887.
116. R. Crabb and A. Hubel, *Tissue Eng., Part A*, 2008, **14**, 173.
117. L. J. Bray, K. A. George, D. W. Hutmacher, T. V. Chirila and D. G. Harkin, *Biomaterials*, 2012, **33**, 3529.
118. A. M. A. Shadforth, K. A. George, A. S. Kwan, T. V. Chirila and D. G. Harkin, *Biomaterials*, 2012, **33**, 4110.
119. D. G. Harkin, K. A. George, P. W. Madden, I. R. Schwab, D. W. Hutmacher and T. V. Chirila, *Biomaterials*, 2011, **32**, 2445.
120. B. D. Lawrence, Z. Pan, A. Liu, D. L. Kaplan and M. I. Rosenblatt, *Acta Biomater.*, 2012, **8**, 3732.
121. M. W. Grinstaff, *Biomaterials*, 2007, **28**, 5205.
122. P. C. Kang, M. A. Carnahan, M. Wathier, M. W. Grinstaff and T. Kim, *J. Cataract Refract. Surg.*, 2005, **31**, 1208.
123. T. Goda and K. Ishihara, *Expert Rev. Med. Devices*, 2006, **3**, 167.
124. S. L. Willis, J. L. Court, R. P. Redman, J.-H. Wang, S. W. Leppard, V. J. O'Byrne, S. A. Small, A. L. Lewis, S. A. Jones and P. W. Stratford, *Biomaterials*, 2001, **22**, 3261.
125. T. Uchiyama, Y. Kiritoshi, J. Watanabe and K. Ishihara, *Biomaterials*, 2003, **24**, 5183.
126. Y. Kiritoshi and K. Ishihara, *J. Biomater. Sci., Polym. Ed.*, 2002, **13**, 213.
127. Y. Kiritoshi and K. Ishihara, *Polymer*, 2004, **45**, 7499.
128. T. Goda, J. Watanabe, M. Takai and K. Ishihara, *Polymer*, 2006, **47**, 1390.
129. M. X. Chu, K. Miyajima, D. Takahashi, T. Arakawa, K. Sano, S. Sawada, H. Kudo, Y. Iwasaki, K. Akiyoshi, M. Mochizuki and K. Mitsubayashi, *Talanta*, 2011, **83**, 960.
130. M. X. Chu, T. Shirai, D. Takahashi, T. Arakawa, H. Kudo, K. Sano, S. Sawada, K. Yano, Y. Iwasaki, K. Akiyoshi, M. Mochizuki and K. Mitsubayashi, *Biomed. Microdevices*, 2011, **13**, 603.
131. X.-D. Huang, K. Yao, H. Zhang, X.-J. Huang and Z.-K. Xu, *Clin. Exp. Ophthalmol.*, 2007, **35**, 462.
132. X.-D. Huang, K. Yao, Z. Zhang, Y. Zhang and Y. Wang, *J. Cataract Refract. Surg.*, 2010, **36**, 290.

133. K. Yao, X.-D. Huang, X.-J. Huang and Z.-K. Xu, *J. Biomed. Mater. Res., A*, 2006, **78A**, 684.
134. Y. Okajima, S. Kobayakawa, A. Tsuji and T. Tochikubo, *Invest. Ophthalmol. Vis. Sci.*, 2006, **47**, 2971.
135. Y. Okajima, S. Saika and M. Sawa, *J. Cataract Refract. Surg.*, 2006, **32**, 666.
136. M. Shigeta, T. Tanaka, N. Koike, N. Yamakawa and M. Usui, *J. Cataract Refract. Surg.*, 2006, **32**, 859.
137. T. Tanaka, M. Shigeta, N. Yamakawa and M. Usui, *J. Cataract Refract. Surg.*, 2005, **31**, 1648.
138. S. F. Long, S. Clarke, M. C. Davies, A. L. Lewis, G. W. Hanlon and A. W. Lloyd, *Biomaterials*, 2003, **24**, 4115.
139. J. M. Hackett, N. Lagali, K. Merrett, H. Edelhauser, Y. Sun, L. Gan, M. Griffith and P. Fagerholm, *Invest. Ophthalmol. Vis. Sci.*, 2011, **52**, 651.
140. C. R. McLaughlin, M. C. Acosta, C. Luna, W. Liu, C. Belmonte, M. Griffith and J. Gallar, *Biomaterials*, 2010, **31**, 2770.
141. W. Liu, C. Deng, C. R. McLaughlin, P. Fagerholm, N. S. Lagali, B. Heyne, J. C. Scaiano, M. A. Watsky, Y. Kato, R. Munger, N. Shinozaki, F. Li and M. Griffith, *Biomaterials*, 2009, **30**, 1551.
142. M. Ayaki, A. Iwasawa and Y. Niwano, *Jpn. J. Ophthalmol.*, 2011, **55**, 541.
143. A. S. Hoffman, *J. Controlled Release*, 2008, **132**, 153.
144. B. P. Timko, K. Whitehead, W. Gao, D. S. Kohane, O. Farokhzad, D. Anderson and R. Langer, *Annu. Rev. Mater. Res.*, 2011, **41**, 1.
145. R. Lehner, X. Wang, S. Marsch and P. Hunziker, *Nanomedicine*, 2013, **9**, 742, DOI: 10.1016/j.nano.2013.01.012.
146. J. B. Wolinsky, Y. L. Colson and M. W. Grinstaff, *J. Controlled Release*, 2012, **159**, 14.
147. M. Ahmed and R. Narain, *Prog. Polym. Sci.*, 2013, **38**, 767.
148. D. J. Siegwart, J. K. Oh and K. Matyjaszewski, *Prog. Polym. Sci.*, 2012, **37**, 18.
149. E. De Giglio, S. Cometa, M. A. Ricci, A. Zizzi, D. Cafagna, S. Manzotti, L. Sabbatini and M. Mattioli-Belmonte, *Acta Biomater.*, 2010, **6**, 282.
150. B. J. Jeon, S. Y. Jeong, A. N. Koo, B.-C. Kim, Y.-S. Hwang and S. C. Lee, *Macromol. Res.*, 2012, **20**, 715.
151. S. Jang, S. Lee, H. Kim, J. Ham, J.-H. Seo, Y. Mok, M. Noh and Y. Lee, *Polymer*, 2012, **53**, 4678.
152. W. Khan, S. Farah and A. Domb, *J. Controlled Release*, 2012, **161**, 703.
153. T. Htay and M. W. Liu, *Vasc. Health Risk Manage.*, 2005, **1**, 263.
154. D. Martin and F. Boyle, *Med. Eng. Phys.*, 2011, **33**, 148.
155. J.-P. Xu, J. Ji, W.-D. Chen, D.-Z. Fan, Y.-F. Sun and J.-C. Shen, *Eur. Polym. J.*, 2004, **40**, 291.
156. V. Bhatia, R. Bhatia and M. Dhindsa, *Postgrad. Med. J.*, 2004, **80**, 13.
157. K. Enomoto, T. Hasebe, R. Asakawa, A. Kamijo, Y. Yoshimoto, T. Suzuki, K. Takahashi and A. Hotta, *Diamond Relat. Mater.*, 2010, **19**, 806.
158. J. S. Lee and J. Feijen, *J. Controlled Release*, 2012, **161**, 473.

159. O. Onaca, R. Enea, D. W. Hughes and W. Meier, *Macromol. Biosci.*, 2009, **9**, 129.
160. D. M. Webster, P. Sundaram and M. E. Byrne, *Eur. J. Pharm. Biopharm.*, 2013, **84**, 1.
161. R. Matsuno and K. Ishihara, *Macromol. Symp.*, 2009, **279**, 125.
162. M. Wada, H. Jinno, M. Ueda, T. Ikeda, M. Kitajima, T. Konno, J. Watanabe and K. Ishihara, *Anticancer Res.*, 2007, **27**, 1431.
163. A. Ishikawa, M. Fujii, K. Morimoto, T. Yamada, N. Koizumi, M. Kondoh and Y. Watanabe, *Results Pharma Sci.*, 2012, **2**, 16.
164. R. Miyata, M. Ueda, H. Jinno, T. Konno, K. Ishihara, N. Ando and Y. Kitagawa, *Int. J. Cancer*, 2009, **124**, 2460.
165. N. Kameyama, S. Matsuda, O. Itano, A. Ito, T. Konno, T. Arai, K. Ishihara, M. Ueda and Y. Kitagawa, *Cancer Biother. Radiopharm.*, 2011, **26**, 697.
166. M. Licciardi, E. F. Craparo, G. Giammona, S. P. Armes, Y. Tang and A. L. Lewis, *Macromol. Biosci.*, 2008, **8**, 615.
167. H. Lomas, J. Du, I. Canton, J. Madsen, N. Warren, S. P. Armes, A. L. Lewis and G. Battaglia, *Macromol. Biosci.*, 2010, **10**, 513.
168. I. Canton, M. Massignani, N. Patikarnmonthon, L. Chierico, J. Robertson, S. A. Renshaw, N. J. Warren, J. P. Madsen, S. P. Armes, A. L. Lewis and G. Battaglia, *FASEB J.*, 2013, **27**, 98.
169. N. Bhuchar, R. Sunasee, K. Ishihara, T. Thundat and R. Narain, *Bioconjugate Chem.*, 2012, **23**, 75.
170. G.-Y. Liu, C.-J. Chen and J. Ji, *Soft Matter*, 2012, **8**, 8811.
171. G.-Y. Liu, L.-P. Lv, C.-J. Chen, X.-F. Hu and J. Ji, *Macromol. Chem. Phys.*, 2011, **212**, 643.
172. G.-Y. Liu, L.-P. Lv, C.-J. Chen, X.-S. Liu, X.-F. Hu and J. Ji, *Soft Matter*, 2011, **7**, 6629.
173. J. Du and S. P. Armes, *Langmuir*, 2009, **25**, 9564.
174. S. Tu, Y.-W. Chen, Y.-B. Qiu, K. Zhu and X.-L. Luo, *Macromol. Biosci.*, 2011, **11**, 1416.
175. C. Liu, L. Long, Z. Li, B. He, L. Wang, J. Wang, X. Yuan and J. Sheng, *J. Microencapsulation*, 2012, **29**, 242.
176. G. Liu, Q. Jin, X. Liu, L. Lv, C. Chen and J. Ji, *Soft Matter*, 2011, **7**, 662.
177. S. Höbel, A. Loos, D. Appelhans, S. Schwarz, J. Seidel, B. Voit and A. Aigner, *J. Controlled Release*, 2011, **149**, 146.
178. S. Srinivasachari, Y. Liu, L. Prevette and T. Reineke, *Biomaterials*, 2007, **28**, 2885.
179. M. Ahmed and R. Narain, *Biomaterials*, 2011, **32**, 5279.
180. M. Ahmed and R. Narain, *Biomaterials*, 2012, **33**, 3990.
181. M. Ahmed, N. Bhuchar, K. Ishihara and R. Narain, *Bioconjugate Chem.*, 2011, **22**, 1228.
182. M. Ahmed, Z. Deng, S. Liu, R. Lafrenie, A. Kumar and R. Narain, *Bioconjugate Chem.*, 2009, **20**, 2169.
183. M. Ahmed, X. Jiang, Z. Deng and R. Narain, *Bioconjugate Chem.*, 2009, **20**, 2017.

184. M. Ahmed, M. Jawanda, K. Ishihara and R. Narain, *Biomaterials*, 2012, **33**, 7858.
185. S. Hemp, A. Smith, J. Bryson, M. J. Allen and T. Long, *Biomacromolecules*, 2012, **13**, 2439.
186. S.-W. Huang and R.-X. Zhuo, *Phosphorus, Sulfur Silicon Relat. Elem.*, 2008, **183**, 340.
187. D. Armstrong, G. Fleming, M. Markman and H. Bailey, *Gynecol. Oncol.*, 2006, **103**, 391.
188. F. Wang, Y.-C. Wang, L.-F. Yan and J. Wang, *Polymer*, 2009, **50**, 5048.
189. Y.-C. Wang, H. Xia, X.-Z. Yang and J. Wang, *J. Polym. Sci., Part A: Polym. Chem.*, 2009, **47**, 6168.
190. J. Wu, X.-Q. Liu, Y.-C. Wang and J. Wang, *J. Mater. Chem.*, 2009, **19**, 7856.
191. N. Cuong, M. Hsieh, Y. Chen and I. Liau, *J. Biomater. Sci., Polym. Ed.*, 2011, **22**, 1409.
192. Y.-C. Wang, Y. Li, T.-M. Sun, M.-H. Xiong, J. Wu, Y.-Y. Yang and J. Wang, *Macromol. Rapid Commun.*, 2010, **31**, 1201.
193. J.-Z. Du, X.-J. Du, C.-Q. Mao and J. Wang, *J. Am. Chem. Soc.*, 2011, **133**, 17560.
194. M.-H. Xiong, Y. Bao, X.-Z. Yang, Y.-C. Wang, B. Sun and J. Wang, *J. Am. Chem. Soc.*, 2012, **134**, 4355.
195. J. He, P. Ni, S. Wang, H. Shao, M. Zhang and X. Zhu, *J. Polym. Sci., Part A: Polym. Chem.*, 2010, **48**, 1919.
196. J. He, M. Zhang and P. Ni, *Soft Matter*, 2012, **8**, 6033.
197. R. Ikeuchi and Y. Iwasaki, *J. Biomed. Mater. Res., A*, 2013, **101A**, 318.
198. X. Liu, P. Ni, J. He and M. Zhang, *Macromolecules*, 2010, **43**, 4771.
199. Y. Hao, J. He, M. Zhang, Y. Tao, J. Liu and P. Ni, *J. Polym. Sci., Part A: Polym. Chem.*, 2013, **51**, 2150.
200. E. Johnson and P. Soucacos, *Injury*, 2008, **39S**, S30.
201. S. Wang, A. Wan, X. Xu, S. Gao, H. Mao, K. Leong and H. Yu, *Biomaterials*, 2001, **22**, 1157.
202. A. Wan, H. Mao, S. Wang, K. Leong, L. Ong and H. Yu, *Biomaterials*, 2001, **22**, 1147.
203. Q. Li, J. Wang, S. Shahani, D. Sun, B. Sharma, J. Elisseeff and K. Leong, *Biomaterials*, 2006, **27**, 1027.
204. A. N. Koo, I. K. Kwon, S. C. Lee, S.-K. Lee, H.-S. Kim, Y.-H. Woo, S.-H. Jeon, J.-H. Chae and K.-W. Kang, *Macromol. Res.*, 2010, **18**, 1030.
205. S. C. Lee, H. W. Choi, H. J. Lee, K. J. Kim, J. H. Chang, S. Y. Kim, J. Choi, K.-S. Oh and Y.-K. Jeong, *J. Mater. Chem.*, 2007, **17**, 174.
206. A. Brunsen, C. Diaz, L. I. Pietrasanta, B. Yameen, M. Ceolin, G. J. A. A. Soler-Illia and O. Azzaroni, *Langmuir*, 2012, **28**, 3583.
207. H. Shao, K. N. Bachus and R. J. Stewart, *Macromol. Biosci.*, 2009, **9**, 464.
208. H. Shao and R. J. Stewart, *Adv. Mater. (Weinheim, Ger.)*, 2010, **22**, 729.
209. S. Kaur, G. Weerasekare and R. Stewart, *ACS Appl. Mater. Interfaces*, 2011, **3**, 941.

210. T. Miyata, K. Nakamae, A. S. Hoffman and Y. Kanzaki, *Macromol. Chem. Phys.*, 1994, **195**, 1111.
211. K. Nakamae, T. Miyata and A. S. Hoffman, *Makromol. Chem.*, 1992, **193**, 983.
212. K. A. George, E. Wentrup-Byrne, D. J. T. Hill and A. K. Whittaker, *Biomacromolecules*, 2004, **5**, 1194.
213. G. Ilia, *Polym. Adv. Technol.*, 2009, **20**, 707.
214. J. Madsen, S. P. Armes, K. Bertal, H. Lomas, S. MacNeil and A. L. Lewis, *Biomacromolecules*, 2008, **9**, 2265.
215. C. Wachiralarpphaithoon, Y. Iwasaki and K. Akiyoshi, *Biomaterials*, 2006, **28**, 984.
216. J.-Z. Du, T.-M. Sun, S.-Q. Weng, X.-S. Chen and J. Wang, *Biomacromolecules*, 2007, **8**, 3375.
217. C. Salinas and K. Anseth, *J. Dent. Res.*, 2009, **88**, 681.
218. A. Engler, S. Sen, H. Sweeney and D. Discher, *Cell*, 2006, **126**, 677.
219. C. R. Nuttelman, D. S. W. Benoit, M. C. Tripodi and K. S. Anseth, *Biomaterials*, 2006, **27**, 1377.
220. K. M. Bratlie, R. L. York, M. A. Invernale, R. Langer and D. G. Anderson, *Adv. Healthcare Mater.*, 2012, **1**, 267.
221. A. Mang, J. Pill, N. Gretz, B. Kränzlin, H. Buck, M. Schoemaker and W. Petrich, *Diabetes Technol. Ther.*, 2005, **7**, 163.
222. T. Uchiyama, J. Watanabe and K. Ishihara, *J. Membr. Sci.*, 2002, **208**, 39.
223. T. Uchiyama, J. Watanabe and K. Ishihara, *Trans. Mater. Res. Soc. Jpn.*, 2002, **27**, 415.
224. T. Nowak, K. Nishida, S. Shimoda, Y. Konno, K. Ichinose, M. Sakakida, M. Shichiri, N. Nakabayashi and K. Ishihara, *J. Artif. Organs*, 2000, **3**, 39.
225. M. Shichiri, M. Sakakida, K. Nishida and S. Shimoda, *Artif. Organs*, 1998, **22**, 32.
226. C. Wang, A. Javadi, M. Ghaffari and S. Gong, *Biomaterials*, 2010, **31**, 4944.
227. A. Sambanis, in *Comprehensive Biotechnology*, ed. M. Moo-Young, Pergamon, Oxford, 2nd edn, 2011, vol. 5, pp. 699–711.

CHAPTER 10

Complexation with Metals: Anticorrosion Phosphorus-Containing Polymer Coatings

GHISLAIN DAVID* AND CLAIRE NEGRELL-GUIRAO

Institut Charles Gerhardt (ICGM)/Equipe Ingénierie et Architectures Macromoléculaires (IAM), ENSCM, 8 rue de l'école normale, 34296 Montpellier, France
*Email: ghislain.david@enscm.fr

10.1   Introduction

Corrosion is a broad scientific/technological area that is truly an interdisciplinary mixture of several sciences, such as electrochemistry, organic chemistry, materials science, metallurgy, *etc*. According to several reports, the general cost of corrosion is estimated to be about 1–5% of the national gross domestic product.[1] Corrosion control and management can be based on several strategies, one of which uses chemical additives in contact with the metallic surface to be protected. These chemical additives are known as corrosion inhibitors, *i.e.* which delay or even stop metallic corrosion.[2–4] There is a wide range of effective corrosion inhibitors acting over a broad range of pH conditions. Phosphonate-type inhibitors are one of those operating by complexation with metal ions.[5,6] The complexation of phosphonate/phosphonic acids to metals has been extensively studied and the connectivity of PO_3 groups can significantly vary according to the nature and oxidation state of the metal atom, as well as the reaction conditions

RSC Polymer Chemistry Series No. 11
Phosphorus-Based Polymers: From Synthesis to Applications
Edited by Sophie Monge and Ghislain David
© The Royal Society of Chemistry 2014
Published by the Royal Society of Chemistry, www.rsc.org

(temperature, pH, *etc.*) and bulkiness of the organic component bound to the phosphorus groups.[7] Thus in the presence of metal cations, phosphonic acids in their deprotonated form give soluble compounds that precipitate to provide a 3D protective thin film. According to water-related technologies, a wide range of phosphonate inhibitors are available and their activity as corrosion inhibitors was recently reviewed in an excellent paper by Demadis *et al.*[8] Nevertheless, a high concentration of inhibitor is generally required.

Sheffer *et al.*[9] reported the enhanced corrosion protection of sol–gel films on an aluminum substrate by entrapping the phosphonate group in organosilanes. However, phosphonate groups which remain as additives in a physical mixture can lead to dynamic phenomena such as aggregation, phase separation, or leaching by solvent or water with time. Such disadvantages can be overcome if they form a part of polymer network structure. Kannan *et al.*[10,11] reported the synthesis of a highly networked methacrylate-phospho-silicate hybrid by copolymerizing 2-(methacryloyloxy)ethyl phosphate (EGMP), containing a polymerizable methacrylate group and functional phosphate group, with [3-(methacryloyloxy)propyl]-trimethoxysilane (MEMO), which possesses a polymerizable methacryloxy group at one end and alkoxysilane groups capable of forming inorganic networks *via* sol–gel route at the other end. Another strategy consists of embedding phosphonate into polymers in order to reduce the inhibitor concentration. Research in this area is also stimulated by the need to develop inhibitor formulations free from chromates, nitrates, and inorganic phosphorus. In this chapter we will focus on the development of organic coatings containing phosphonic acid moieties, able to act as anticorrosion coatings.

10.2 Coatings Made from Blend Polymers

One solution to protect galvanized steel-plate against corrosion is a single-coat system consisting of a polymer matrix mixture with a corrosion inhibitor. In our case, we propose blends made with fluorinated polymers for the matrix and phosphonated (co)polymers for the inhibitor part. The proportion of this type of additive favoring adhesion and anticorrosion properties is used in a few percent of cases.

Commercial anticorrosion phosphonated polymer compounds are generally formed from Sipomer® or Phosmer® monomers, which are *phosphate*-type (meth)acrylates and can be readily polymerized *via* emulsion or solution.[12,13] Polymers with some phosphonate functionality have long been established as excellent adhesives and anticorrosion compounds;[14–21] however, there has been very little investigation into the use of *phosphonate*-type methacrylates for the same purpose.[12,13]

Dimethyl [2-(methacryloyloxy)ethyl]phosphonate (MAPC2) (Scheme 10.1) has been successfully copolymerized with methyl methacrylate (MMA) and used as an additive with poly(vinylidene fluoride) [poly(VDF)].[22,23] The incorporation of a phosphonic component results in a copolymer with highly

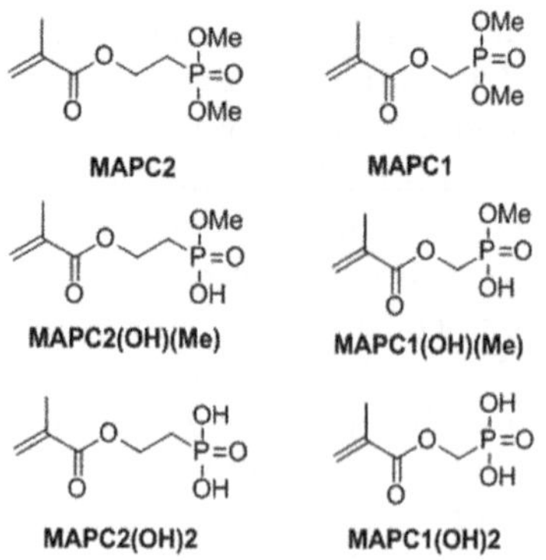

MAPC2   **MAPC1**

MAPC2(OH)(Me)   **MAPC1(OH)(Me)**

MAPC2(OH)2   **MAPC1(OH)2**

Scheme 10.1   Structures of phosphonated methacrylate monomers.

Table 10.1   Salt spray test on PVDF blend coatings with MMA/MAPC2 copolymers as inhibitor compounds with molar ratios of each monomer (10% w/w of additive compared to PVDF).

Coating	Sample	Wt% of phosphorus into the blend coating	Salt spray test percentage of corroded surface (%)
A	PVDF	0	Delamination
B	PVDF + 10 % Copolymer [MMA-MAPC2(OH)$_2$ – 90/10]	0.24	99
C	PVDF + 5% Copolymer [MMA-MAPC2(OH)$_x$(OMe)$_{2-x}$ – 90/10]	0.13	84
D	PVDF + 10% Copolymer [MMA-MAPC2(OH)$_x$(OMe)$_{2-x}$ – 90/10]	0.25	31
E	PVDF + 10% Copolymer [MMA-MAPC2(OH)$_1$(OMe)$_1$ – 90/10]	0.22	96
F	PVDF + 10% Copolymer [MMA-MAPC2(OH)$_1$(OMe)$_1$ – 80/20]	0.44	99

enhanced adhesion onto the metallic surface, as alkylphosphonic acids are well known to form resonance-stabilized phosphonate complexes with a wide range of metal alloys,[22] while remaining soluble in the poly(VDF) matrix which inhibits water penetration and thus corrosion. These authors tested different coatings with a varied content of phosphorus in the blends. The additives (10% w/w compared to PVDF) were different copolymers of MMA/MAPC2 with a monoalkyl phosphonate group, MMA/MAPC2(OH)$_1$(OMe)$_1$, a diphosphonic acid group, MMA/MAPC2(OH)$_2$, or a mixture of three phosphonate functions [25% P(O)(OMe)$_2$, 50% P(O)(OMe)(OH), and 25% P(O)(OH)^2], MMA/MAPC2(OH)$_x$(OMe)$_{2-x}$. The anticorrosion properties of the blends with good adhesion were evaluated with a salt spray test on galvanized steel plates coated with a 20–25 μm film (well established as the best standard by which to measure the adhesion/anticorrosive properties of a surface coating).[24] The results, listed in Table 10.1, showed that the

Table 10.2 Salt spray test on PVDF blend coatings with MMA/MAPC1 copolymers as inhibitor compounds with molar ratios of each monomer (10% w/w of additive compared to PVDF).

Coating	Sample	Wt% of phosphorus into the blend coating	Salt spray test percentage of corroded surface (%)
G	PVDF + Copolymer [MMA-MAPC1(OH)$_2$ – 90/10]	0.24	99
H	PVDF + Copolymer [MMA-MAPC1(OH)$_1$(OMe)$_1$ – 90/10]	0.22	40
I	PVDF + Copolymer [MMA-MAPC1(OH)$_1$(OMe)$_1$ –80/20]	0.44	99

percentage of corroded surfaces increased when the percentage of phosphorus in the blends increased. They showed also that the phosphonic acid groups introduced into PVDF from statistical copolymers maintain some level of adhesion, resulting in prevention of the spread of corrosion. Coating D shows the best results in term of corrosion resistance. The authors ascribed this result to high wet adhesion properties.

More recently, El Asri et al.[25] also copolymerized dimethyl [(methacryloyloxy)methyl]phosphonate (MAPC1) (Scheme 10.1) with MMA, leading to statistical copolymers. They prepared three different additives, one based on MMA/MAPC1(OH)$_2$ copolymer [initial molar ratios of 10% for MAPC1(OH)$_2$)] and two others based on MAPC1(OH)(OMe)/MMA [10% and 20% for MAPC1(OH)(OMe)].

The anticorrosive behavior of these copolymers was evaluated using the salt spray test during 14 days; the results are gathered in Table 10.2. Coating H showed only 40% of corroded surface due to the high level of adhesion maintained by the phosphonic hemi-acid group. These adhesion/anticorrosive properties should be enhanced by favoring migration of the phosphonic groups towards the metallic surface. Gradient copolymers [used as additives in the poly(VDF) matrix] increase migration of phosphonic groups, as previously demonstrated by the work of Rixens et al.[26] Compared to gradient copolymers, diblock copolymers should be better candidates to favor migration of phosphonic groups and promote better adhesion/anticorrosive properties.

David et al.[16] have tested these new diblock copolymers as additives for adhesion/anticorrosive properties of metals. In order to protect against corrosion, a coating system was prepared consisting of a blend of poly(VDF) (inhibiting water penetration) with poly(MMA)-*b*-poly(monophosphonic acrylate) diblock copolymer with 10 wt% [compared to poly(VDF)]. This diblock copolymer was obtained by atom transfer radical polymerization using post-functionalization and hydrolysis to introduce the phosphonic acid function onto the polymeric backbone. The anticorrosive properties of virgin poly(VDF) were also determined without any additive to be compared

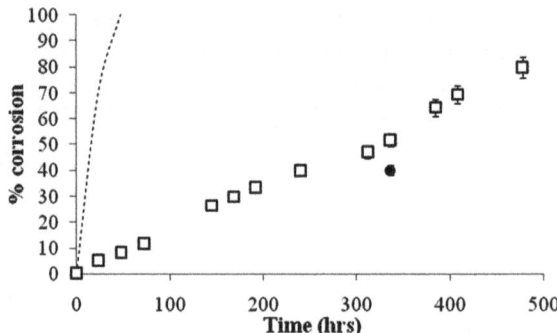

Figure 10.1 Graphical result of corrosion after being subjected to the salt spray test (0.5 mol L^{-1} NaCl in water at 35 °C), where (—) = virgin poly(VDF), (□) = diblock copolymer, and (●) = statistical copolymer.

Table 10.3 Properties of poly(VDF) blended with hemi-phosphonic statistical and diblock copolymer additives [10% w/w compared to poly(VDF)]; (●) = MMA unit; (P) = hemi-phosphonic acid unit.

Blends (weight %)	Structure of the monophosphonic additive	Ia of blend (mg KOH/g)	% of corroded surface (after 14 days)
Virgin PVDF	/	0	100 (after 2 days)
PVDF	Statistical copolymer	4.6	40
PVDF	Diblock copolymer	2.7	48

with the blends. The anticorrosive behavior of these systems was evaluated using the salt spray test.

Figure 10.1 shows an almost linear trend for the evolution of corrosion as a function of time when the diblock copolymer is used as additive. As expected, virgin poly(VDF) alone is not able to prevent the metal from corrosion as the surface is almost completely corroded after only 48 h. Figure 10.1 also shows a high improvement when poly(VDF) is blended with poly(MMA)-b-poly(phosphonate methacrylate) diblock copolymer additive, although the statistical copolymer seems to afford better adhesion towards the metallic surface and hence better anticorrosion properties.

Table 10.3 gathers characteristics of both blends, i.e. acid value and weight % of additive, and gives the content of corroded surface after being submitted for 14 days to the salt spray test. The blend made from the diblock copolymer additive and poly(VDF) brings high efficiency towards corrosion [compared to virgin poly(VDF)], as about 50% of the metallic surface is corroded after 14 days submitted to the salt spray test. However, compared to the blend made of statistical copolymer additive and poly(VDF), no improvement is clearly observed (40% of corroded surface after 14 days). Two main differences arise from these two blends, the first one being the monophosphonic additive structure and the second one being the acid value I_a. To be highly efficient against corrosion, the additive must ensure high adhesion at the interface between the metal and the poly(VDF) matrix. Promoting adhesion induces migration of the phosphonic group at the metal surface. This migration occurs when phase segregation is obtained, i.e. phosphonic acid nodules into a PMMA matrix.[26] Noteworthy is that segregation is favored by both the structure of the copolymer and the content of the phosphonic acid units. The anticorrosion tests shown in Table 10.3 prove that the acid content in the copolymers is the main parameter that controls phase segregation, from which high adhesion towards the metal should be obtained.

10.3   Phosphorus-Containing UV Polymer Coatings

We have shown above that copolymers of MMA with methacrylate-containing phosphonate moieties, when blended with poly(VDF), were able to enhance the anticorrosive properties of the resulting coatings. It is however necessary to afford polymer coatings with the use of a hydrophobic polymer matrix such as poly(VDF). Photopolymerization allows for a more efficient polymerization due to better activation of the radical species, especially with regard to the high surface to volume ratio as is the case for surface coatings. In addition, photopolymerization proceeds without any solvent, which is of increasing importance due to the environmental impact of industrial organic solvents. We have performed a series of UV photopolymerization reactions, which proceed by first mixing (meth)acrylates with pentaerythritol tetrakis(3-mercaptopropionate) as a thiol allowing reticulation and Darocur® as a photoinitiator. The mixture of (meth)acrylate is composed of tripropylene glycol diacrylate and hexanediol diacrylate, which allows creation of the polymer network. In this (meth)acrylate mixture, only a small amount of phosphonic acid-containing methacrylate is added, thus reacting with diacrylate monomers during the crosslinking reaction. Dimethyl [(methacryloyloxy)methyl]phosphonate (MAPC1)[25] under its ester or acid form was first added to the acrylate formulations, with mass content ranging from about 1 to 10 wt% (Table 10.4).[27]

The coatings were applied on a steel substrate and the corrosion resistance was characterized by electrochemical impedance spectroscopy. By following the evolution of the impedance response of the steel substrate/coating/electrolyte system over an extended period of time, the degradation of the coating material and the onset of the corrosion processes can be monitored.[28–32] In Figure 10.2 we show the impedance diagrams obtained after 24 h immersion in the 0.1 M NaCl solution for the formulations F1 and F2.

Table 10.4 Composition of the different formulations prepared for the electrochemical characterization. In addition to the contents listed here, each formula contains 6% Darocur.

Formulation	Acrylates (w %)	MAPC1 (w %)	MAPC1(OH)₂ (w %)	Oleic acid phosphonic acid (w %)
1	92.7	1.3	/	/
2	92.7	/	1.3	/
3	91.5	/	2.5	/
4	89	/	5	/
5	84	/	10	/
6	89	/	2.5	2.5

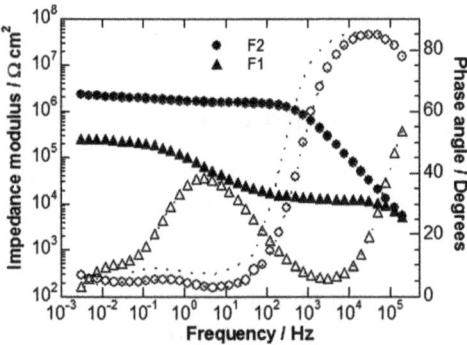

Figure 10.2 Electrochemical impedance diagrams (Bode representation) obtained for the F1 and F2 samples after 24 h immersion in 0.1 M NaCl solution.

The diagrams are characterized by two time constants: the high-frequency (HF) part of the diagrams (from 10^5 to 1 Hz) is related to the coating and attributed to the barrier properties of the film, while the low-frequency (LF) part (from 1 to 10^{-3} Hz) corresponds to the reactions occurring at the metal/coating interface through defects and pores in the coating.[33,34] In Figure 10.2 it can be seen that for the F1 coating the barrier effect (HF range) is significantly lower than for the F2 coating. The impedance value is only 2×10^4 Ω cm² (plateau in the HF range). The impedance modulus in the low-frequency range is low, which indicates that the F1 coating is poorly efficient to protect the steel surface and which is also a consequence of the poor barrier effect. The diagram for the F2 coating revealed a high impedance modulus in the HF range (2×10^6 Ω cm²), indicating that the barrier effect is significant. The barrier effect leads to a decrease of the surface area in contact with the electrolyte (high impedance modulus in the LF range) and as a consequence this coating is more protective. This quantitative study clearly demonstrates that phosphonic acid-containing methacrylate leads to better corrosion protection, compared to its ester form. In Figure 10.3 the impedance modulus at low frequencies ($|Z|_{3\ \text{mHz}}$) is shown for the four formulations with different

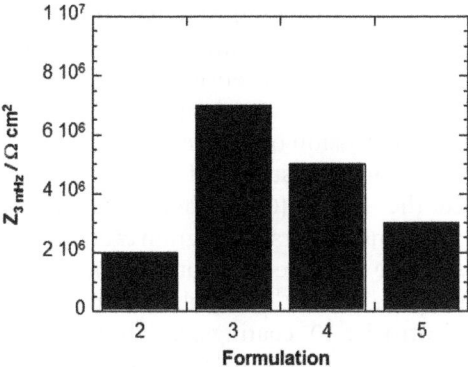

Figure 10.3 $|Z|_{3 \text{ mHz}}$ after 24 h of immersion in 0.1 M NaCl solution for the formulations with different MAPC1(OH)$_2$ concentrations.

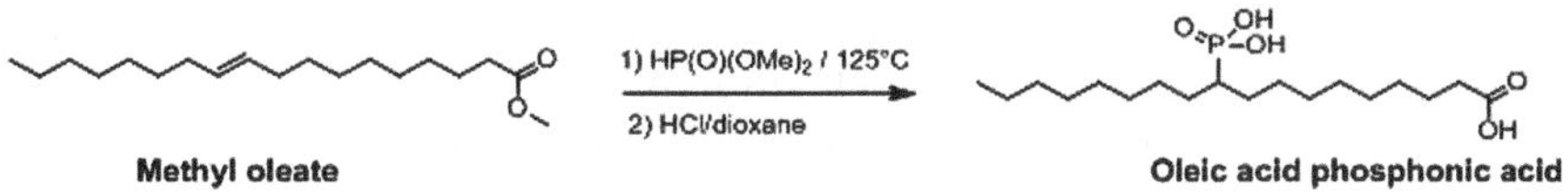

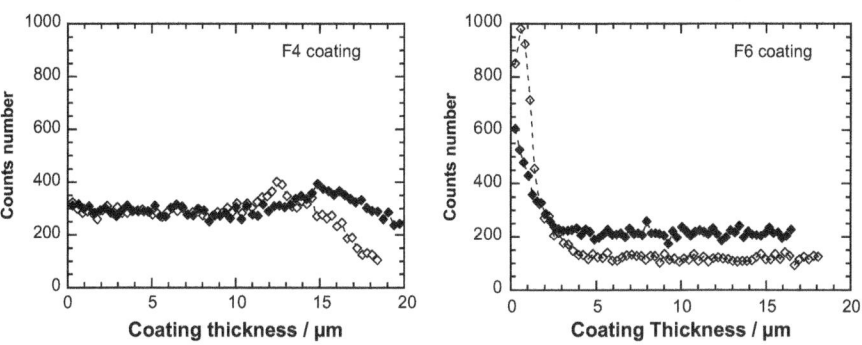

Scheme 10.2 Synthesis of 10-phosphonooleic acid.

Figure 10.4 Phosphorus profiles along the coating thickness for F4 (*left*) and F6 (*right*) coatings (♦, ◇: results of two independent measurements on the same sample).

MAPC1(OH)$_2$ contents and a minimum efficient MAPC1(OH)$_2$ concentration of around 2.5% can be seen.

Interestingly, 10-phosphonooleic acid ("oleic acid–phosphonic acid"), which was obtained *via* a two-step reaction (Scheme 10.2), was added to the methacrylate formulations (F6). It was shown that the addition of the 10-phosphonooleic acid significantly improved the barrier properties of the coating and as a consequence increased the corrosion protection.[27]

To analyze the role of the 10-phosphonooleic acid, SEM/EDX analyses were performed on both F4 and F6 coatings (Figure 10.4). From the F4 coating SEM the detected phosphorus quantity is constant through the whole coating thickness, whereas from the F6 coating SEM we noticed a phosphorus profile significantly modified at the metal/coating interface. Indeed, about 3 μm of the internal zone of the coating was enriched in phosphorus, which means that the 10-phosphonooleic acid can migrate through the film and reach the substrate to inhibit the corrosion of the steel.[35,36]

To improve the corrosion resistance of such methacrylate coatings, it is possible to replace the MAPC1(OH)$_2$ monomer by bisphosphonic acid-containing methacrylate monomers. Chougrani *et al.* have developed a series of new aminobisphosphonic acid methacrylate derivatives as well as their *gem*-bisphosphonic acid equivalents.[37,38] They also compared the efficiency of three monomers[39] as anticorrosive UV coatings, *i.e.* [3-(2-methylacryloyloxy)propyl]-phosphonic acid (MAC$_3$P), [({3-[(2-methylacryloyl)oxy]propyl}imino)dimethane-diyl]bis(phosphonic acid) (MAC$_3$NP$_2$), and (2-{2-[(2-methylacryloyl)oxy]-ethoxy}ethane-1,1-diyl)bis(phosphonic acid) (MAC$_3$P$_2$) (Scheme 10.3).

The mass content of such monomers in the UV formulations is similar to that of MAPC1(OH)$_2$ shown in Table 10.4, *i.e.* less than 3 wt%. For the methacrylate coatings, salt spray tests were conducted and the resistance to corrosion was assessed over time (Figure 10.5).

First from Figure 10.5 it can be observed that the adhesion properties are similar for the formula with either the monophosphonic methacrylate additive MAC$_3$P or the bisphosphonic MAC$_3$NP$_2$, the N denoting the inclusion of an amine group. Aminophosphonic acids are highly efficient at

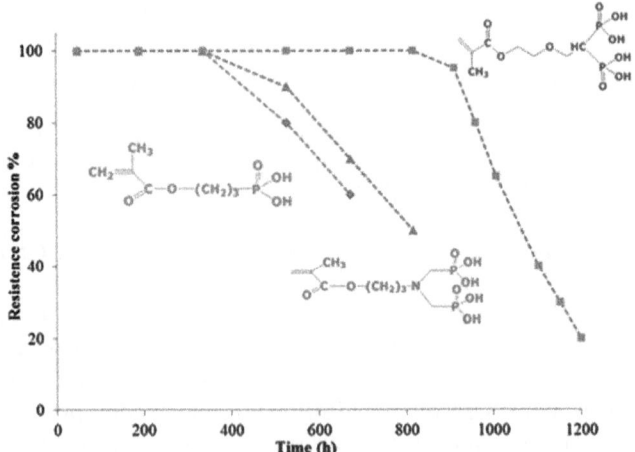

Scheme 10.3   Structures of MAC$_3$P (*left*), MAC$_3$NP$_2$ (*middle*), and MAC$_3$P$_2$ (*right*).

Figure 10.5   Graphical result of the resistance to corrosion after being subject to the salt spray test (0.5 mol L^{-1} NaCl in water at 35 °C).

forming adhesive complexes with calcium-based substrates,[40,41] or forming micelles with calcium salts in aqueous medium,[42] but as yet there have been no investigations into how they will adhere to metal or metal alloy surfaces. It is known that tertiary amines will form a quaternary ammonium complex when in aqueous solution; thus an assumption could be made that the reason behind the reduction in adhesive properties for the aminophosphonic acid-based formulas is due to the formation of an ammonium complex. Therefore these monomers show little promise as adhesion promoters. On the other hand, the system based on MAC_3P_2 (and at only 1% w/w concentration), having two phosphonic functional groups and no amine function, shows significantly better adhesion that the other formulas since no corrosion occurs until 900 h when subjected to the salt spray test. This indicates that bisphosphonic functional groups show high capacity as adhesion promoters and could be further optimized for industrial applications.

Following the same strategy, Lam *et al.*[43] suggested adding oligomer species into the UV acrylate formulations. Such oligomers were made from chain transfer reactions of phosphonic acid-containing methacrylate with MMA and *n*-butyl methacrylate, this last monomer aimed at increasing the hydrophobicity of the coating to avoid water incorporation between the coating and the metallic surface (Scheme 10.4). The degree of polymerization was about 30 and the mass content of the phosphonic acid was about 8 wt%.

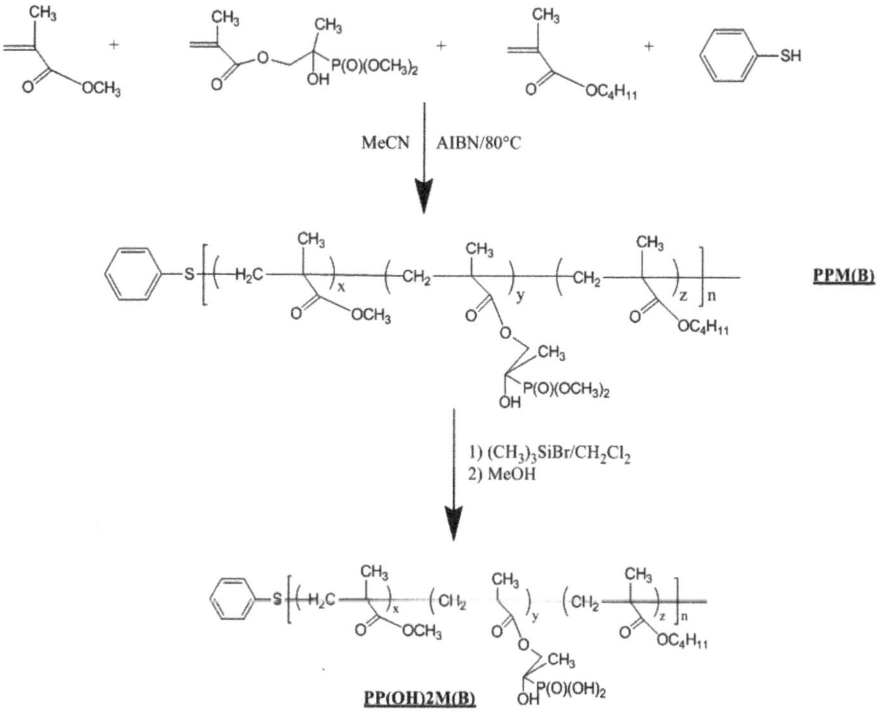

Scheme 10.4 Synthesis of oligomers containing phosphonic acid moieties and butyl methacrylate.

When the PPM(B) terpolymer (phosphonate groups) is added to the acrylate UV coating (*i.e.* composed of tripropylene glycol diacrylate and hexanediol diacrylate), an improvement is observed due to migration of the phosphonate groups towards the metallic surface, but all the metallic surface is corroded after 500 h. Migration of phosphonate coatings towards the metallic surface was evidenced by Rixens *et al.*[26] However, adhesion of the coating onto the metallic surface is increased by using phosphonic groups instead. This is indeed confirmed by the resistance to the corrosion of formulation 3, *i.e.* an acrylate coating with PP(OH)$_2$M(B) terpolymer. Even after 1000 h subject to the salt spray test, no corrosion was observed, and only 10% of the surface was corroded after 1150 h. The coating obtained from formulation 3 shows outstanding resistance to corrosion despite the low content of acid groups; the acid index I_a was evaluated at about 1.38 mequiv g^{-1} KOH. For comparison, the coatings based on bisphosphonic acid methacrylates, with an acid index I_a ranging from about 9 to 28 mequiv g^{-1} KOH, showed a maximum 100% resistance to corrosion after 900 h when subject to the salt spray test, for the best coatings.

10.4   Phosphonation of Pre-Made Polymers

The relatively good adhesion to the metallic surface of methacrylate coatings is limited, however, by hydrolysis of the methacrylate function.[17,18] Since methacrylates undergo hydrolysis, some authors proposed the use of (meth)acrylamide-bearing phosphonic acid groups. The resulting cross-linked polymers improved adhesion and were efficiently used in dental applications[44–47] (see Chapter 8). The post-phosphonation of pre-made polymers, which are highly stable towards hydrolysis, is another strategy to introduce phosphonic acid groups. David *et al.*[48] have recently developed the chemical grafting of phosphonic acid ester groups onto poly(vinyl acetal) pre-made polymers, and especially poly(vinyl butyral) (PVB). PVB is obtained by acetalization of poly(vinyl alcohol) (PVA) with butyraldehyde, using either a solvent process or an aqueous process.[49–53] PVB is extensively used in the manufacture of textiles, safety glass, adhesives, engineering plastics, and binders.[54–57] Interestingly, PVB can be also used as metal primers and coatings due to its good adhesion to metallic substrates.[58] David *et al.* carried out the acetalization of PVA with both (3-diethylphosphonate)propanal and butyraldehyde (Scheme 10.5).

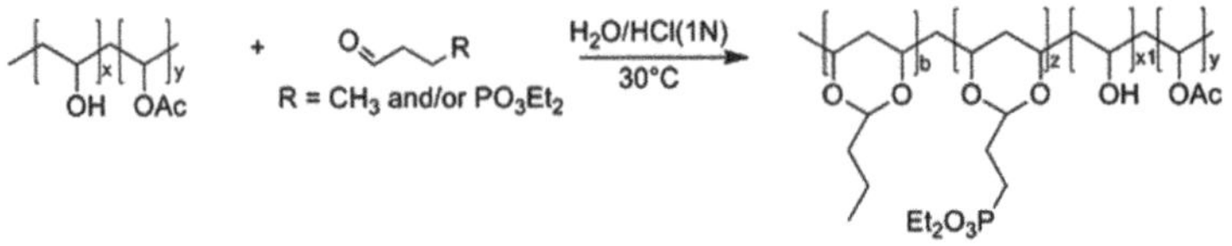

Scheme 10.5   Acetalization of PVA with both (3-diethylphosphonate)propanal and butyraldehyde.

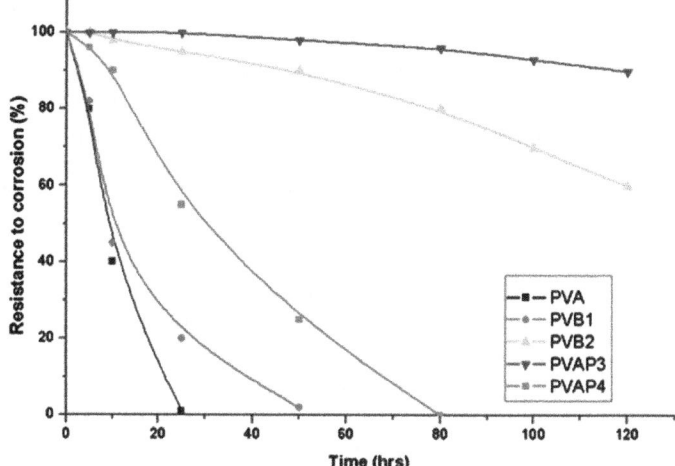

Figure 10.6 Graphical result of the resistance to corrosion after being subject to the salt spray test (0.5 mol L^{-1} NaCl in water at 35 °C) for PVA, PVB1 (48 mol% butyral), PVB2 (73 mol% butyral), PVAP3 (60 mol% butyral and 20 mol% phosphonate), and PVAP4 (50 mol% phosphonate).

Four different PVBs, *i.e.* PVB1 (48 mol% butyral), PVB2 (73 mol% butyral), PVAP3 (60 mol% butyral and 20 mol% phosphonate), and PVAP4 (50 mol% phosphonate), as well as poly(vinyl alcohol), were subjected to the salt spray test (Figure 10.6). As expected, PVA did not show any resistance against corrosion since the whole surface was corroded after 20 h of exposure. The resistance to corrosion for PVB1 was poorly improved compared to the pristine PVA. Hence, the high vinyl alcohol molar content seems to be detrimental for resistance to corrosion; as an example, rapid corrosion of the plate was also observed when coated with PVAP4 (having about 50 mol% of vinyl alcohol). The resistance to corrosion was clearly enhanced for the coating based on PVB2, *i.e.* having more than 70 mol% of butyral units. Finally, the PVAP3-based coating showed very good resistance against corrosion. Indeed, after 80 h subject to the salt spray test the plate did not show any corrosion. After 90 h, the corrosion slowly spread out. From the empirical data presented it is apparent that the presence of the phosphonate group is directly responsible for the excellent adhesion and anticorrosive properties.

10.5 Conclusions

In this chapter, we have summarized three different strategies to use phosphonated copolymers as corrosion inhibitors to protect metals. In the first part, the blends of organophosphorus monomers copolymerized with MMA in a PVDF matrix showed relatively good efficiency. These studies also showed phosphonic acid groups as good adhesion/corrosion promoters, but the macromolecular structure (statistical or diblock) was not the main parameter, unlike the phosphonic acid content. In second part, we showed alternative solutions dealing with UV coatings; the photopolymerization process is interesting for a good environmental impact. Formulations with phosphonated monomers and a crosslinking agent were applied on a steel substrate and the corrosion resistance was characterized by electrochemical impedance spectroscopy, showing good results principally with the phosphonic acid form. Moreover, the addition of 10-phosphonooleic acid in these coatings increased the efficiency of the protection against corrosion. New monomers with a higher ratio of phosphorus on the unit like bisphosphonic functional groups showed high capacity as adhesion promoters with no corrosion occurring until 900 h when subjected to the salt spray test. The last solution to UV coating was the use of a phosphonic terpolymer since only 10% of the surface was corroded after 1150 h. Finally, for avoiding the side reaction of methacrylate hydrolysis, PVA copolymers were tested as coatings. After 90 h the corrosion slowly spread out, providing little interest in this type of polymer for anticorrosion applications; meanwhile the efficiency of phosphonic functions was clearly shown for adhesive and anticorrosive properties.

References

1. V. S. Sastri, J. R. Perumareddi and M. Elboujdaini, in *Environmentally Induced Cracking of Metals*, eds. E. Mimoun, G. Edward and Z. Wenyue, *Proceedings of the International Symposium*, Ottawa, ON, Canada, 2000, vol. 2000, pp. 167–176.
2. R. Javaherdashti, *Mater. Perform.*, 2002, **41**, 30–32.
3. R. Javaherdashti, *Corros. Rev.*, 2003, **21**, 311–325.
4. R. Javaherdashti, *Mater. Perform.*, 2006, **45**, 52–54.
5. Y. I. Kuznetsov, *Zashch. Met.*, 1990, **26**, 954–964, *Chem. Abstr.*, 1991, **CAN114**, 70948.
6. Y. I. Kuznetsov and Y. A. Popkov, *Zh. Prikl. Khim. (Leningrad)*, 1990, **63**, 1042–1049, *Chem. Abstr.*, 1990, **CAN113**, 158355.
7. C. Queffelec, M. Petit, P. Janvier, D. A. Knight and B. Bujoli, *Chem. Rev.*, 2012, **112**, 3777–3807.
8. K. D. Demadis, M. Papadaki and D. Varouchas, *Green Corros. Chem. Eng.*, 2012, 243–296.
9. M. Sheffer, A. Groysman, D. Starosvetsky, N. Savchenko and D. Mandler, *Corros. Sci.*, 2004, **46**, 2975–2985.
10. A. G. Kannan, N. R. Choudhury and N. K. Dutta, *Polymer*, 2007, **48**, 7078–7086.
11. A. G. Kannan, N. R. Choudhury and N. K. Dutta, *J. Electroanal. Chem.*, 2010, **641**, 28–34.
12. M. Zakikhani and J. Davis, (Albright & Wilson), *Eur. Pat.*, 765 889, 1997.
13. T. Okamoto, H. Mori and H. Matsuda, (Okura Industrial), *U.S. Pat.*, 4 433 124, 1984.

14. W. Herbst, H. Ludwig, F. Roehlitz and H. Vilcsek, (Metallgesellschaft), *Ger. Pat.*, 1 187 100, 1965.
15. N. Moszner, F. Zeuner, U. K. Fischer and V. Rheinberger, *Macromol. Chem. Phys.*, 1999, **200**, 1062–1067.
16. N. Moszner, U. Salz and J. Zimmermann, *Dent. Mater.*, 2005, **21**, 895–910.
17. U. Salz, J. Zimmermann, F. Zeuner and N. Moszner, *Polym. Preprints.*, 2004, **45**, 325–326.
18. U. Salz, J. Zimmermann, F. Zeuner and N. Moszner, *J. Adhes. Dent.*, 2005, 7, 107–116.
19. F. Zeuner, N. Moszner, T. Volkel, K. Vogel and V. Rheinberger, *Phosphorus, Sulfur Silicon Relat. Elem.*, 1999, **144/146**, 133–136.
20. F. Zeuner, N. Moszner, M. Drache and V. Rheinberger, *Phosphorus, Sulfur Silicon Relat. Elem.*, 2002, **177**, 2263.
21. O. Senhaji, J. J. Robin, M. Achchoubi and B. Boutevin, *Macromol. Chem. Phys.*, 2004, **205**, 1039–1050.
22. C. Bressy-Brondino, B. Boutevin, Y. Hervaud and M. Gaboyard, *J. Appl. Polym. Sci.*, 2002, **83**, 2277–2287.
23. C. Brondino, B. Boutevin, J.-P. Parisi and J. Schrynemackers, *J. Appl. Polym. Sci.*, 1999, **72**, 611–620.
24. A. Shida, H. Sugimura, M. Futsuhara and O. Takai, *Surf. Coat. Technol.*, 2003, **169/170**, 686–690.
25. Z. El Asri, K. Chougrani, C. Negrell-Guirao, G. David, B. Boutevin and C. Loubat, *J. Polym. Sci., Part A: Polym. Chem.*, 2008, **46**, 4794–4803.
26. B. Rixens, R. Severac, B. Boutevin and P. Lacroix-Desmazes, *J. Polym. Sci., Part A: Polym. Chem.*, 2005, **44**, 13–24.
27. F. Millet, R. Auvergne, S. Caillol, G. David, A. Manseri and N. Pebere, *Prog. Org. Coat.*, 2014, **77**, 285–291.
28. J. E. O. Mayne, *J. Soc. Chem. Ind., London*, 1947, **66**, 93–95.
29. L. Beaunier, I. Epelboin, J. C. Lestrade and H. Takenouti, *Surf. Technol.*, 1976, **4**, 237–254.
30. F. Mansfeld, M. W. Kendig and S. Tsai, *Corrosion (Houston)*, 1982, **38**, 478–485.
31. F. Mansfeld, *Electrochim. Acta*, 1990, **35**, 1533–1544.
32. P. L. Bonora, F. Deflorian and L. Fedrizzi, *Electrochim. Acta*, 1996, **41**, 1073–1082.
33. N. Pebere, T. Picaud, M. Duprat and F. Dabosi, *Corros. Sci.*, 1989, **29**, 1073–1086.
34. C. Le Pen, C. Lacabanne and N. Pebere, *Prog. Org. Coat.*, 2000, **39**, 167–175.
35. Y. Gonzalez, M. C. Lafont, N. Pebere, G. Chatainier, J. Roy and T. Bouissou, *Corros. Sci.*, 1995, **37**, 1823–1837.
36. Y. Gonzalez, M. C. Lafont, N. Pebere and F. Moran, *J. Appl. Electrochem.*, 1996, **26**, 1259–1265.
37. K. Chougrani, B. Boutevin, G. David and G. Boutevin, *Eur. Polym. J.*, 2008, **44**, 1771–1781.
38. K. Chougrani, G. Niel, B. Boutevin and G. David, *Beilstein J. Org. Chem.*, 2011, 7, 364.

39. K. Chougrani, B. Boutevin, G. David, S. Seabrook and C. Loubat, *J. Polym. Sci., Part A: Polym. Chem.*, 2008, **46**, 7972–7984.
40. J. J. Vepsalainen, *Curr. Med. Chem.*, 2002, **9**, 1201–1208.
41. A. D. Geddes, S. M. D'Souza, F. H. Ebetino and K. J. Ibbotson, *Bone Mineral Res.*, 1994, **8**, 265–306.
42. N. J. Alfano and D. M. Shenberger, (Calgon Corp.), *U.S. Pat.*, 5 454 954, 1995.
43. O. A. Lam, G. David, Y. Hervaud and B. Boutevin, *J. Polym. Sci., Part A: Polym. Chem.*, 2009, **47**, 5090–5100.
44. N. Nishiyama, K. Suzuki, H. Yoshida, H. Teshima and K. Nemoto, *Biomaterials*, 2003, **25**, 965–969.
45. X. Xu, R. Wang, L. Ling and J. O. Burgess, *J. Polym. Sci., Part A: Polym. Chem.*, 2006, **45**, 99–110.
46. Y. Catel, M. Degrange, L. Le Pluart, P.-J. Madec, T.-N. Pham and L. Picton, *J. Polym. Sci., Part A: Polym. Chem.*, 2008, **46**, 7074–7090.
47. N. Moszner, J. Pavlinec, I. Lamparth, F. Zeuner and J. Angermann, *Macromol. Rapid Commun.*, 2006, **27**, 1115–1120.
48. G. David, E. Ortega, K. Chougrani, A. Manseri and B. Boutevin, *React. Funct. Polym.*, 2011, **71**, 599–606.
49. G. S. Stamatoff, (E. I. du Pont de Nemours), *U.S. Pat.*, 2 697 053, 1954.
50. G. S. Stamatoff, (E. I. du Pont de Nemours), *U.S. Pat.*, 789 055, 1949.
51. G. S. Stamatoff, (E. I. du Pont de Nemours), *U.S. Pat.*, 2 358 355, 1944.
52. M. D. Fernandez, M. J. Fernandez and P. Hoces, *J. Appl. Polym. Sci.*, 2006, **102**, 5007–5017.
53. M. Hajian, G. A. Koohmareh and M. Rastgoo, *J. Appl. Polym. Sci.*, 2010, **115**, 3592–3597.
54. H. Takehara, M. Ichitani and Y. Ochiai, (Sekisui Chemical), *Jpn. Pat.*, 2007 112 902, 2007.
55. W. I. Patnode, E. J. Flynn and J. A. Weh, *J. Ind. Eng. Chem.*, 1939, **31**, 1063–1071; CAN142:114851 (2004).
56. H. Yuan, P. Ma, J. Wang, J. Li and X. Guo, *Gongcheng Suliao Yingyong*, 2004, **32**, 43–46, *Chem. Abstr.*, 2004, **CAN142**, 114851.
57. D. S. Plumb, *Text. Age*, 1947, **11**, 64, 66, 68–69.
58. H. Z. Zhang, B. L. Liu, R. Luo, Y. Wu and D. Lei, *Polym. Degrad. Stab.*, 2006, **91**, 1740–1746.

CHAPTER 11

Use of Phosphorus-Containing Polymers for the Removal of Metal Ions from Wastewater

LAVINIA LUPA,[a] ADRIANA POPA*[b] AND GHEORGHE ILIA[b]

[a] Politehnica University of Timisoara, Faculty of Industrial Chemistry and Environmental Engineering, B-dul Vasile Parvan, no. 6, 300223, Timisoara, Romania; [b] Institute of Chemistry Timisoara of Romanian Academy, B-dul Mihai Viteazu 24, 300223 Timisoara, Romania
*Email: apopa_ro@yahoo.com; apopa@acad-icht.tm.edu.ro

11.1 Introduction

Metals in the environment are a major concern due to their toxicity to many life forms. They will not degrade into harmless end products like organic pollutants, which in the majority are susceptible to biological degradation. Heavy metals can be found in a variety of industries, in particular mining, metal processing, finishing, and plating. Effluents from these industries contain heavy metals, the concentrations of which are regulated by environmental agencies, so it is often necessary to remove heavy metals before discharging them. The main techniques employed for their removal from wastewaters include precipitation, reduction, ion exchange, and adsorption, or alternate inexpensive methods using industrial waste materials (discarded tires, fly ash, sludge), agricultural products and byproducts (starch, tree barks, onion skin, coconut shell, palm pressed fibers, lignin),

RSC Polymer Chemistry Series No. 11
Phosphorus-Based Polymers: From Synthesis to Applications
Edited by Sophie Monge and Ghislain David
© The Royal Society of Chemistry 2014
Published by the Royal Society of Chemistry, www.rsc.org

and naturally occurring minerals (coal, peat moss, pyrite). One of the most investigated kinds of material for the removal of metal ions are polymers. Insoluble polymeric matrixes are commonly used in the recovery of metal ions from dilute solutions.[1] Nowadays, increasing interest has been put on the minimization of the environmental impact and in reducing costs. For this reason, researchers have focused on the use of low-cost materials which are also eco-friendly and available in abundance, such as natural polysaccharides. Cellulose and chitosan (natural polysaccharides) have some unique properties, such as nontoxicity, biocompatibility, biodegradability, hydrophilicity, and good adsorption properties.[2-4] Therefore these materials are preferred for use in the process of obtaining new materials which can be used in various fields, especially in the adsorptive removal process of various metal ions. Chitosan is obtained from the deacetylation of the natural biopolymer chitin, found in crustaceous shells, insects, and fungal cell walls. Chitosan consists of β-(1,4)-2-acetamido-2-deoxy-β-D-glucose and β-(1,4)-2-amino-2-deoxy-β-D-glucose units and contains high contents of amino and hydroxyl groups, which favor the modification of this biopolymer and the introduction of new functional groups.[5-7] In order to improve the efficiency of metallic ion fixation onto cellulose and chitosan, they are subjected to a chemical modification process. From all the studied chemical modification processes, the use of phosphorylated agents has proved to be the most efficient for obtaining the desirable properties.[2-4]

Another class of polymer used in removal of metal ions form wastewater is synthetic chelating resin containing phosphorus moieties. The chelating resins are a group of products having chelating groups on their surface. They are obtained by the chemical modification of their surface with different phosphorus pendant groups.[8] The use of chelating resins in the extraction of metal ions has several advantages[9] over conventional methods. Resins are principally useful in metal ion recuperation, when the concentration of ions in solution is small and therefore the utilization of large volumes of solution is inevitable. Selective determination of metal ions will be possible using a chelating resin that has high selectivity with the ligand for the targeted metal ions. The technique of chelation is an economic method because it uses only a small amount of ligand and extraction solvent and this also increases the sensitivity of the system. Trace amounts of metal ions (with extremely low ppb levels) can be removed because the target ion is rich in the solid phase resin.[9]

The efficiency of metal ion removal methods depends on two sets of parameters:

1. The process parameters: type and charge of target ion(s), type, charge and concentration of ballast ion(s), and pH of the solution.[10]
2. The properties of the ion-exchange/coordination resin, such as degree of crosslinking, swelling, and the nature and structure of the ligands present in the structure of the resin.[10]

11.2 Natural Polymers

11.2.1 Cellulose

11.2.1.1 Type of Phosphorus Pendant Groups Grafted onto Cellulose

In research, various sources of cellulose are used: wood sawdust,[4] pure cotton,[2,4] or agricultural waste (*e.g.* corn cobs).[5-7] The cellulose is used in its pure form or is treated with some bacterial culture (*e.g.* a kind of *Acetobacter*), obtaining a bacterial cellulose.[2] For the purpose of modifying cellulose, various phosphorylating agents were studied: mono- and dibasic ammonium phosphates, alkali metal phosphates, phosphonates [*e.g.* 1-hydroxyethylidene-1,1-diphosphonic acid (HEDP), di-2-ethylhexyl phosphoric acid (DEHPA), chloromethylphosphonic acid, phosphoric acid, phenylphosphonic acid (PPA), 2-carboxyethylphosphonic acid (CEPA), and *N*-(phosphonomethyl)iminodiacetic acid (PMIDA)].[2-4,11-13] The phosphorylation process is influenced by the morphology of cellulose; therefore before the phosphorylation process with any agent the cellulose must be washed in order to remove all the impurities, and it must be dried. The drying process of the starting cellulose influences the fibrous structure of the cellulose.[2] Oshima *et al.* demonstrated that lyophilization is a more efficient drying method than either drying in an oven or the alcohol substitution method, because it prevents aggregation of the microfibrils. In most of the studies, either urea alone was used to promote phosphorylation of cellulose or urea along with *N,N*-dimethylformamide (DMF)[2] or dicyandiamide.[3] A representative synthetic route to phosphorylated bacterial cellulose (PBC) is presented in Scheme 11.1.[2]

Nada *et al.* studied the influence of the corn cobs treatment, before phosphorylation, with different concentrations of the NaOH solution upon the phosphorylation efficiency and upon the metal ions adsorption. They demonstrated that the use of a 7.5% NaOH solution lead to the highest level of phosphate groups which were incorporated into the treated corn cobs. The modified cellulose was washed several times with distilled water (and maybe acetone, or an alcohol), dried, and then could be used in further experiments. In the phosphorylation process the phosphoric group should be introduced at the primary carbon with hydroxyl groups and the phosphorylation is complete when all the C–OH groups are substituted.[2]

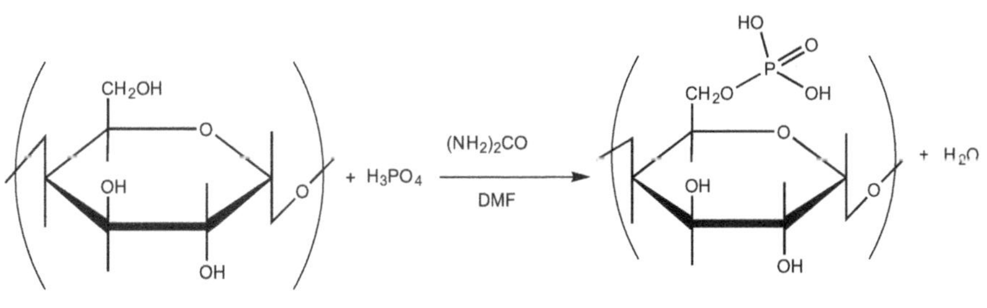

Scheme 11.1 The phosphorylation route of bacterial cellulose.

11.2.1.2   Metal Ion Adsorption onto Cellulose Grafted with
Different Phosphorylated Groups

The cellulose phosphate was widely applied in the removal process of di-valent and trivalent (Cu^{2+}, Zn^{2+}, Co^{2+}, Ni^{2+}, Pb^{2+}, Mn^{2+}, Fe^{2+}, Fe^{3+}, Cr^{3+}) ions because of the rapid adsorption of the metals ions in comparison with the synthetic polymers. Also it was observed that the cellulose phosphate functions as an ion exchanger for lanthanide ion removal from aqueous solution. In order to evaluate the adsorbent characteristics of each support in the removal process of metal ions from aqueous solution, the most widely used method by researchers has been the batchwise one. In any adsorption process the most important parameter is the pH of the medium, which depends on the nature of the support used and also on which metal ions are to be removed. In most cases, researchers worked with solutions that have an acid pH value in order to avoid the possible precipitation of the studied metal ions.[12] A complex study of the pH influence upon the efficiency of Fe^{3+}, Cu^{2+}, Mn^{2+}, Zn^{2+}, Co^{2+}, and lanthanide(III) ions removal from aqueous solutions with PBC was made by Oshima *et al.* Their conclusion was that the lanthanide ions are adsorbed when the pH value of the aqueous solution is less than 3, and in the case of transition metal ions the adsorption percentage increases with increasing aqueous pH and reaches over 90% at a pH value of around 4.5.

In the experiments carried out by Buasri *et al.* on the removal of Cu^{2+} and Zn^{2+} ions from aqueous solution onto modified cellulose and corn cob, respectively, with phosphoric acid, the influence of the different initial biomass concentration was studied. The number of sites available for biosorption depends upon the amount of the biosorbent. In all cases it was observed that an increase of the biosorbent concentration lead to an increase in the metal ion uptake. They concluded that the optimum S : L ratio was 3 g/100 mL. Beyond this dosage there is a rise in the biomass surface area and in the number of potential binding sites.[11,13] In the same studies the influence of the temperature on the adsorption process was evaluated. An increase of temperature leads to activation of the metal ions and their binding by the adsorbent coordinating site is enhanced, so the metal uptake is favored. In these studies the maximum adsorption capacities for Zn^{2+} and Cu^{2+} onto the modified cellulose and corn cob, respectively, with phosphoric acid, was obtained at 70 °C when 1 g of biomass was used for an initial concentration of 500 ppm of metal ions.[11,13]

Another parameter, and perhaps the most important one because it offers information about the maximum adsorption capacities of the studied materials, is the influence of the initial concentration of the studied metal ions. In all the studies the experimental data regarding the variation of the initial concentration of the metal ions are fitted with the help of isotherm models. The most widely used isotherm and which is perfectly fitted in all cases is the Langmuir isotherm. In Table 11.1 are summarized the maximum adsorption capacities obtained from the Langmuir isotherm in cases of various removal

Table 11.1 The maximum adsorption capacities of various phosphorylated cellulose in the removal process of different metal ions from aqueous solutions.

Cellulose source	Phosphorylation conditions	Adsorbed metal ions	pH	q_{max} obtained from the Langmuir plot	Ref.
Bacterial cellulose	With phosphoric acid after promoted with urea and DMF	La^{3+}	3	0.91 mmol g^{-1}	2
Corn cob	POCl$_3$ after treatment with 7.5% NaOH solution	Zn^{2+}	$-^a$	100 µmol g^{-1b}	12
		Fe^{2+}		160 µmol g^{-1b}	
		Pb^{2+}		85 µmol g^{-1b}	
		Ni^{2+}		40 µmol g^{-1b}	
		Cr^{3+}		180 µmol g^{-1b}	
Corn cob	POCl$_3$ after treatment with 7.5 % NaOH solution. The corn cob was crosslinked using epichlorohydrin [2-(chloromethyl)oxirane]	Zn^{2+}	$-^a$	120 µmol g^{-1b}	12
		Fe^{2+}		180 µmol g^{-1b}	
		Pb^{2+}		90 µmol g^{-1b}	
		Ni^{2+}		50 µmol g^{-1b}	
		Cr^{3+}		190 µmol g^{-1b}	
Cellulose	With phosphoric acid	Cu^{2+}	3	11.54 mg g^{-1}	11
		Zn^{2+}		17.67 mg g^{-1}	
Corn cob	With phosphoric acid	Zn^{2+}	3	14.60 mg g^{-1}	13

^{a}The pH value is not mentioned.
^{b}The values of the adsorption capacities are obtained from the experimental data; the Langmuir isotherm was not studied in this article.

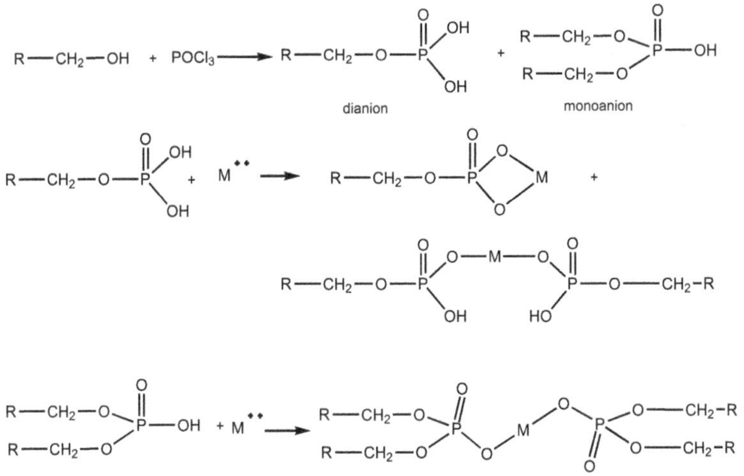

Scheme 11.2 The mechanism of metal ions adsorption onto phosphorylated cellulose.

processes of metal ions from aqueous solutions by different sources of cellulose phosphorylated under different conditions and at specific pH values of the medium. Nada *et al.* proposed a mechanism of metal ion adsorption onto phosphorylated cellulose[12] (Scheme 11.2).

11.2.2   Chitin and Chitosan

11.2.2.1   *Type of Phosphorus Pendant Groups Grafted onto Chitin and Chitosan*

Compared to synthetically substituted cellulose, chitin and chitosan contain a higher percentage of nitrogen (6.89%) and therefore present a commercial interest.[14] These two abundant natural polymers also have some limited reactivity and processability due to their solubility.[5,14] In order to eliminate this drawback, a lot of researchers have studied the effects of chemical modification of these natural polymers on their solubility and the application of these new derivatives. Chemical modification of chitin and chitosan to generate new biofunctional materials is of primary interest because, depending on the nature of the group introduced, such a procedure would not change the fundamental skeleton of chitin and chitosan and they could retain their original physicochemical and biochemical properties.[14–23] It was observed that the chelating properties of chitin and chitosan were increased and their solubility was modified by the introduction of phosphonic acid or phosphonate groups, respectively, onto these molecules.[5,14–23] Therefore, owing to the advantages of chitin and chitosan phosphorylation, a review in 2006 by Jayakumar *et al.* systematized the related issues regarding the various methods of preparation of phosphorylated chitin and chitosan and the potential applications of these compounds in various fields such as metal ion adsorption, drug delivery, fuel cells, blood compatibility, food, and tissue engineering.[14]

There are several techniques to obtain phosphate derivatives of chitin and chitosan. Some researchers studied the ionic interaction between the positively charged amino groups and a negatively charged counter ion, tripolyphosphate (TPP), to prepare chitosan beads.[5,15,16] By using TPP as a possible crosslinking agent for the natural polymer phosphorylation, there is obtained nontoxic, porous chitosan–TPP beads. TPP interacts with chitosan through electrostatic forces. The protonated amine groups in chitosan will interact with the negatively charged counter ion, TPP, through an ionic interaction and create ionic crosslinked networks. This interaction can be controlled by the charge density of TPP and chitosan, which is dependent on the pH of the solution.[5,15–20] Laus *et al.* obtained crosslinked chitosan by the phase inversion method. Before the ionic crosslinking of chitosan with triphosphate, they incorporated in the chitosan molecule epichlorohydrin [2-(chloromethyl)oxirane], *via* a covalent crosslinking reaction.[5] Kim *et al.* studied in this work the phosphorylation of chitin using urea and phosphoric acid.[21] This method is also mentioned in the review written by Varma in 2004.[22] Varma also stated that Nishi *et al.* achieved in 1986 the phosphorylation of chitosan through reaction with phosphorus pentoxide in methanesulfonic acid. Jayakumar *et al.* studied in 2009 the alginate/phosphorylated chitin blend film in the removal process of Ni^{2+}, Zn^{2+}, and Cu^{2+} from aqueous solutions. They prepared the alginate/P-chitin in two steps.

In the first step the phosphorylation of chitin was achieved with a mixture of P_2O_5, H_3PO_4, Et_3PO_4, and hexanol, and then the P-chitin was mixed with alginate in the weight ratio $1:1$.[23]

11.2.2.2 Metal Ion Adsorption onto Chitin and Chitosan Grafted with Different Phosphorylated Groups

All researchers mention in their work that the pH of the aqueous solution is an important parameter in the adsorption processes; therefore the influence of the pH on the adsorption process of the metal ions was studied. In studies of lead removal from aqueous solutions through adsorption onto phosphorylated chitin and chitosan, pH values around 5 were always used, because at higher pH values the lead ions gradually precipitate.[5,16,21] The effects of pH upon copper ion adsorption onto various phosphorylated chitin and chitosan was studied by Ngah and Fatinathan, Laus *et al.*, and Jayakumar *et al.*[5,16,23] Ngah and Fatinathan worked at a pH value around 5, but the other researchers stated that the optimum pH for Cu^{2+} removal was almost 6. In the case of Cd^{2+} ion removal, Laus *et al.* found that the optimum pH value was 7.[5] Besides Cu^{2+} ion removal, Jayakumar *et al.* studied the removal of Ni^{2+} and Zn^{2+} ions through adsorption onto alginate/P-chitin. For both the studied metals ions the optimum pH was around 7. Therefore it can be concluded that in all the adsorption processes of metal ions from aqueous solutions through adsorption onto phosphorylated chitin and chitosan, the metal uptake increases with an increase of the pH of the solutions, but this increase should not reach the pH value at which precipitation of the metal ions occurs.

Different equilibrium times were achieved in the removal process of Cu^{2+} ions from aqueous solutions. When chitosan (CTS)–epichlorohydrin (ECH)–TPP is used as adsorbent the equilibrium between the adsorbent and Cu^{2+} ions is achieved in 30 h.[5] In the removal processes of Cu^{2+} ions through adsorption onto chitosan–tripolyphosphate beads the equilibrium time was achieved in 24 h when the initial concentration of Cu^{2+} was at 300 mg L^{-1}.[18] Jayakumar *et al.* achieved a lower equilibrium time (5 h) in studying the process of Cu^{2+} ion removal onto alginate/P-chitin blend film.[23] The most efficient equilibrium time from the economic point of view (100 min) was obtained by Ngah *et al.* when chitosan tripolyphosphate (CTPP) was used as adsorbent for Cu^{2+} and Pb^{2+} ions.[16] For lead ions, almost the same equilibrium time was obtained by Kim *et al.* through adsorption onto chitin phosphorylated with urea and phosphoric acid,[21] in comparison with the highest equilibrium time of 72 h obtained by Laus through adsorption of Pb^{2+} onto CTS–ECH–TPP. In the same study an equilibrium time of 10 h was obtained when Cd^{2+} ions were in contact with CTS–ECH–TPP.[5] The uptake of Ni^{2+} and Zn^{2+} ions onto alginate/P-chitin increases with contact time until 4 h of contact, when equilibrium is reached.[23]

Kinetic models were used to determine the adsorption kinetics and rate-limiting steps for the adsorption of Pb^{2+}, Cu^{2+}, and Cd^{2+} onto CTPP beads and CTS–ECH–TPP.[5,15,16] In all the studies the pseudo-first-order and pseudo-second-order kinetic models were used. It was observed that in the case of the pseudo-second-order kinetic model the obtained correlation co-efficients are close to unity and the calculated equilibrium adsorption capacity values are very close to the values obtained experimentally.[5,15,16] Therefore it can be concluded that the adsorption of metal ions onto the phosphorylated chitin and chitosan is actually a chemisorption, which involves valence forces through sharing or exchange of electrons between the metal ions and the adsorbent. These two kinetic models cannot predict or confirm any particular adsorption mechanism. Therefore the intraparticle diffusion model was studied, owing to the fact that it is well known that the adsorption process of the metal ions onto a porous material is mainly controlled by a multi-step process. These steps involve the transport of the metal ions from the bulk solution, film diffusion, intraparticle diffusion, and finally metal ion adsorption onto the sites. The intraparticle diffusion model plot of metal ion adsorption onto CTPP showed two interdependent linear lines. The first initial linear portion represents the intraparticle diffusion while the second one corresponds to the equilibrium state. The straight line does not pass through the origin, indicating that intraparticle diffusion is not the sole rate-limiting step in the removal process of metals ions through adsorption onto CTPP beads.[15,16]

Owing to the fact that CTPP was so efficient in the removal process of Pb^{2+} and Cu^{2+} ions from aqueous solution, Ngah and Fatinathan also studied the effect of different temperatures on the adsorption process, as the thermo-dynamic parameters would be helpful in the practical application of the process. It was observed that the adsorption process of Pb^{2+} and Cu^{2+} ions onto CTPP beads was spontaneous and endothermic in nature. Kim *et al.* and Wu *et al.* also studied the temperature dependence of the equilibrium con-stant. In their studies, the same negative values of the Gibbs free energy (ΔG°) were obtained, indicating that the adsorption process is both thermo-dynamically favorable and spontaneous under experimental conditions. Also, positive values for the enthalpy change (ΔH°), indicating the endothermic nature of the adsorption process of metal ions onto the studied materials, and the obtained positive values of the entropy change (ΔS°) show the affinity of the studied materials for the metal ions due to an irregular increase in ran-domness at the adsorbent/solution surface.[15,21] These thermodynamic par-ameters are practical indicators for the applicability of an adsorption process.

Beside kinetic and thermodynamic studies, equilibrium studies were done in order to determine the influence of the initial concentration of the metals ions upon the efficiency of the adsorption process and in order to determine the maximum adsorption capacities of the phosphorylated chitin and chit-osan materials in the removal process of various metal ions from aqueous solutions. The experimental equilibrium data were fitted to the Langmuir and Freundlich isotherms. In all the cases the experimental results showed a better fit to the Langmuir than to the Freundlich equation. The maximum adsorption capacities obtained from the Langmuir isotherm plot in the

Table 11.2 The maximum adsorption capacities of various phosphorylated chitins and chitosans in the removal process of different metal ions from aqueous solutions.

Material	Phosphorylation conditions	Adsorbed metal ions	pH	q_{max} obtained from the Langmuir plot	Ref.
Phosphorylated chitin	With urea and phosphoric acid	Pb^{2+}	4	248 mg g^{-1}	21
Alginate/ phosphorylated chitin blend film	With P_2O_5, H_3PO_4, Et_3PO_4, and hexanol mixture	Ni^{2+} Zn^{2+}	7 7	5.67 mg g^{-1a} 2.85 mg g^{-1a}	23
	The crosslinking between P-chitin and alginate was realized with $CaCl_2$ solution	Cu^{2+}	6	11.7 mg g^{-1a}	
Chitosan– tripolyphosphate beads (CTPP)	With sodium tripolyphosphate	Pb^{2+} Cu^{2+}	4.5	57.33 mg g^{-1} 26.06 mg g^{-1}	16
Porous chitosan– tripolyphosphate beads	With sodium tripolyphosphate solution	Cu^{2+}	5	208.3 mg g^{-1}	15
Non-porous chitosan– tripolyphosphate beads	With sodium tripolyphosphate solution	Cu^{2+}	5	196.1 mg g^{-1}	15
Chitosan crosslinked with epichlorohydrin– triphosphate (CTS– ECH–TPP)	Phase inversion method	Cu^{2+} Cd^{2+} Pb^{2+}	6 7 5	130.72 mg g^{-1} 83.75 mg g^{-1} 166.94 mg g^{-1}	5

^{a}The adsorption capacities are obtained from the values of the experimental data; the Langmuir isotherm was not studied in this article.

various removal processes of metal ions from aqueous solutions through adsorption onto phosphorylated chitin and chitosan derivatives at specific pH values of the medium are summarized in Table 11.2. One may notice that the maximum adsorption capacities are obtained in the removal process of metal ions from aqueous solutions, through adsorption onto phosphorylated chitin and chitosan, prepared through the simple phosphorylation method with urea and phosphoric acid or tripolyphosphate solutions. Ngha *et al.* have proposed a binding mechanism for the adsorption of metal ions onto CTPP beads (Scheme 11.3).[16]

In order to recover the adsorbed metal ions and to regenerate the adsorbent for subsequent materials, various eluents at different concentrations were used for the desorption studies. The eluents were H_2O, HNO_3, H_2SO_4, HCl, KCl, NH_4Cl, and EDTA solutions. Ngah and Fatinathan[16] showed that Pb^{2+} and Cu^{2+} ions adsorbed on CTPP beads can be easily and effectively desorbed using 0.1, 0.01, or 0.001 M EDTA solutions. Laus *et al.* demonstrated that the best desorption performance of Cu^{2+}, Cd^{2+}, and Pb^{2+} from CTS–ECH–TPP was obtained with HNO_3 or HCl as eluent.[5]

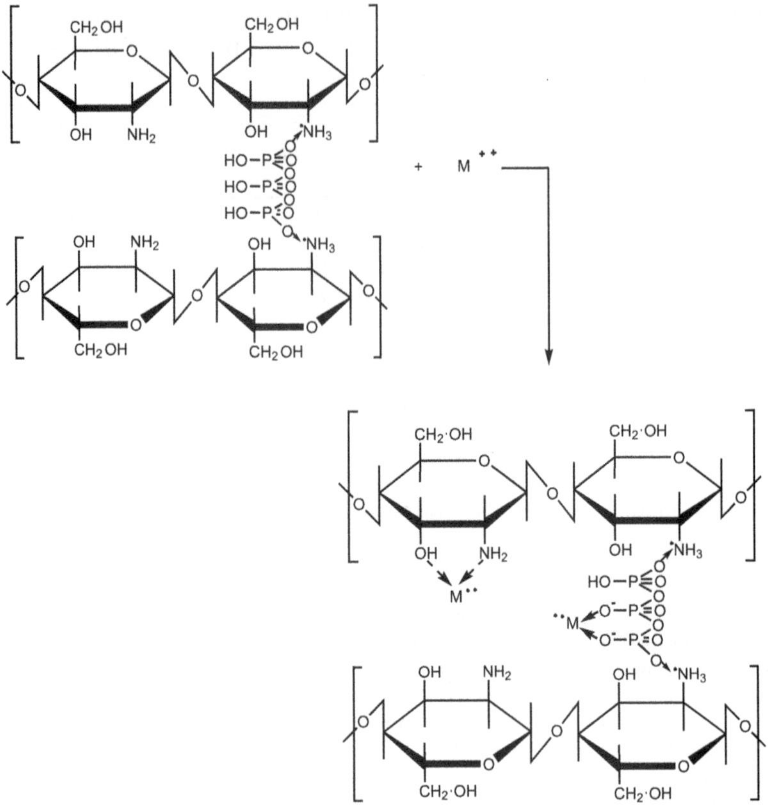

Scheme 11.3 The binding mechanism for metal ion adsorption onto chitosan-tripolyphosphate beads.

11.3 Synthetic Polymers

11.3.1 Styrene–Divinylbenzene Copolymer

11.3.1.1 Type of Phosphorus Pendant Groups Grafted onto Styrene–Divinylbenzene Copolymer

For the removal of metal ions from aqueous solutions, styrene–divinylbenzene copolymers (St-DVB) have been also used as a solid support. In order to enhance the adsorption capacities of St-DVB in the removal process of metal ions from aqueous solutions, it was grafted with different phosphorus pendant groups alone or in combination with other functionalized groups.

Polymers bearing a functional group based on diisobutylphosphine sulfide and containing different spacer arms between the polymeric matrix and the phosphorus functional group were synthesized by Sanchez and co-workers[24–27] from chloromethylated polystyrene (crosslinked with 4% DVB)

(Scheme 11.4). This was done in order to determine the sorbents' selectivity towards gold and palladium from other noble (Pt, Rh, and Ir) and base (Fe, Cu, Ni, and Zn) metals.

Another way to introduce a phosphorus pendant group onto the St-DVB copolymer was with the use of the Michael reaction (Scheme 11.5), when different kinds of substituted phenylphosphonic acid ligands were obtained.[10,28]

Trochimczuk obtained several phosphorus pendant groups, *i.e.* phosphonate esters, phosphinate esters, and phosphine oxide ligands,

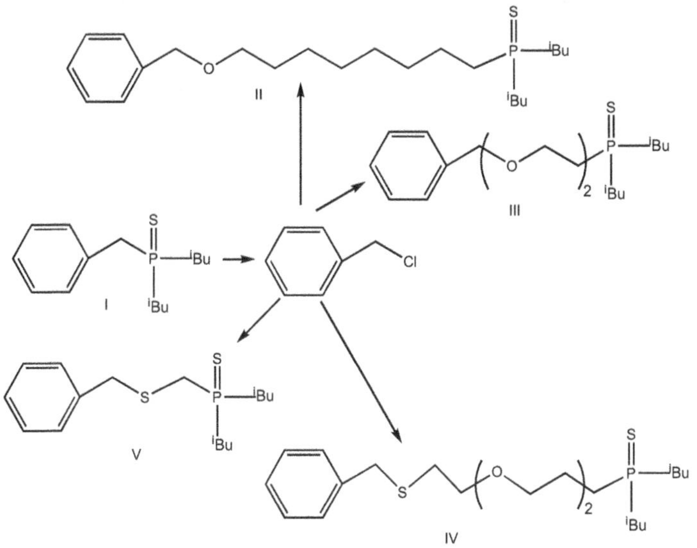

Scheme 11.4 Polymers with the diisobutylphosphine sulfide group and containing different spacer arms.

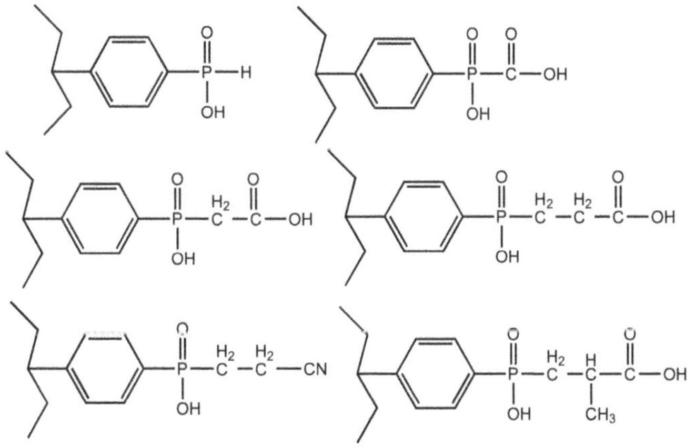

Scheme 11.5 Functionalized phenylphosphinic acid resins.

functionalized on polymeric resins through the Arbuzov reaction.[29] The Arbuzov-type reaction was also studied by Popa *et al.*[30] They prepared some chelating resins through the reaction between chloromethyl polystyrene–divinylbenzene copolymers and triethyl phosphite, yielding the phosphonate ester resin (a) (Scheme 11.6). This was further hydrolyzed by HCl to yield the phosphonate/phosphonic acid resin (b).[30] Polymeric resins with phosphonate esters (a) were also synthesized *via* the Michaelis–Becker reaction and used in the removal of Cu^{2+} from aqueous solutions.[31]

Alexandratos and Zhu used as phosphorus pendant groups a phosphonate diester (PDE), a phosphorylated ethylene glycol (PEG), and a phosphorylated pentaerythritol (PPE) in order to functionalize the polystyrene resins (Scheme 11.7).[32]

They also studied the effect of glycol units, $-(CH_2CH_2O)_{1-4}-$, *versus* phosphate ligands immobilized on crosslinked polystyrene beads.[33] β-Ketophosphonic acid, phosphonoacetic acid, and phosphonic acid polymers (Scheme 11.8) with moieties between carbonyl and phosphoryl groups were studied.[34]

Another possible source of a phosphorus pendant group was [(2-dihydroxyarsinoylphenylamino)methyl]phosphonic acid, which was used to modify the Amberlite XAD-16 polymeric matrix.[35] Subramanian *et al.* also modified the polystyrene–divinylbenzene copolymer beads (Amberlite XAD-16) by anchoring 6,6,6-trifluoro-2,5-dioxo-4-(thiophene-2-carbonyl)hexyl-phosphinic acid in its matrix in order to obtain a new chelating ion-exchange multidentate grafted polymer (Scheme 11.9).[36]

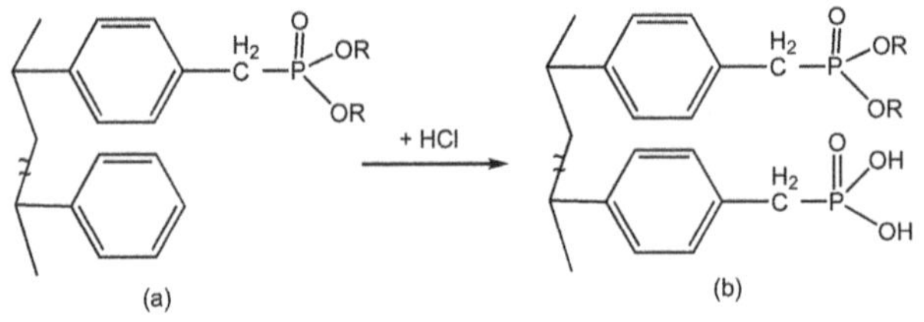

Scheme 11.6 Polymers with phosphonate/phosphonic groups.

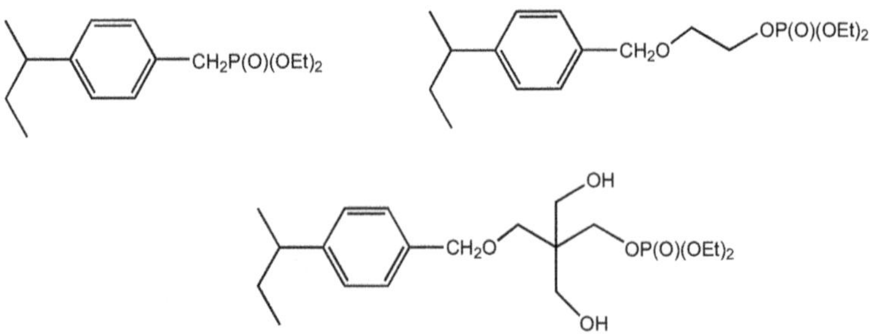

Scheme 11.7 Polymers functionalized with ion-complexing reagents.

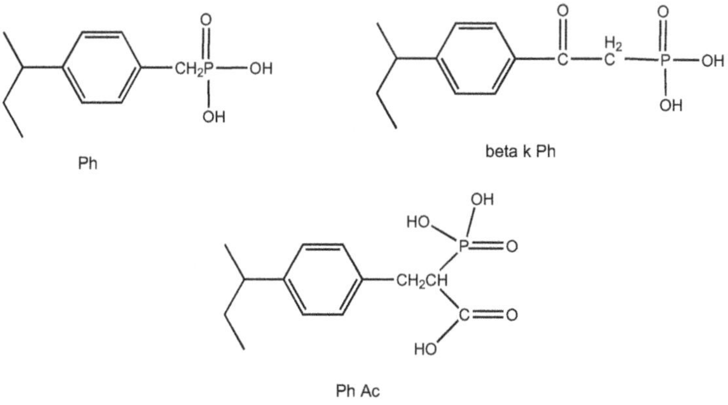

Scheme 11.8 Polymers with carbonyl and phosphoryl groups.

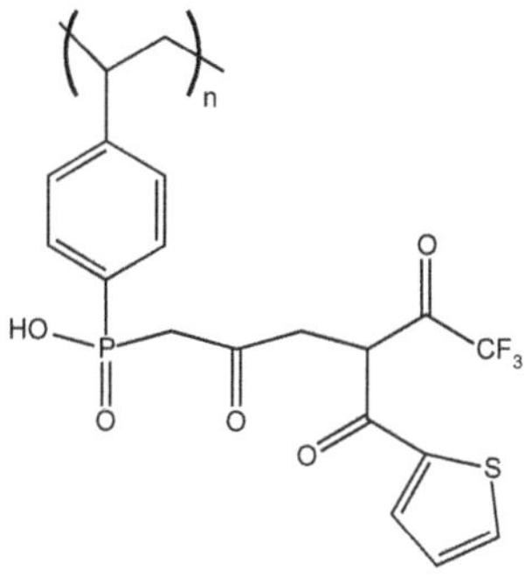

Scheme 11.9 Multidentate grafted polymer with 6,6,6-trifluoro-2,5-dioxo-4-(thiophene-2-carbonyl)hexylphosphinic acid.

Also, they prepared a new material by grafting Merrifield chloromethylated resin with octyl(phenyl)[(N,N-diisobutylcarbamoyl)methyl]phosphine oxide (Scheme 11.10).[37]

A new solid-phase extractant was obtained using Amberlite XAD-16 functionalized with [(3,4-dihydroxybenzoyl)methyl]phosphonic acid (Scheme 11.11).[38]

Also, a new chelating ion-exchange polymer was obtained using Amberlite XAD-16 as the inert polymeric support chemically modified with [(N,N-dihexylcarbamoyl)methyl]phosphonic acid by Maheswari *et al.* (Scheme 11.12).[39]

Bifunctional groups were also used to functionalize the St-DVB copolymer. The bifunctional polymers with two different ligands functionalized on the support were synthesized[40] in order to determine whether synergistic interactions between the ligands could lead to enhanced metal ion affinities and selectivities. These polymers were named "dual mechanism bifunctional polymers" (DMBPs) (Scheme 11.13).[40]

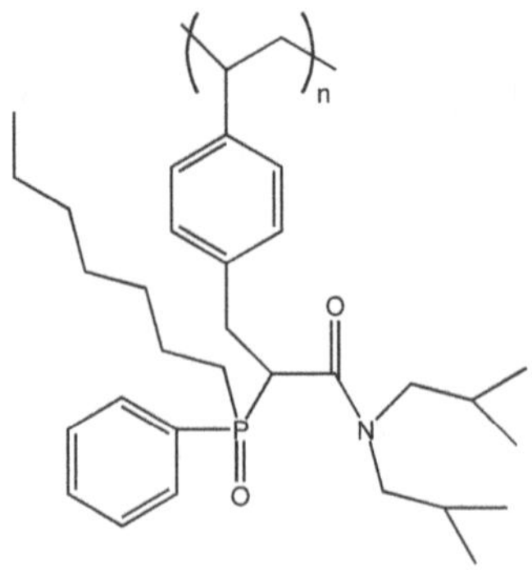

Scheme 11.10   The chloromethylated resin with octyl(phenyl)[(*N,N*-diisobutylcar-bamoyl)methyl]phosphine oxide.

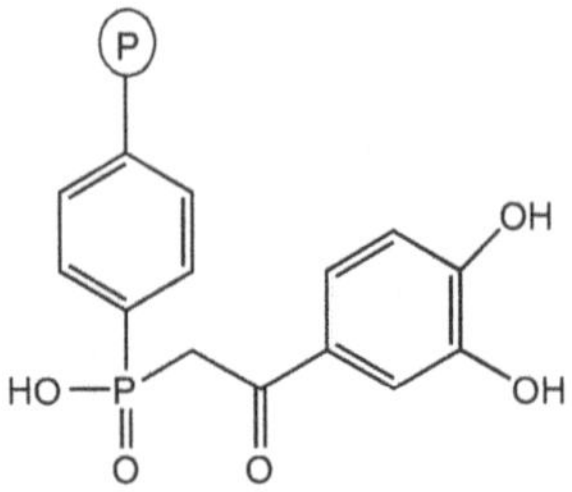

Scheme 11.11   Amberlite functionalized with [(3,4-dihydroxybenzoyl)methyl]phos-phonic acid.

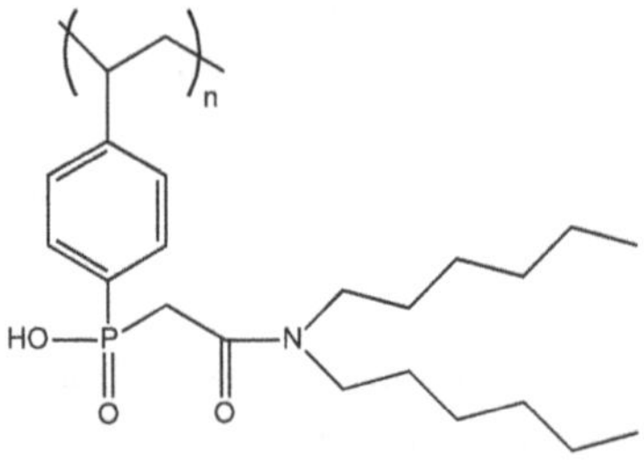

Scheme 11.12   Polymer support modified with [(*N,N*-dihexylcarbamoyl)methyl]-phosphonic acid.

Chelating/ion-exchange resins, having the functional groups of both phosphonic acid and an N-substituted amide of a carboxylic acid, were synthesized using amination of the styrene–divinylbenzene copolymer bearing (ethoxycarbonylmethyl)phosphonate groups with ethylenediamine and diethylenetriamine.[41] A series of bifunctional aminomethylphosphonic acid resins was obtained *via* reaction with formaldehyde and phosphorous acid,[42] and Alexandratos *et al.* have used the Mannich reaction for the

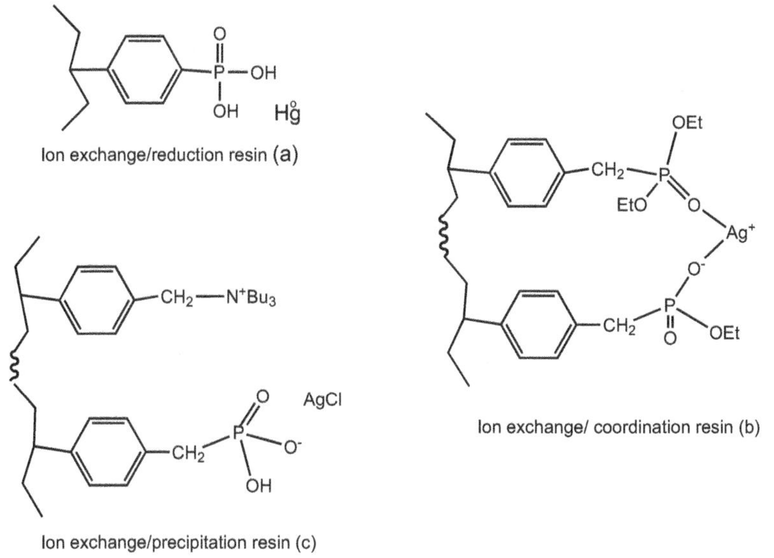

Ion exchange/reduction resin (a)

Ion exchange/coordination resin (b)

Ion exchange/precipitation resin (c)

Scheme 11.13   Bifunctional polymers.

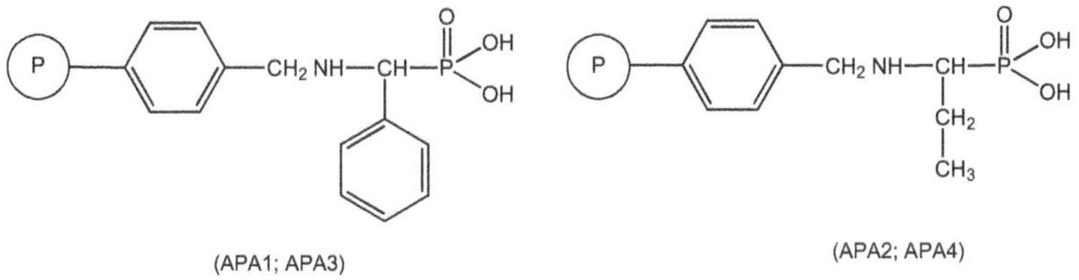

(APA1; APA3)

(APA2; APA4)

Scheme 11.14   Styrene–divinylbenzene copolymers functionalized with aminophosphonic acid pendant groups.

phosphorylation of amine resins with monoamine, ethylenediamine, diethylenetriamine, and tetraethylenepentamine ligands. The aminomethylphosphonic acid ligand was found to be a bifunctional ion exchange/coordination polymer showing a dual mechanism.[42] Jyo et al. have prepared a bifunctional chelating resin containing aminomethylphosphonate and sulfonate groups and a monofunctional aminomethylphosphonate resin by the Mannich condensation reaction and also by eliminating the sulfonation step.[43] The synthesis of poly[styrene-*co*-(1% or 15%)divinylbenzene] functionalized with aminophosphonic acid groups by "one-pot" reactions (Scheme 11.14) was described in 2013 by Popa et al.[44] They tested the aminophosphonic acid grafted on St-DVB copolymer as adsorbents in the removal process of Cu^{2+} ions from aqueous solutions.[45]

Also, they introduced the aminophosphinic acid ligands onto polystyrene through "one-pot" reactions (Scheme 11.15).[46]

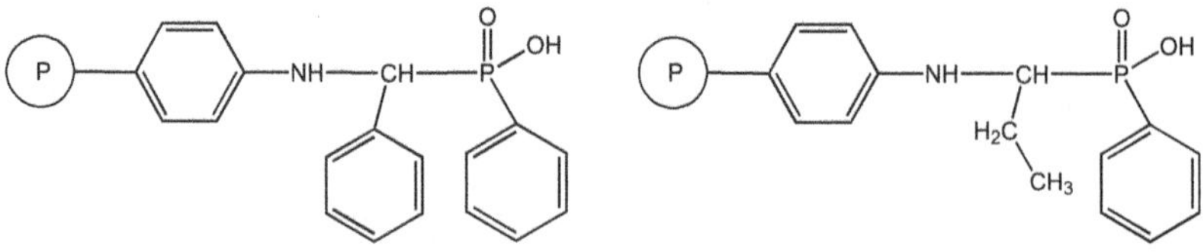

S2   S4

Scheme 11.15   The aminophosphinic acid groups grafted on a St-DVB copolymer.

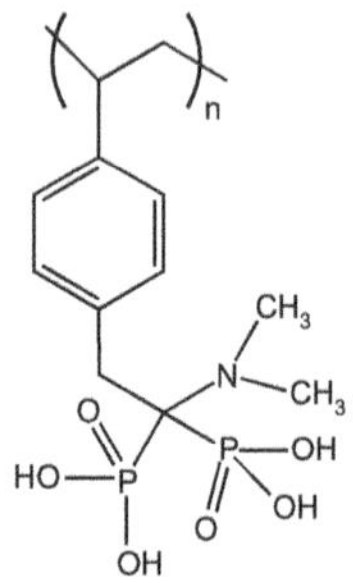

Scheme 11.16   Amberlite XAD-16 functionalized with DAPPA.

Raju *et al.* have synthesized a polyfunctional polymer containing diphos-phonic acid groups and dimethyl amino groups by grafting them onto the Merrifield chloromethylated resin (Scheme 11.16).[47]

11.3.1.2   Adsorption of Metal Ions onto Styrene–Divinylbenzene Copolymer Grafted with Different Phosphorus Pendant Groups

Sanchez *et al.* showed that the functionalization of different water-insoluble macroreticular polystyrene–divinylbenzene polymers with diisobutylphos-phine sulfide groups increased their selectivity towards Pd^{2+} and Au^{3+} ions.[24] The adsorption capacity of the studied materials depended on the length of the spacer arm and the presence of the heteroatoms (O or S) in these arms. Also, an increase in hydrophilicity leads to an increase in the chelating capacity of the resin. It allows better contact between the metal ions in solution and the chelating groups in the resin, and it sometimes affects the kinetics of the adsorption process, the resin capacity, and the selectivity of the process.[24] The experimental capacities obtained for Au^{3+} and Pd^{2+} in batch adsorption studies for the resin synthesized with diiso-butylphosphine sulfide groups are presented in Table 11.3 (below). Sánchez *et al.* have observed the best performance characteristics, in terms of affinity, optimum quantities, and selectivity to gold and palladium, where there are resins bearing spacer arms containing O {*e.g.* 2-[2-(2-ethoxy)ethoxy]ethane}

Table 11.3 The maximum adsorption capacities of various phosphorylated styrene–divinylbenzene copolymers in the removal of different metals ions from aqueous solutions.

Support material	Type of phosphorus groups grafted onto St-DVB	Adsorbed metal ions	pH	q_{max} obtained from the Langmuir plot	Ref.
DVB	Diisobutylphosphine sulfide	Pd^{2+}	$-^a$	0.7 mmol$\,g^{-1b}$	24
Polystyrene		Au^{3+}		6.5 mmol$\,g^{-1b}$	
VBC/DVB	$CH_2P(O)(OEt)_2$	Au^{3+}	$-^a$	120 mg$\,g^{-1b}$	30
	$CH_2P(O)(OBu)_2$			110 mg$\,g^{-1b}$	
VBC/St/DVB	1,1-Dicarboxylate-2-ethanephosphonate	Pb^{2+}	$-^a$	4.0 meq$\,g^{-1b}$	49
		Cu^{2+}		1.7 meq$\,g^{-1b}$	
		Cd^{2+}		1.1 meq$\,g^{-1b}$	
		Zn^{2+}		1.1 meq$\,g^{-1b}$	
		Ni^{2+}		1.0 meq$\,g^{-1b}$	
VBC/DVB	$(EtO)_2P(O)CH_2CH_2COOEt$	Cu^{2+}	2	0.51 mg$\,g^{-1b}$	50
	$(HO)_2P(O)CH_2CH_2COOEt$			1.00 mg$\,g^{-1b}$	
	$(EtO)_2P(O)CH_2CH_2COOH$			0.54 mg$\,g^{-1b}$	
	$(HO)_2P(O)CH_2CH_2COOH$			1.45 mg$\,g^{-1b}$	
	$(HO)_2P(O)CH_2CH_2P(O)(OH)_2$			>2.10 mg$\,g^{-1b}$	
St-1%DVB-gel type	Aminophosphonic acid group	Ni^{2+}	7.3	1.24 mg$\,g^{-1}$	38
St-15%DVB	Aminophosphonic acid group			1.52 mg$\,g^{-1}$	
Poly St-DVB	Aminophosphonic acid group	Cu^{2+}	$-^a$	0.97 mg$\,g^{-1}$	39
St-7%DVB	Triethyl phosphite	Ni^{2+}	7	19 mg$\,g^{-1b}$	36
Amberlite XAD-16	[(N,N-Dihexylcarbamoyl)methyl]phosphonic acid	U^{6+}	6.5	1.429 mmol$\,g^{-1b}$	46
		Th^{4+}	4	1.228 mmol$\,g^{-1b}$	
		La^{3+}	5	1.356 mmol$\,g^{-1b}$	
Amberlite XAD-16	[(2-Dihydroxyarsinoylphenyl-amino)methyl]phosphonic acid	U^{6+}	5	1.49 mmol$\,g^{-1b}$	59
		Th^{4+}	4	1.40 mmol$\,g^{-1b}$	
		La^{3+}	1–3	1.12 mmol$\,g^{-1b}$	
Merrifield resin	Phosphonic acid	U^{6+}	3.6	161.55 mg$\,g^{-1}$	60

^{a}The pH value is not mentioned.
^{b}The adsorption capacities are obtained from the experimental values; the Langmuir isotherm was not studied in this article.

or S and O heteroatoms in the spacer groups {*e.g.* 2 [2 (2 thioethyl)ethoxy]ethane}.[25] The resins synthesized by Sanchez *et al.* have a great affinity towards gold and palladium in hydrochloric acid media and no affinity for other noble metals. The authors showed the influence of different parameters such as flow rate and temperature under dynamic conditions and took into account the previous results obtained in batch conditions. From the experimental data presented in Table 11.3 it can be observed that the selectivity is higher in the case of gold than for palladium.[26] The polymer which has O atoms in the side chain is characterized by a lower ion exchange capacity (0.5 ± 0.1 mmol g^{-1} polymer) towards Pd^{2+} ions than the polymer which contains S atoms (0.7 ± 0.1 mmol g^{-1} polymer).[26,27,48]

Sánchez *et al.* observed that the increase of temperature from 298 to 333 K resulted in an increase in the sorption of Au^{3+} ions compared to that of Pd^{2+} ions. Thiourea and sodium nitrate were used as eluents. Sodium nitrate (2 mol L^{-1}, pH 4.7) desorbed 75% of Pd^{2+} and did not elute Au^{3+}, but thiourea (0.5 mol L^{-1}, $[H^+] = 1.0$ mol L^{-1}) allowed for almost complete recovery of Au^{3+} containing trace amounts of Pd^{2+} ions.[26,27,48]

Trochimczuk obtained an adsorption capacity of Au^{3+} from hydrochloric acid solutions of 110–120 mg of Au^{3+} per g of resin, using as adsorbent polymeric resins with phosphonate esters, phosphinate esters, and phosphine oxide ligands synthesized *via* the Arbuzov reaction.[29]

Phosphonated vinylbenzyl chloride (VBC)/styrene (St)/divinylbenzene (DVB) copolymers with functional acidic ligands were tested by Trochimczuk. They were used in nitric acid solutions in the removal process of Cu^{2+}, Ni^{2+}, Pb^{2+}, Zn^{2+}, and Cd^{2+} ions. Of these two resins, the one obtained with 1,1-diphosphonate-2-ethanecarboxylate (which has a higher content of phosphonate groups in its structure) presented the highest distribution coefficient. This is in accordance with the fact that this ligand has the most acidic character. The resins presented the following order of affinity for the studied metals ions: $Pb^{2+} > Cu^{2+} > Cd^{2+} > Zn^{2+} > Ni^{2+}$. In the removal process of the divalent ions, the highest adsorption capacity was obtained in the case of the 1,1-dicarboxylate-2-ethanephosphonate immobilized onto VBC/St/DVB. The results are presented in Table 11.3 (below). Because the studied resins showed high affinity towards divalent metal cations the authors investigated the efficiency of the studied copolymer in the removal process of divalent cations under dynamic conditions. It was observed that the second resin displayed a higher affinity for Pb^{2+}. The authors determined the selectivity and capacity of their resin towards divalent cations through a column study. These resins can be used in more cycles of adsorption of divalent cations and elution of the adsorbed divalent cations, because using 1 M nitric acid the striping of the metal ions in the resin is completed within 10 bed volumes. Therefore these two resins can be used in the removal of metal cations from aqueous solutions and in their preconcentration.[49]

Also, Trochimczuk *et al.* studied the effect of close proximity of acidic and basic ligand sites on chelating/ion-exchange of Cu^{2+}, Cd^{2+}, Ni^{2+}, and Zn^{2+} metal cations.[41] The obtained resins have the same kind of functional groups (phosphonic acid and *N*-substituted amide of a carboxylic acid) but deposited randomly within the polymer material. They found that close proximity of both acidic and basic sites, interacting with each other, gave antagonistic effect at pH 3.7 and 5.6 and suppressed uptake of Cu^{2+}, Cd^{2+}, Ni^{2+}, and Zn^{2+} from weakly acidic solutions. The resins $(EtO)_2$-$P(O)CH(CH_2)CH_2COOEt$ and $(HO)_2P(O)CH(CH_2)CH_2COOEt$ with carbamyl-ethylenephosphonates displayed at pH 3.7 a log K_d for Cu^{2+} equal to 2.10 and 3.06, whereas the reference resins $(EtO)_2P(O)CH(CH_2)CH_2COOH$ and $(HO)_2P(O)CH(CH_2)CH_2COOH$ had a log K_d for Cu^{2+} of 2.83 and 4.50, respectively.[41] The resin obtained by the reaction of $(EtO)_2P(O)CH_2CH_2COOEt$ with VBC/DVB was selectively hydrolyzed in order to obtain a series of chelating/cation-exchange resins. These resins had different combinations of oxygen-containing different groups (phosphonic acid, ethylphosphonate, carboxylic acid, and carboxylate ester) and the resulting materials were used in the process of removing Cu^{2+} ions from different concentrations of nitric acid solutions. It can be noticed that the type of ligand functionality has a strong influence on the amount of Cu^{2+} ions complexed and/or exchanged by the ligand.[50] The Cu^{2+} uptake from 0.01 M nitric acid solutions by selectively hydrolyzed resins is presented in Table 11.3 (below).

Resins with carboxyl groups in the α and β positions display higher divalent metal uptake when the pH of the solution is above 1.5. Resins with either a carboxyl or a nitrile group in the γ position are less effective in metal ion uptake than the parent, phenylphosphinic, polymer.[28] It was found that resins in which the phosphinic functionality or both phosphinic and carboxylic functionalities were in acid form displayed a high affinity towards In^{3+} and Ga^{3+}.[10]

Alexandratos proposed the use of dual mechanism bifunctional polymers (DMBPs), which actually is a support functionalized with two different ligands, in order to increase the ionic selectivity and adsorption velocity in the removal process of metals ions from aqueous solutions.[40] An efficient bifunctional polymer used in the removal process of U^{6+} ions from 1 M nitric acid solution is a resin which contains both diphosphonic and sulfonic acids. This resin showed a distribution coefficient more than 100 times higher than the distribution coefficient of the corresponding monofunctional polymer. This bifunctional polymer has been studied intensively and commercialized (as DIPHONIX resin). In a review in 1997, owing to its higher efficiency, Chiariza *et al.* emphasized its properties and applications in the removal process of metal cations from a number of aqueous solutions. They also emphasized its application in the actinide separation field, in drinkable water purification, in the preparation of special purity reagents for high-tech applications, and in hydrometallurgy and electro-wining processes.[51] A paper by Alexandratos and Natesa explains the fact that bifunctionality is an important alternative for enhancing the metal ion separation process by a macroporous resin.[52] They made a comparison between the efficiency of the bifunctional phosphonic/sulfonic acid resin and the efficiency of the monofunctional phosphonic acid resin in the removal process of Eu^{3+} and Fe^{3+} from nitric acid solutions, explaining the principle of bifunctionality.[52] Also, Eu^{3+} ions were removed by a bifunctional polymer obtained through the immobilization of the sulfonic acid group onto polystyrene which was previously grafted with phosphonic acid ligands through the Arbuzov reaction,[53] or onto VBC/St/DVB functionalized with the sodium salt of tetra-isopropyl methylenediphosphonate (R_4MDA).[54] In order to obtain a maximized metal ion uptake, it is important to have an equilibrium between the chemical interaction and physical parameters, but it was observed that the bifunctional resin complexes a two to eight times higher amount of Eu^{3+} compared to the hydrolyzed resins.[53,54] Another bifunctional polymer which was studied is the aminomethylphosphonic acid resin, where the ligand was immobilized *via* the Mannich reaction. The metal ion complexing abilities of this bifunctional resin were compared with those of amine and phosphonic acid resins in the removal process of Cu^{2+}, Cd^{2+}, Pb^{2+}, and Eu^{3+} from nitric acid solutions. These resins were evaluated using buffered solutions at pH 5, in order to obtain reproducibility in the amount of complexed metal ion. A synergistic action is observed when both groups operating together complex more metal ions than either one alone. The ion exchange action of the phosphonate segment and the coordination action of the amine segment work together as a dual mechanism for metal ion complexation.[42]

The phosphonic and phosphoric acid ligands have a high affinity for metal ions in acidic solutions due to ion exchange, but when the phosphorylation of a resin is realized using the phosphate diester ligand, the affinity for metal ions in acid solutions is very low. This is because the interaction between the metal ion and the diester ligand is realized through coordination at the phosphoryl oxygen.[33] Therefore Alexandratos and Zhu studied how the ion-complexing ability of the phosphate diester ligand could be increased by adding glycol units, which have a far greater affinity for metal ions than diester ligands alone.[33]

Popa *et al.* studied the adsorption of Ni^{2+} and Cu^{2+} onto poly(styrene-*co*-divinylbenzene) functionalized with aminophosphonic acid groups.[44,45] They conducted the study at a pH value lower than the precipitation pH value of the studied divalent ions. The effects of time, temperature, and initial ion concentration on the adsorption efficiency were investigated. They made a comparison between the adsorption efficiency of aminophosphonic acid groups grafted onto two synthesized macroporous copolymers *versus* two commercial copolymers having a gel structure, in the removal process of Cu^{2+} and Ni^{2+} from aqueous solutions. By applying kinetic models to the experimental data it was found that the adsorption of Cu^{2+} and Ni^{2+} ions onto the studied materials is described by the pseudo-second-order kinetic model. The temperature dependence of sorption reveals the increase in sorption performance of the sorbent with temperature. For both metal ions which were studied, the equilibrium sorption data were modeled using Freundlich and Langmuir isotherms. The Langmuir model provided an excellent fit of the equilibrium adsorption data. Also it was observed that by the functionalization of the macroporous copolymer, the maximum adsorption capacities obtained were close to (or even higher than) those achieved by the functionalized commercial copolymer in the removal process of Ni^{2+} and Cu^{2+} ions from aqueous solutions (Table 11.3).[44,45] Therefore they also studied these synthesized macroporous copolymers in the removal process of the same divalent ions from aqueous solutions, but this time using for the functionalization the aminophosphinic acid ligands.[46] The adsorption experiments were conducted in the same way, and it was concluded that the aminophosphinic acid polymeric resins synthesized readily by "one-pot" polymer-analogous reactions are highly effective in the removal of Cu^{2+} and Ni^{2+} ions. Thus this resin format has good potential for technological use.[46] The same group of researchers studied the removal of

Cu^{2+}, Ca^{2+}, and Ni^{2+} ions from aqueous solutions using a simple phosphonate resin and bifunctional phosphonate/phosphonic acid resins, which were functionalized this time with triethyl phosphite. It was observed that the studied divalent ions were retained by the bifunctionalized resin at pH 7.[30]

Jyo *et al.* highlighted in their work the influence of the type of the group grafted (functionalized) on a macroporous polymer upon the metal ion selectivity.[55] Therefore when the phosphonic acid group was attached to phenyl groups of the copolymer matrix they obtained the following overall metal ion selectivity sequences: $Mo^{6+} \approx Fe^{3+} \approx U^{6+} \approx Bi^{3+} \approx Lu^{3+} > Al^{3+} \approx Gd^{3+} > La^{3+} > Cr^{3+} > Pb^{2+} > Mn^{2+} \approx Cd^{2+} \approx Cu^{2+} > Ca^{2+} \approx Co^{2+} \approx Zn^{2+} \approx Ba^{2+} \approx Sr^{2+} > Ni^{2+} \approx Mg^{2+}$.[55] When a macroreticular methylphosphonic acid resin was used as the adsorbent for metal ions removal from a nitric solution, the following decreasing order of metal ion affinity was achieved: $Fe^{3+} \approx U^{6+} \approx Mo^{6+} > Bi^{3+} > Al^{3+} > Gd^{3+} > La^{3+} \approx V^{5+} > Pb^{2+} > Cd^{2+} > Cu^{2+} \approx Ca^{2+} \approx Ba^{2+} \approx Zn^{2+} > Mg^{2+} \approx Co^{2+} \approx Ni^{2+}$.[56] They also made a comparison between a bifunctional chelating resin containing aminomethylphosphonate and sulfonate groups with a monofunctional aminomethylphosphonate resin. For the bifunctional polymer the increasing order of metal ion selectivity was as follows: $Ni^{2+} < Zn^{2+} < Cu^{2+} \approx Mg^{2+} < Ca^{2+} < Pb^{2+} < La^{3+} < Gd^{3+} < Lu^{3+} < Fe^{3+}$ and for the monofunctional polymer the increasing selectivity of the metal ions was: $Zn^{2+} < Ni^{2+} < Ca^{2+} < Mg^{2+} < Pb^{2+} < Cu^{2+} < La^{3+} < Gd^{3+} < Lu^{3+} < Fe^{3+}$.[43] In order to show the influence of the type of the phosphorylated group which is grafted on a solid support they studied three macroreticular chelating resins containing phosphinic, phosphonic, or methylene phosphonic groups in the removal process of Cr^{3+} from nitric acid. The highest efficiency was developed by the resin grafted with methylenephosphonic groups, followed by the resin containing phosphonic groups, and the least efficient was the resin containing phosphinic group. These results were in agreement with the cation exchange capacity of the resins.[57]

Souza *et al.* studied the influence of the degree of porosity of different sorbents based on styrene–divinylbenzene copolymers containing a phosphoryl group in the removal process of Pb^{2+} from aqueous solutions. The authors explained that the porous structure and the swelling capacity of a polymer influence the phosphorylation process of the resin matrix and therefore influence the accessibility of the metal ions to the introduced groups.[58] The authors obtained different degrees of porosity in the copolymers of styrene and DVB by their treatment with varied mixtures of toluene and *n*-heptane. It was observed that increasing the *n*-heptane portion lead to a slight increase in the degree of porosity of the copolymers, and further the increase in porosity lead to an increase in the incorporation of phosphorus into the network structure. Also the increase in the porosity provoked a decrease in the complexation capacity, which can be explained by the fact that the introduction of the phosphoryl groups into St-DVB copolymers increases the swelling capacity in water.[58]

The maximum adsorption capacities obtained by the styrene–divinylbenzene copolymer grafted with different functionalized groups in the removal process of various metal ions from aqueous solutions are presented in Table 11.3.

11.3.2   Other Types of Polymers

Abderrahim *et al.* studied the selectivity of a polymer functionalized with [(polyethylenimine)methylene]phosphonic acid (which possessed phosphonic and amine moieties as chelating groups) (Scheme 11.17).[61] They tested its selectivity/capacity to recover various metal ions from a solid waste which had resulted from the electrolysis of zinc. The solid waste was dissolved in nitric acid, at different pH levels and then the polymer was added.

The selectivity of the resin is strongly influenced by the initial pH of the medium and the results varied according to the equilibrium pH as follows: $Cu^{2+} < Cd^{2+} < Pb^{2+} < Ni^{2+} < Fe^{3+}$ at $pH_e = 2.4$, $Ni^{2+} < Co^{2+} < Zn^{2+} < Cd^{2+} < Cu^{2+} < Pb^{2+}$ at $pH_e = 3.6$. The studied functionalized resin presented the highest selectivity for Fe^{3+} ions at $pH_i < 3.7$, and the lowest selectivity in the case of Co^{2+} and Zn^{2+} ions at $pH_i < 5$.[61] Because of the efficiency of this polymer when functionalized with methylenephosphonic acid, the authors made a more detailed study regarding the removal of copper ions from aqueous solutions. In this work the following factors were studied: the effect of the agitation time, the solution pH, the effect of the initial Cu^{2+} concentration, the effect of ionic strength, and the effect of the copper salt on the adsorption kinetics.[62] Based on the reported results the [(polyethylenimine)methylene]phosphonic acid can be used as an effective extractant polymer for removing Cu^{2+} from dilute contaminated water.[62] The same detailed study was carried out by the same authors on the removal of U^{4+} from a nitrate aqueous media[60] and on the extraction of Cd^{2+} in a batch process[63] through adsorption onto a Merrifield resin grafted with phosphonic acid. In this case, a phosphonic group, triethyl phosphite, was used as a source, and in the second step the phosphonate-grafted polystyrene resin beads were treated with hydrobromic acid.[60,63] The presented results indicate that the sorption of metal ions depends on the operational parameters. However, the studied Merrifield resin grafted with phosphonic acid was found to be feasible for the removal of metal ions from wastewaters.

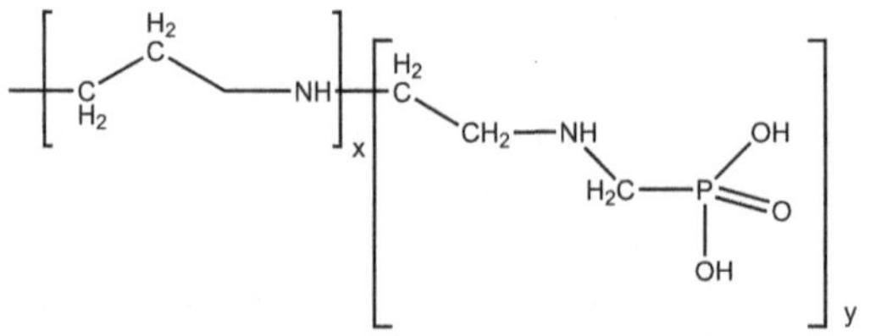

Scheme 11.17   The [(polyethylenimine)methylene]phosphonic acid.

Ali *et al.* synthesized a crosslinked polyzwitterionic acid through cyclocopolymerization of diallylaminomethylphosphonic acid and 1,1,4,4-tetraallylpiperazinium dichloride (10 mol%).[64] The polyzwitterionic acid was transformed into a crosslinked anionic polyelectrolyte (CAPE). The adsorption capacity of Pb^{2+} (7.50 mmol g^{-1}) was greater than that of Cu^{2+} (6.05 mmol g^{-1}). The experimental data for the adsorption of Pb^{2+} and Cu^{2+} on CAPE fitted the Lagergren second-order kinetic model as well as the Langmuir and Freundlich isotherm models. The adsorption process was spontaneous and endothermic in nature, with negative and positive values for ΔG and ΔH, respectively.[64]

Crosslinked styrene (St)/maleic anhydride (MA) copolymers have been synthesized, hydrolyzed with dicarboxylic acid, and converted to bear dihydroxyphosphino functionalities.[65] On the dihydroxyphosphino-functionalized styrene–methyl acrylate (20% divinylbenzene) copolymer, the adsorption of Pb^{2+} showed a linear relationship with the concentrations and fitted the Langmuir isotherm. The kinetics of Pb^{2+} adsorption on this dihydroxyphosphino-functionalized copolymer were studied. The metal-ion adsorption kinetics of this copolymer appeared to show particle diffusion. They explained that the moving boundary advanced from the surface of the molecule towards the center.[65]

Nonaca *et al.* prepared copolymer beads having phosphinic acid groups by suspension copolymerization of [(acryloyloxy)propyl](*n*-octyl)phosphinic acid, *N*-isopropylacrylamide, and tetraethylene glycol dimethacrylate (Scheme 11.18).[66]

The copolymer beads having phosphinic acid groups showed higher adsorption capacity for lanthanide metal ions (Eu^{3+}, Sm^{3+}, Nd^{3+}, or La^{3+}) than for main transition metal ions (Cu^{2+}, Ni^{2+}, or Co^{2+}). In each single metal ion solution the pH was measured. Values ranged from 3.5 to 6.5 at 20 °C. The selective adsorption for lanthanide metal ions was as follows: $Eu^{3+} > Sm^{3+} > Nd^{3+} > La^{3+}$.[66]

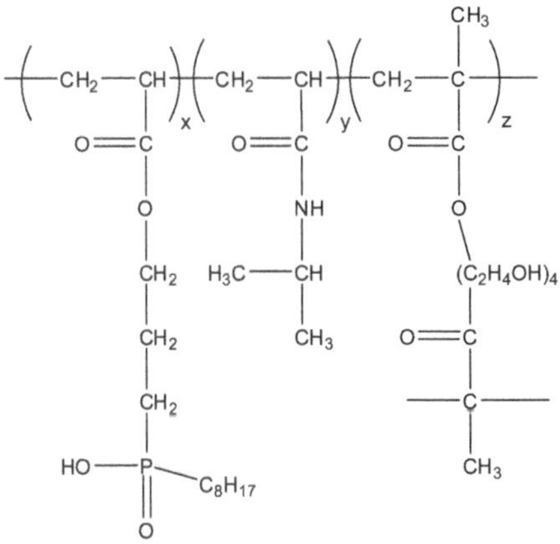

Scheme 11.18 Copolymer beads having phosphinic acid groups.

Hwang *et al.* synthesized poly(vinylphosphonic acid-*co*-methacrylic acid) microbeads by suspension polymerization, and they investigated the adsorption properties with indium.[67] The maximum adsorption predicted by the Langmuir adsorption isotherm model was greatest at a 0.5 mol ratio of vinylphosphonic acid.[67] The adsorption capacity of indium was 0.70 mmol g^{-1}. They calculated the thermodynamic parameters ($\Delta G°$, $\Delta H°$, and $\Delta S°$) from the temperature-dependent adsorption isotherms and indicated that the adsorption was spontaneous and endothermic.[67]

11.4 Conclusions

One may notice that the modification of the polymeric matrix through phosphorylation of its surface with different phosphorus pendant groups leads to an increase in the adsorption efficiency of the polymer in the removal process of metal ions from various aqueous solutions. It was observed that the selectivity for various metal ions and the maximum adsorption capacity of the functionalized polymer depended on the following: the type of functionalized groups grafted onto the polymeric matrix; the properties of the ion-exchange/coordination resin; and the phosphorylation conditions. Additionally, the adsorption efficiency of the polymer in adsorbing the different metal ions is influenced by the various experimental parameters. The most important, and perhaps the one with the highest influence, is the medium pH. The effects of time, temperature, and initial ion concentration on the adsorption efficiency were also investigated. The kinetics of the adsorption of metal ions onto functionalized resin follows the pseudo-second-order rate. The equilibrium isotherm for sorption of the investigated metal ions has been modeled successfully using the Langmuir isotherm. Various thermodynamic parameters such as ΔG, ΔH, and ΔS were calculated. In all the cases the thermodynamics parameters indicated a spontaneous and endothermic nature of the adsorption process of metal ions onto a functionalized resin.

In conclusion, the use of a functionalized polymer with various types of phosphorus groups showed a high feasibility in the removal process of metal ions from different aqueous solutions.

References

1. L. R. Bernabé, S. Villegas, B. Ruf and I. M. Peric, *J. Chil. Chem. Soc.*, 2007, **52**, 1164.
2. T. Oshima, K. Kondo, K. Ohto, K. Inoue and Y. Baba, *React. Funct. Polym.*, 2008, **68**, 376.
3. E. J. Blanchard and E. E. Graves, *Textile Res. J.*, 2003, **73**, 22.
4. S. Elbariji, A. Petrissans, M. Elamine, K. Ouzaouit, H. Kabli, A. Albourine and P. Gerardine, *Part. Sci. Technol.*, 2011, **29**, 320.
5. R. Laus, T. G. Costa, B. Szpoganicz and V. T. Favere, *J. Hazard. Mater.*, 2010, **183**, 233.

6. W. S. W. Ngah, A. Kamari and Y. J. Koay, *Int. J. Biol. Macromol.*, 2004, **34**, 155.

7. A. O. Martins, E. L. Silva, E. Carasek, N. S. Gonçalves, M. C. M. Laranjeira and V. T. Favere, *Anal. Chim. Acta*, 2004, **521**, 157.

8. A. Popa, *Functionalized Polymers with Phosphorus and Nitrogen Groups*, Vest University Publisher, Timisoara, 2009.

9. B. S. Garg, R. K. Sharma, N. Bhojak and S. Mittal, *Microchem. J.*, 1999, **61**, 94.

10. A. W. Trochimczuk and S. Czerwinka, *React. Funct. Polym.*, 2005, **63**, 215.

11. A. Buasri, N. Chaiyuti, K. Tapang, S. Jaroensim and S. Panphrom, *Civil Environ. Res.*, 2012, **2**, 17.

12. A. M. A. Nada, A. A. Mahdy and A. A. El-Gendy, *BioResources*, 2009, **4**, 1017.

13. A. Buasri, N. Chaiyuta, K. Tapanga, S. Jaroensina and S. Panphroma, *APCBEE Procedia*, 2012, **3**, 60.

14. R. Jayakumar, R. L. Reis and J. F. Mano, *e-Polym.*, 2006, **35**, 1.

15. S. J. Wu, T. H. Liou, C. H. Yeh, F. L. Mi and T. K. Lin, *J. Appl. Polym. Sci.*, 2013, **127**, 4573.

16. W. S. W. Ngah and S. Fatinathan, *J. Environ. Manage.*, 2010, **91**, 958.

17. M. S. Chiou and H. Y. Li, *Chemosphere*, 2003, **50**, 1095.

18. S. T. Lee, F. L. Mi, Y. J. Shen and S. S. Shyu, *Polymer*, 2001, **42**, 1879.

19. A. K. Anal, W. F. Stevens and C. R. Lopez, *Int. J. Pharm.*, 2006, **312**, 166.

20. J. A. Ko, H. J. Park, S. J. Hwang, J. B. Park and J. S. Lee, *Int. J. Pharm.*, 2002, **249**, 165.

21. S. H. Kim, H. Song, G. M. Nisola, J. Ahn, M. M. Galera, C. H. Lee and W. J. Chung, *J. Ind. Eng. Chem.*, 2006, **12**, 469.

22. A. J. Varma, S. V. Deshpande and J. F. Kennedy, *Carbohydr. Polym.*, 2004, **55**, 77.

23. R. Jayakumar, M. Rajkumar, H. Freitas, N. Selvamurugan, S. V. Nair, T. Furuike and H. Tamura, *Int. J. Biol. Macromol.*, 2009, **44**, 107.

24. J. M. Sánchez, M. Hidalgo, M. Valiente and V. Salvadó, *J. Polym. Sci., Part A: Polym. Chem.*, 2000, **38**, 269.

25. J. M. Sánchez, M. Hidalgo and V. Salvadó, *React. Funct. Polym.*, 2001, **49**, 215.

26. J. M. Sánchez, M. Hidalgo and V. Salvadó, *React. Funct. Polym.*, 2001, **46**, 283.

27. J. M. Sanchez, M. Hidalgo and V. Salvado, *Solvent Extr. Ion Exch.*, 2004, **22**, 285.

28. A. W. Trochimczuk, *React. Funct. Polym.*, 2000, **44**, 9.

29. A. W. Trochimczuk, *Sep. Sci. Technol.*, 2002, **37**, 3201.

30. A. Popa, C.-M. Davidescu, P. Negrea, G. Ilia, A. Katsaros and K. D. Demadis, *Ind. Eng. Chem. Res.*, 2008, **47**, 2010.

31. C. M. Davidescu, A. Popa, R. Ardelean, P. Negrea, G. Bandur, G. Ilia and G. Rusu, *Chem. Bull. "Politehnica" Univ. (Timisoara)*, 2010, **55**(69), 32.

32. S. D. Alexandratos and X. Zhu, *React. Funct. Polym.*, 2007, **67**, 375.

33. S. D. Alexandratos and X. Zhu, *Sep. Sci. Technol.*, 2008, **43**, 1296.

34. S. D. Alexandratos and S. D. Smith, *J. Appl. Polym. Sci.*, 2004, **91**, 463.
35. T. P. Rao, P. Metilda and J. M. Gladis, *Talanta*, 2006, **68**, 1047.
36. M. A. Maheswari and M. S. Subramanian, *Anal. Lett.*, 2005, **38**, 1331.
37. C. S. K. Raju and M. S. Subramanian, *J. Hazard. Mater.*, 2007, **145**, 315.
38. M. A. Maheswari and M. S. Subramanian, *React. Funct. Polym.*, 2005, **62**, 105.
39. M. A. Maheswari and M. S. Subramanian, *Talanta*, 2004, **64**, 202.
40. S. D. Alexandratos, *J. Hazard. Mater.*, 2007, **A139**, 467.
41. A. W. Trochimczuka and J. Jezierska, *Polymer.*, 2000, **41**, 3463.
42. S. D. Alexandratos and M.-J. Hong, *Sep. Sci. Technol*, 2002, **37**, 2587.
43. Y. Hamabe, Y. Hirashima, J. Izumi, K. Yamabe and A. Jyo, *React. Funct. Polym.*, 2009, **69**, 828.
44. C.-M. Davidescu, M. Ciopec, A. Negrea, A. Popa, L. Lupa, E.-S. Dragan, R. Ardelean, G. Ilia and S. Iliescu, *Polym. Bull.*, 2013, **70**, 277.
45. M. Ciopec, L. Lupa, A. Negrea, C.-M. Davidescu, A. Popa, P. Negrea, M. Motoc, D. David and D. Kaycsa, *Rev. Chim. (Bucharest)*, 2012, **63**, 49.
46. M. Ciopec, C.-M. Davidescu, A. Negrea, L. Lupa, A. Popa, C. Muntean, R. Ardelean and G. Ilia, *Polym. Eng. Sci.*, 2012, **53**, 1117.
47. C. S. K. Raju and M. S. Subramanian, *Talanta*, 2005, **67**, 81.
48. Z. Hubicki, M. Wawrzkiewicz and A. Wolowicz, *Chem. Anal. (Warsaw)*, 2008, **53**, 759.
49. A. W. Trochimczuk, *Eur. Polym. J.*, 1999, **35**, 1457.
50. A. W. Trochimczuk and J. Jezierska, *Polymer*, 1997, **38**, 2431.
51. R. Chiariza, E. P. Horwitz, S. D. Aleandratos and M. J. Gula, *Sep. Sci. Technol.*, 1997, **32**, 1.
52. S. D. Alexandratos and S. Natea, *Eur. Polym. J.*, 1999, **35**, 431.
53. A. W. Trochimczuk and S. Alexandratos, *J. Appl. Polym. Sci.*, 1994, **52**, 1273.
54. S. D. Alexandraos, A. W. Trochimczul, E. P. Horwitz and R. Gartone, *J. Appl. Polym. Sci.*, 1996, **61**, 273.
55. K. Yamabe, T. Ihara and A. Jyo, *Sep. Sci. Technol.*, 2001, **36**, 3511.
56. A. Jyo, K. Yamabe and H. Egawa, *Sep. Sci. Technol.*, 1997, **32**, 1099.
57. N. Kabay, N. Gizli, M. Demircioğlu, M. Yuksel, A. Jyo, K. Yamabe and T. Shuto, *Chem. Eng. Commun.*, 2003, **190**, 813.
58. M. A. V. Souza, L. Claudio de Santa Maria, M. A. S. Costa, W. S. Hui, L. C. Costa, H. C. A. Filho and S. C. Amico, *Polym. Bull.*, 2011, **67**, 237.
59. D. Prabhakaran and M. S. Subramanian, *Anal. Bioanal. Chem.*, 2004, **379**, 519.
60. N. Ferrah, O. Abderrahim, M. A. Didi and D. Villemin, *J. Radional. Nucl. Chem*, 2011, **289**, 721.
61. O. Abderrahim, M. A. Didi, B. Moreau and D. Villemin, *Solvent Extr. Ion Exch.*, 2006, **24**, 943.
62. N. Ferrah, O. Abderrahim, M. A. Didi and D. Villemin, *Desalination*, 2011, **269**, 17.
63. N. Ferrah, O. Abderrahim, M. A. Didi and D. Villemin, *J. Chem.*, 2013, Article ID 980825, 10; http://dx.doi.org/10.1155/2013/980825.

64. O. C. S. Al Hamouz and S. A. Ali, *Sep. Purif. Technol.*, 2012, **98**, 94.
65. A. R. Mcclain and Y.-L. Hsieh, *J. Polym. Sci., Part A: Polym. Chem.*, 2004, **42**, 92.
66. T. Nonaka, A. Yasunaga, T. Ogata and S. Kurihara, *J. Appl. Polym. Sci.*, 2006, **99**, 449.
67. N.-S. Kwak, Y. Baek and T. S. Hwang, *J. Hazard. Mater.*, 2012, **203/ 204**, 213.

CHAPTER 12

Flame Retardancy of Phosphorus-Containing Polymers

RODOLPHE SONNIER,* LAURENT FERRY AND
JOSÉ-MARIE LOPEZ-CUESTA

Centre des Matériaux (C2MA), Ecole des Mines d'Alès, 30100 Alès, France
*Email: rodolphe.sonnier@mines-ales.fr

12.1 Introduction

The chemical modification of a polymer through phosphorus incorporation may impart new properties to the material, like increased solubility, heavy metal ion adsorption, modified glass transition temperature, and flame retardancy. Much work has been done over several decades and especially in the past 15 years to overcome the poor fire behavior of polymers by incorporating phosphorus into the macromolecular structure through covalent bonding.

The choice of phosphorus-containing groups to incorporate is very large, even if the studied groups in the literature are mainly the following four: phosphonate, phosphate, phosphine oxide, and DOPO (9,10-dihydro-9-oxa-10-phosphaphenanthrene 10-oxide). Table 12.1 shows several phosphorus-containing groups used as flame retardant groups.

Other comprehensive reviews can be found elsewhere (for example, the reader will find many other references not discussed in this chapter, in particular about polyphosphazenes).[1] In the present chapter, the effect of

RSC Polymer Chemistry Series No. 11
Phosphorus-Based Polymers: From Synthesis to Applications
Edited by Sophie Monge and Ghislain David
© The Royal Society of Chemistry 2014
Published by the Royal Society of Chemistry, www.rsc.org

Table 12.1 Various phosphorus-containing groups discussed in this chapter.

O=P---	O=P-O---	$\begin{array}{c} O \\ \| \\ O=P--- \\ \| \\ O \end{array}$	$\begin{array}{c} O \\ \| \\ O=P--O--- \\ \| \\ O \end{array}$	DOPO structure
Phosphine oxide	Phosphinate	Phosphonate	Phosphate	DOPO

phosphorus on the different parameters which control the response of a polymer upon a fire will be reviewed. The objective is to highlight in a few pages some general rules that allow the best phosphorus-containing structures for flame retardancy to be determined.

12.2 Thermal Stability and Charring Promotion

Phosphorus is known as an effective char promoter. Charring limits the release of fuels to the flame and therefore reduces the heat released. Moreover, the char accumulates on the surface of the material and can act as a protective layer which limits the heat transfer from the flame to the condensed phase and the gas transfer from the pyrolysis zone to the flame. This phenomenon is called the barrier effect. Nevertheless, it should be noted that the presence of char is not a sufficient condition to observe an effective barrier effect. Its structure (cohesion, porosity, thickness) is also very important but rarely studied. Thermal stability is another important parameter to assess the reaction to fire of a material. Indeed, the higher is the degradation temperature of a polymer, the greater is the heat required to start its pyrolysis. Table 12.2 lists the effects of phosphorus-containing groups on the thermal stability and charring for a variety of polymers.

According to Table 12.2, no clear tendency could be drawn. Most generally, thermal stability in the presence of phosphorus is lowering due to the low energy of P–C and P–O bonds (respectively 264 and 335 kJ mol^{-1}). Charring is quite often promoted (Figure 12.1), but to a varied extent. Some articles report even a decrease of char content when incorporating phosphorus.[25] Phosphorus compounds could degrade into phosphorus acids. Such acids (particularly phosphoric acid) are believed to act through a dehydrating action, generating double bonds in the polymer and finally crosslinking. Therefore, even if lowering the thermal stability is a drawback, this is generally considered as necessary to make phosphorus efficient as a char promoter.

12.3 Combustion

Heat release is generally evaluated either using the well-known cone calorimeter or using a pyrolysis combustion flow calorimeter (PCFC). The PCFC is a useful tool to study the flammability of very small polymeric samples (2–5 mg).[29,30] The sample is pyrolyzed under nitrogen flow at a

Table 12.2 Influence of phosphorus on charring and thermal stability from various studies.

Modified polymer	Maximum phosphorus content (wt%)	Phosphorus group	Thermal stability	Charring increase	Ref.
Epoxy resin	1.5	DOPO with nitrogen atoms	Decreased (argon)	Slightly	2
Epoxy resin	2.7	Phosphate, phosphonate, phosphinate, phosphine oxide	Decreased (nitrogen)	Yes	3
Polyacrylonitrile	1.4	Phosphorylamino group	Decreased (nitrogen)	Yes	4
Aromatic polyamide	8.1	DOPO (pendant)	Decreased (nitrogen)	Lower charring	5
Polyester (PET)	4	Phosphonate (backbone)	Decreased (nitrogen)	Lower charring	6
Polyester (PET)	3	DOPO (pendant)	Slightly decreased (nitrogen)	No	6
Polyester (PET)	1.2	Phosphinate (backbone)	No change (air)	No	7
Polyester (PBT)	5.9	DOPO (pendant)	Enhanced (nitrogen)	Yes	8
Polyester (PTT)	2.25	DOPO (pendant)	Slightly enhanced (nitrogen)	Yes	9
Aromatic polyester	6.8	DOPO (pendant)	Slightly enhanced (air)	Yes	10
Poly(ester-imide)	5.2	DOPO (pendant)	Slightly decreased (air)	Yes	10
Poly(ester-imide)	5.0	DOPO (pendant)	No change (air)	Yes	11
Polyimide	5.8	Phosphate (pendant)	Decreased (air and nitrogen)	Yes	12
Polyimide	8.3	Phenoxaphosphine oxide rings	Decreased (air and nitrogen)	Yes	12
Poly(methyl methacrylate)	1.6	Phosphonate (pendant) with silicon atoms	Decreased (nitrogen)	Slightly	13
Poly(methyl methacrylate)	10	Phosphonate (MAPC1)	Slightly enhanced (nitrogen)	Yes	14
Poly(methyl methacrylate)	13.7	Aminobisphophonate	Decreased (nitrogen)	Yes	15

Polymer					
Poly(methyl methacrylate)	3.5	Acrylate/methacrylate phosphate/phosphonate	Enhanced (nitrogen)	Slightly	16
Poly(methyl methacrylate)	3.5	Methacrylate phosphonate	No change (nitrogen)	Slightly	17
Polystyrene	15.6	Phosphonate or phosphoramidate (pendant)	Variable (nitrogen)	Yes	18
Polystyrene	3.1	Phosphate and phosphonate (pendant)	Variable (nitrogen)	Slightly	19
Polystyrene	10.3	Phosphonate (pendant) with nitrogen atoms	Decreased (nitrogen)	Slightly	20
Poly(4-hydroxystyrene)	12.8	Phosphate (pendant)	Decreased (air and nitrogen)	Yes	21
Chloromethylated polysulfone	6.0	DOPO (pendant)	Enhanced (nitrogen)	No	22
Polyurethane	6.5	Phenyl phosphonate and phenyl phosphine oxide (backbone)	Decreased (air and nitrogen)	Yes	23
Polyurethane	1.9	Phosphonate (backbone)	Decreased (nitrogen)	Slightly	24
Polyurethane	2	Phosphate (pendant)	No change (air)	Lower charring	25
Unsaturated polyester (UPR)	5.2	Phosphonate	Decreased (nitrogen)	Yes	26
Vinyl acetate/butyl acrylate copclymer	5	Phosphonate and phosphate	Decreased (air)	Yes	27

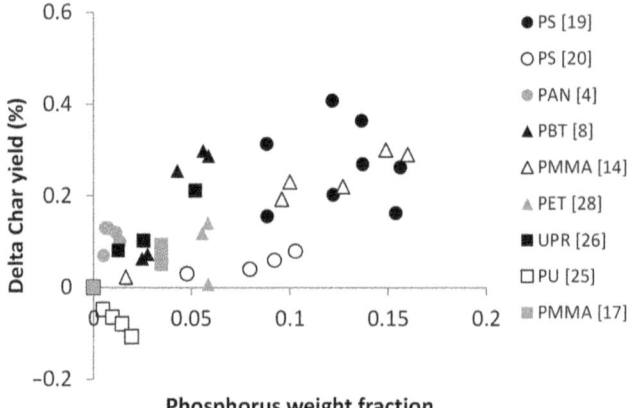

Figure 12.1 Increase in char yield (calculated as the difference between the char yield of the phosphorus-containing polymer and those of the phosphorus-free polymer) *versus* the phosphorus content, from various studies.[4,8,14,17,19,20,25,26,28]

constant heating rate as in thermogravimetric analysis (TGA), but the heating rate is generally higher (typically 1 K s^{-1}). The gases released are sent to a combustor in the presence of an excess of oxygen at high temperature (typically 20% of oxygen and 900 °C) to ensure complete combustion. Heat release is calculated from the oxygen consumption according to Huggett's relation.[31]

Phosphorus is believed to act mainly as a char promoter. Therefore, a fraction of the fuel is trapped in the condensed phase and the total heat release (THR) is lowered.[8,9,19,20,26] The decrease of the peak of the heat release rate (pHRR) in the PCFC is quite often accompanied by other minor peaks and/or by a decrease of the peak temperature.[4,18,19,20] Therefore the PCFC results do not bring a definite conclusion about flame retardancy improvement.

There are literature claims that phosphorus acts not only as a char promoter but also in the vapor phase, leading to a decrease in the effective heat of combustion (EHC). This activity could be either flame inhibition (*i.e.* incomplete combustion due to scavenging of radicals like H· or HO· by phosphorus radicals in the vapor phase) or a change in the pyrolytic gases. If the decrease in the EHC is generally well observed, very few articles provide evidence of flame inhibition.

First of all, it is important to evaluate how the phosphorus is allocated between the condensed and vapor phases during the thermal degradation of the polymer, even if the volatilization of phosphorus does not show that it acts in the vapor phase. Vahabi *et al.* have proposed an original approach to study the efficiency of phosphorus in both condensed and vapor phases.[14] Two indices (called respectively charring efficiency and efficiency in the vapor phase) were calculated to evaluate the influence of phosphorus to improve the charring and to decrease the EHC (measured from TGA and PCFC). These calculations are based on the measurement by elemental analysis of phosphorus remaining in the condensed phase after degradation. They also analyzed pyrolytic gases by TGA-Fourier transformed infrared spectroscopy coupling and measured the phosphorus content in the residue after isothermal degradation in TGA at various temperatures. They found strong differences between the studied copolymers. Indeed, the release of phosphorus in the vapor phase occurs over a wider range of temperatures for a copolymer containing aminobisphosphonated groups than for a copolymer containing monophosphonated groups.

Price *et al.* have measured the P content in the char for copolymers synthesized from methyl methacrylate (MMA) and various phosphonate- or phosphate-containing groups.[17] Char yield ranged from 6.5 to 10.8 wt%, while the P content in the char was strongly changing: from 1 wt% (for an acrylate phosphonate, allowing the lowest char yield) to 11.3 wt% for an acrylate phosphate, leading to the highest char yield). Tibiletti *et al.* have also calculated the partition of phosphorus between condensed and vapor phases after the cone calorimeter test for unsaturated polyester resins modified with dimethyl vinylbenzylphosphonate.[26] Interestingly, the higher is the initial phosphorus content in the resin, the higher is the percentage of phosphorus remaining in the condensed phase. The authors explain that phosphonate groups could interact at a greater extent when the initial phosphorus content is high enough, allowing a higher fraction of the phosphorus to remain in the condensed phase. Moreover, they noticed a decrease in the EHC at low phosphorus content when the phosphorus is mainly released in the vapor phase.

Gases released during pyrolysis could change in the presence of phosphorus due to the release of phosphorus-containing gases or because the fraction of fuels which is trapped in the condensed phase may be not representative of the whole chemical structure. Therefore the EHC of such gases could be modified. In general, the EHC decreases when incorporating phosphorus, as seen in Figure 12.2, even if it is not systematic. Sablong *et al.*

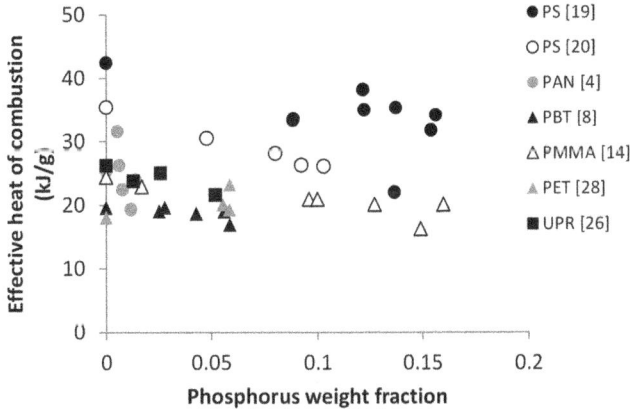

Figure 12.2 Change in effective heat of combustion *versus* phosphorus weight fraction from various studies.

incorporated a phosphorus-containing aliphatic-aromatic diol into poly-(butylene terephthalate) (PBT).[8] The EHC calculated from the PCFC results is stable and close to 17–20 kJ g^{-1} despite a quite high phosphorus content (higher than 5 wt%).

Flame inhibition is the main mode-of-action of halogenated flame retardants. Even if phosphorus compounds promote charring, phosphorus is considered a better flame inhibitor than halogenated elements.[32] A recent method to study the flame inhibition effect was proposed by monitoring the combustor temperature in a PCFC.[33,34] This method consists in measuring the combustion efficiency at different temperatures of combustion: the ratio of the total heat released at each temperature to the total heat released at 900 °C (combustion is believed to be complete at 900 °C). The combustion efficiency should decrease faster with the temperature of combustion in the case of flame inhibition. However, poly(diethyl vinylbenzylphosphonate) showed a higher combustion efficiency than polystyrene at low temperature (600–650 °C).[33] This may be due to particular conditions of combustion in a PCFC.

Hergenrother et al. have calculated the flaming combustion efficiency of epoxy resins prepared from various epoxy- or amine-functional organophosphorus compounds by dividing the EHC measured in fire calorimetry tests (Ohio State University apparatus and cone) by the effective heat of complete combustion measured in a PCFC.[35] The combustion efficiency ranged roughly from 0.6 to 0.8 (versus 0.7 for the phosphorus-free epoxy resin) whatever the oxidation state of phosphorus and other parameters. The authors concluded that "there is no discernible effect [...] on the combustion efficiency in the flame".

Nevertheless, some authors have provided some evidence of a flame inhibition effect of phosphorus. Schartel et al. have measured the EHC of epoxy resins in a PCFC (i.e. corresponding to complete combustion) and in a cone calorimeter.[36] Epoxy resins containing DOPO covalently linked to the network have been studied. All thermosets exhibited the same EHC in a PCFC (approximately 24 kJ g^{-1}), but in cone calorimeter tests the composites prepared from the phosphorus-containing resins (with 60 vol% of carbon fibers) exhibited a significantly lower EHC than the composite prepared from the phosphorus-free epoxy resin. In this study, a clear flame inhibition effect was evidenced.

12.4 Calculation of the Phosphorus Contribution to Flammability

Lyon proposed to calculate the contributions of chemical groups to different flammability properties (HRC, EHC, THR, char yield) using Van Krevelen's approach.[37,38] Heat release capacity (HRC) is defined as the peak of heat release rate divided by the heating rate. The contributions to the HRC of two phosphorus-containing groups were found to be very low and even negative (see Table 12.3). Pursuing the preceding approach, Sonnier et al. calculated

Table 12.3 Contributions of some phosphorus-containing groups.

| | Contribution to | | | | |
| | Char yield ($g\,g^{-1}$) | HRC (sum) ($J\,g^{-1}\,K^{-1}$) | EHC ($kJ\,g^{-1}$) | THR ($kJ\,g^{-1}$) | Ref. |
Phosphorus group					
	–	−179	–	–	37
	–	−807	–	–	37
	–	500	–	0.5	39
	–	400	–	20	39
	0.7	−549	4.7	–	14
	0.78	−258	−0.75	–	14

the contributions of two other phosphorus groups from a series of monomers, oligomers, and polymers.[39] Surprisingly, much higher contributions to the HRC were found. Nevertheless, no contribution could be properly calculated when the molecules contained both phosphate and ester or acid groups. In these molecules, the HRC and THR were always significantly lower than expected. This may be due to the presence of interactions between both groups. Interactions indices were proposed and maximal interaction was determined when one phosphonate group was present together with two ester or acid groups.

Finally, Vahabi *et al.* calculated the contributions of the phosphonate group for two chemically modified PMMAs.[14] Significant differences were found between both molecular structures. This result confirms that the contributions of phosphorus-containing groups could hardly be calculated while their effects depend on their chemical environment.

12.5   Limiting Oxygen Index

Neither TGA nor PCFC are suitable to evaluate the overall fire behavior of a material. Tests on larger samples are needed to take into account most of the phenomena that affect this behavior. Even if it is often accepted that char measured in TGA could provide a good insight into the charring during bench-scale fire tests, it is not always the case. For example, in the study of Ebdon *et al.*, no sharp correlation between char yield in nitrogen and char on burning [during the limiting oxygen index (LOI) test] was noticed for various phosphorus-containing PMMAs and PSs (Figure 12.3).[40]

LOI, UL-94 vertical test, and cone calorimeter are the most used methods to study the flame retardancy of a material at a bench scale. At this scale, all modes-of-action of a flame retardant, including the barrier effect, could be effective. Most generally, phosphorus allows improving the performances of polymers in these tests: higher LOI value, better UL-94 rating, decreased peak of heat release rate and total heat release.

Very high LOIs could be obtained when phosphorus is included in macromolecules. Liaw and Shen showed a LOI equal to 60 for an aromatic polyphosphonate.[41] FRX Polymers proposed some phosphorus-containing homopolymers (called Nofia) exhibiting a LOI as high as 65.[42] Even if such

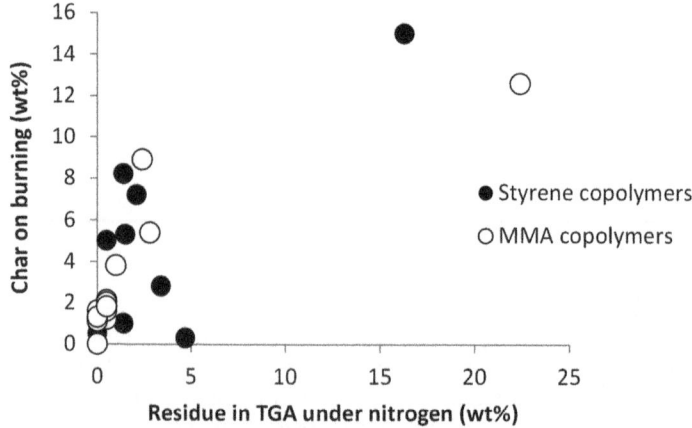

Figure 12.3   Relation between char content in TGA and char on burning (from LOI test) from the study of Ebdon *et al.*[40]

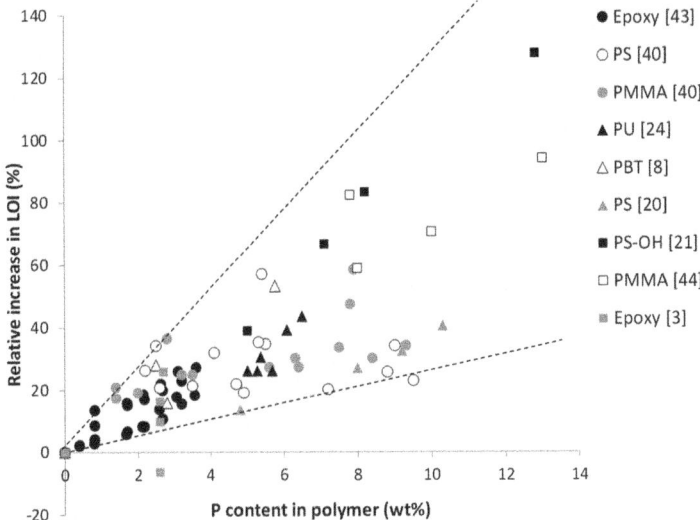

Figure 12.4 Relation between increase in LOI and initial P content from various studies (dotted lines are only drawn for eyes guideline).[3,8,20,21,24,40,43,44]

high values are not always observed, phosphorus leads almost systematically to a significant increase in LOI.

Figure 12.4 shows the relation between the initial phosphorus content in the polymer and the measured increase in LOI for a set of data comprising modified common polymers. An increase of approximately 2.5% in LOI per 1 wt% of phosphorus could be reasonably expected. Only a small decrease was observed by Braun *et al.* for an epoxy resin modified by a phosphate-based hardener.[3] However, the phosphorus epoxy resin was compared to a reference using 4,4'-diaminodiphenyl sulfone (DDS) as hardener. Owing to the presence of sulfur and the aromatic structure of the hardener, a DDS-based resin still has a high LOI value of 31. Therefore, the substitution of DDS by the phosphate-based hardener leads to a lower LOI value.

12.6   Chemical Environment of Phosphorus

According to the results of Ebdon *et al.* obtained from two series of MMA and styrene copolymers, char promotion could not be easily anticipated only from the phosphorus content.[40] This means that a range of parameters modifies the efficiency of phosphorus as a flame retardant: not only the phosphorus content, but also its chemical environment, its position in the macromolecule, and the host polymer. In this short chapter it is not possible to list in detail the influence of all these parameters. Therefore, only the main conclusions will be discussed in the following, from a narrow set of selected articles.

12.6.1   Aromatic Rings

Phenyl groups linked to phosphorus are believed to limit the lowering of the thermal stability. Dumitrascu and Howell have synthesized polystyrenes with various phosphorus-containing pendant groups.[18] The thermal stability was much higher for diphenyl than for diethyl phosphonates or phosphates. Phenyl groups could also enhance char promotion. For example, diglycidyl phenylphosphonate is slightly more efficient than diglycidyl methylphosphonate to increase charring when incorporated as reactive flame retardant groups into phosphorus-containing epoxy resins.[35]

12.6.2   Oxidation State of Phosphorus

Phosphate- and phosphonate-containing copolymers based on MMA were compared by Price's team.[16,17] For modified PMMA, phosphates appear slightly more efficient: higher LOI, higher time-to-ignition, lower pHRR, and higher char yield in cone calorimeter tests. A strong difference concerns the phosphorus content in the char: close to 10 wt% for phosphate-containing PMMA but only 1–2.5 wt% for phosphonate-containing PMMA. This result highlights the good efficiency of phosphate as a char promoter. The authors explain that phosphate decomposition leads to phosphoric acid. Phosphonate decomposes rather to phosphonic acid, which is a weaker acid and less effective to promote the acid-catalyzed crosslinking of the polymer.

Hergenrother *et al.* have compared various phosphorus-containing groups incorporated within the backbone of epoxy resins.[35] Phosphorus was incorporated as phosphate, phosphonate, phosphite, and phosphine oxide and the authors have observed clear tendencies according to the oxidation state. The char content is the highest for phosphate and the lowest for phosphine oxide. In this last case, the char content is even lower than the value for phosphorus-free epoxy resin. Phosphonate and phosphite lead to intermediate charring. While THR and HRC are very well-correlated to char content, phosphate is also the best flame retardant and phosphine oxide the worst.

The results from the study of Braun *et al.* were similar and opposite at the same time.[3] In this work, phosphine oxide, phosphinate, phosphonate, and phosphate were compared. The phosphorus content was fixed to 2.7 wt%. Once again, the results were correlated to the oxidation state of phosphorus. The higher was this parameter, the lower was the thermal stability but higher was the charring. The decomposition temperature was not modified in the presence of phosphine oxide. On the contrary, the degradation occurred more than 50 °C earlier in the presence of other phosphorus-containing groups. Flaming tests were carried out on composites containing carbon fibers. The LOI was a maximum for phosphine oxide (39%) and a minimum for phosphate (29%). Moreover, phosphine oxide allows reaching a UL-94 rating of V1 while other phosphorus-containing resins were unclassified. It could be assumed that phosphine oxide was much more efficient than other phosphorus-containing groups. According to the authors, the char promotion was not really important in the composites due to the high loading of carbon fibers. Therefore, the best flame retardant is one that acts in the vapor phase.

Other results confirm that phosphine oxide is a poor char promoter.[40,45] Phosphorus-containing acrylates were crosslinked and characterized by Canadell *et al.*[45] Phosphate and phosphonate lead to a high char yield in nitrogen (27–28 and 21–25 wt% respectively), while phosphine oxide exhibits a lower char yield than the phosphorus-free acrylate used as reference (10.6–11.5 *vs.* 13.3 wt%). The authors explain that phosphonate and phosphate could undergo elimination of alkene. Therefore, a phosphorus acid derivative is produced and could improve charring through a modification of the degradation pathway of the polymer. On the contrary, phosphine oxide is linked to phenyl groups and could not follow this mechanism.

Phosphine oxide is sometimes believed to act more significantly in the vapor phase. Braun *et al.* noticed that the effective heat of combustion in a cone calorimeter of various phosphorus-containing epoxy composites decreased only when phosphorus was present as phosphine oxide.[3] More phosphorus-containing gases were detected by Fischer *et al.* when phosphine oxide was incorporated into a PET macromolecule in comparison to DOPO groups. Both groups are located on the aromatic ring in the backbone.[28] However, Hergenrother *et al.* observed neither a specific flame inhibition effect nor a lower effective heat of combustion for phosphine oxide when they incorporated various phosphorus-containing groups into epoxy resins.[35]

12.6.3 DOPO

DOPO was extensively used as an efficient flame retardant incorporated into various polymers, as seen in Table 12.2. In particular, DOPO allows maintaining a higher thermal stability than other phosphorus groups.

Chang *et al.* have incorporated two types of phosphorus-containing groups into PET: a phenylphosphonate into the main chain or a DOPO group as pendant group.[6] Thermal stability is slightly modified in the presence of DOPO and more severely reduced when phosphorus is in the backbone. The authors explain that the P–O bond is easily attacked. The cleavage of this bond in pendant groups does not lead to scission of the main chain, and thermal stability is thus little disrupted. Shieh and Wang have incorporated DOPO or phosphate into epoxy resins.[46] DOPO is located as a pendant group while phosphate is in the backbone. DOPO leads to a slight increase in the onset of the degradation temperature in nitrogen and in air. On the other hand, the degradation occurs at lower temperature with phosphate. It is attributed to the higher stable O=P–O bond in DOPO in comparison to the O–P–O bond for phosphate (see Table 12.2). Moreover, the char yield at 700 °C in air and in nitrogen is higher for DOPO-containing resin

(38–45 *vs.* 32–37 wt% for phosphate-containing resin in nitrogen). Finally, the authors measured the UL-94 rating of their materials. A V0 rating was obtained with only 1.1 wt% of phosphorus when DOPO was used. A twice higher content of phosphorus was needed in the case of phosphate. Schafer *et al.* have compared DOPO and DPPO (2,8-dimethylphenoxaphosphin 10-oxide) incorporated into an epoxy resin cured with DDM (4,4′-diamino-diphenylmethane).[47] Two non-heterocyclic groups (diphenyl phosphite and diphenyl phosphate) were also studied. The phosphorus content was 0–1.67 wt%. From the UL-94 and LOI results the authors concluded that DPPO and DOPO have similar efficiencies. V0 could be obtained with a low phosphorus content (0.6–0.8 wt%). Nevertheless, DOPO allows reaching a higher LOI (up to 39.2 wt% *vs.* 31.1 wt% for DPPO). On the other hand, non-heterocyclic phosphorus-containing groups exhibited poor efficiency: no rating in a UL-94 test was achieved and LOI values were not improved. The authors explained these results according to the release of PO radicals. PO radicals could interact with H and OH radicals in the vapor phase. DOPO-containing epoxy resins released more PO radicals than DPPO during thermal degradation. On the other hand, diphenyl phosphite and diphenyl phosphate did not form PO radicals during decomposition.

As seen in the previous reports, DOPO is generally a better flame retardant compound than other phosphorus-containing groups. Nevertheless, Yan *et al.* studied two phosphorus-containing methyl acrylates copolymerized with styrene.[48] Both DOPO and PEPA (1-oxo-2,6,7-trioxa-1-phosphabicy-clo[2.2.2]oct-4-ylmenthol) were located as pendant groups. PEPA contains a non-aromatic heterocyclic group and phosphorus is linked to four oxygen atoms. The thermal stability of the PEPA-containing copolymer was slightly lower but charring was much more promoted: char yields at 500 °C in nitrogen were approximately 40 wt% with PEPA and less than 10 wt% with DOPO. While the DOPO-containing copolymer exhibited only a V2 rating in the UL-94 test due to dripping, PEPA allowed a V0 rating.

12.6.4   Additive *versus* Reactive Approaches

The chemical incorporation of phosphorus into a macromolecule is believed to provide some advantages over the use of additives. Additives (particularly small molecules) could leach over time, leading to a decrease of flame retardancy. They could also plasticize the polymer matrix.[16,49] Mechanical properties are often strongly reduced by the addition of additives, generally at high loading. Transparency is also most often impaired. Nevertheless, additives can be easily incorporated through classic processes whereas the synthesis of new phosphorus-containing polymers is cost-consuming. Moreover, the presence of phosphorus in the macromolecule could also lead to detrimental changes: undesirable modification of glass transition temperature, lower crosslinking, *etc.*[26]

Price *et al.* have extensively compared both approaches for modified PMMA.[16,17,49] They copolymerized various phosphorus-containing comonomers with MMA. They also added into PMMA or PS flame retardant additives with structures close to those of comonomers. The phosphorus content was fixed to 3.5 wt% in all cases. The reactive approach is clearly more efficient for MMA-based polymers: charring is significantly enhanced (char yield in the cone calorimeter test is 6.7–20.6 wt% for the reactive approach *vs.* 2.5–2.7 wt% for the additive approach). The peak of the heat release rate is also lower for phosphorus-containing copolymers. Ignition is also more enhanced for copolymers. These results should be mainly explained by the lower thermal stability of filled PMMA, as shown by a thorough study comparing PMMA filled with diethyl ethylphosphonate and two phosphonate-containing copolymers. The additive volatilizes at lower temperatures than PMMA, leading to a significant decrease in thermal stability, while copolymers exhibit a higher thermal stability than pure PMMA. There is almost no overlapping between the volatilization of the additive and the decomposition of PMMA. Therefore the possibility of the phosphorus-containing additive to interact with the degradation pathway of the polymer is quite limited.

Schartel *et al.* have studied epoxy resins flame-retarded with DOPO groups.[36] DOPO was incorporated into the network (through diamino hardener) or alternatively as an additive analogue. The authors showed that the additive volatilizes at the beginning of the decomposition. On the other hand, most of the phosphorus remained in the condensed phase when it was covalently linked to the network. Therefore, the thermal stability was much more reduced for the former system and the char yield was much higher in the latter system (26.5 *vs.* 12.5 wt%). The HRC and the total heat release in PCFC were also lower for the reactive approach. Composites synthesized from these systems with 60 vol% of carbon fibers were also tested in a cone calorimeter test. Surprisingly, the additive approach performed better: the total heat release and the peak of heat release rate were significantly lower. The authors explain that the char promotion is not an important parameter in these systems because of the high loading in carbon fibers. The additive approach allows a more significant volatilization of phosphorus and therefore leads to flame inhibition more extensively.

12.6.5 Polymeric Additives

To overcome some problems with low molecular additives like leaching or low thermal stability, phosphorus-containing polymeric additives could be used. Phosphorus is incorporated into macromolecules used as an additive to improve the flame retardancy of a polymer.

Morita *et al.* have found a relation between the increase in the LOI and the polymerization rate of phosphorus- and bromine-containing monomers in ethylene propylene diene monomer (EPDM).[50] The polymerization of these comonomers is initiated simultaneously with the crosslinking of the polymer matrix, but the authors did not check if the comonomers are linked to the EPDM macromolecules. They assumed that the phosphorus-containing comonomers are more effective when they are converted into polymers even if it is not the only influential parameter.

Sonnier *et al.* have compared physical polymer blends, block, and random copolymers synthesized from MMA and MAPC1.[51] The content of both comomomers ranged over a large extent. The heat release rate and total heat release measured in a PCFC followed the rule of mixture for physical polymer blends. On the other hand, the block copolymers exhibited lower heat release rate values than expected from the rule of mixture. Similar results and even better performances need to be confirmed for random copolymers. It is noteworthy that to the best of our knowledge the comparison between random and block copolymers was not carried out.

Generally, however, the incorporation as additive of a phosphorus-containing polymer appears to be a valuable approach to impart flame retardancy to a polymer matrix. Chang *et al.* have synthesized a phosphorus-containing polymeric flame retardant.[52] The phosphorus content in the polymeric flame retardant was 13.9 wt%. The LOI increased from 20.4 to 35.4% when the polymeric flame retardant was incorporated into the epoxy resin at a content of 8 wt%. The UL-94 rating was V0, with only 0.7 wt% of phosphorus. Similar results were found for other polymer matrices (PET and unsaturated polyester).

12.6.6 Synergism with Nitrogen

Synergism between phosphorus and nitrogen (or other elements) is often postulated. Synergism could be assessed when the flame retardancy is improved by the combined presence of phosphorus and nitrogen in comparison to the presence alone of phosphorus or nitrogen at the same global amount. However, it seems quite difficult to apply strictly the rule of mixtures to polymers containing phosphorus and/or other elements. A more restrictive definition is that phosphorus and nitrogen interact in a chemical way to impart flame retardancy. Although other elements (Si, S, Br, *etc.*) have been also incorporated together with phosphorus, only some articles are discussed about the possible synergism between phosphorus and nitrogen.

Dumitrascu and Howell have compared polymers synthesized from phosphorus- or phosphorus-nitrogen-containing styrene monomers with various pendant groups (methoxy, ethoxy, phenoxy, DOPO) bonded to the phosphorus.[19] Nitrogen and phosphorus atoms are directly linked through a N–P bond. The onset temperature of the anaerobic pyrolysis changes mainly according to the pendant groups and could be lower or slightly higher than the onset temperature of polystyrene. The char yield at 600 °C depends also on the pendant groups but a clear tendency could be observed: its value ranges from 15.6 to 26.9 wt% for phosphorus-containing polymers and from 26.2 to 40.8 wt% for phosphorus-nitrogen-containing polymers.

Vahabi *et al.* did not report any synergism between phosphorus and nitrogen.[14] The authors copolymerized MMA with monophosphonated or aminobisphosphonated comonomers (respectively called MAPC1 and

MANP2C3). In this study, N and P are linked through N–C–P bonds. The molar ratio between MMA and the corresponding comonomer is 1 : 1. Both copolymers exhibited quite similar properties. The char content of the co-polymer containing MAPC1 was higher (22 *vs.* 18 wt%) despite a slightly lower phosphorus content (10 *vs.* 12.7 wt%). The heat release properties were measured using a PCFC. Both copolymers exhibited close values for all these combustion properties. Finally, the authors attempted to calculate the contributions of the phosphonate group in both copolymers: similar con-tributions to the residue content and to the EHC were found, confirming that no synergism occurred between nitrogen and phosphorus in such polymers. However, the contribution to the heat release rate of the phos-phonate group in the copolymer between MMA and MANP2C3 was significantly lower.

Schafer *et al.* have compared DPPA (5,10-dihydrophenophosphazine 10-oxide) and DOPO incorporated into DGEBA-based epoxy resins.[43] Contrarily to DOPO, DPPA contains both phosphorus and nitrogen without any direct covalent N–P bond. DOPO is slightly more efficient than DPPA: V0 in the UL-94 test could be achieved for a lower phosphorus content and a higher LOI was also obtained for DOPO at a similar phosphorus content.

12.7 Conclusion

From the great number of works presented here, it is possible to draw some general conclusions about the efficiency of phosphorus as a flame retardant chemically incorporated into a macromolecule. Nevertheless, even the most general tendencies could be contradicted in specific systems, as seen in many cases throughout this chapter. In the opinion of the authors, the fol-lowing conclusions can be drawn.

Basically, the incorporation of phosphorus into a polymer leads to two main effects due to great changes in the thermal degradation pathway. The first one is the decrease of thermal stability due to the low energy of P–O and P–C bonds. The second one is the promotion of charring, even if the initial polymer is non-charring. While the first effect is detrimental for flame retardancy, the second is highly interesting. There is no reason for both effects (charring and lowering of thermal stability) to change with phos-phorus content (or other parameters) at the same rate. Therefore, the phosphorus incorporation could be globally detrimental in some cases. Thermal stability could be monitored by the chemical environment of phosphorus. Aromatic rings and DOPO could allow maintaining high ther-mal stability. Synergism with other elements is more often postulated than proved and even less explained.

Phosphorus could act as char promoter, but also in the vapor phase. The partition of phosphorus between condensed and vapor phases and its modes-of-action (as char promoter, flame inhibitor, *etc.*) depend strongly on various parameters. Phosphine oxide seems to be a relatively poor char promoter. Generally, a high oxidation state should allow more efficient charring. One main advantage of the reactive incorporation is that phosphorus remains more significantly in the condensed phase, leading to strong charring. On the other hand, the additive approach allows phosphorus to volatilize at low temperatures and interactions with the polymer could be limited.

On the whole, much work has allowed identifying some parameters which control the phosphorus efficiency in more or less specific cases. Nevertheless, a systematic study about the influence of all these parameters would be very desirable.

References

1. S.-Y. Lu and I. Hamerton, *Prog. Polym. Sci.*, 2002, **27**, 1661.
2. D. Sun and Y. Yao, *Polym. Degrad. Stab.*, 2011, **96**, 1720.
3. U. Braun, A. I. Balabanovich, B. Schartel, U. Knoll, J. Artner, M. Ciesielski, M. Doring, P. Perez, J. K. W. Sandler, V. Altstadt, T. Hoffmann and D. Pospiech, *Polymer*, 2006, **47**, 8495.
4. P. Joseph and S. Tretsiakova-McNally, *Polym. Degrad. Stab.*, 2012, **97**, 2531.
5. Y.-L. Liu and S.-H. Tsai, *Polymer*, 2002, **43**, 5757.
6. S.-J. Chang, Y.-C. Sheen, R.-S. Chang and F.-C. Chang, *Polym. Degrad. Stab.*, 1996, **54**, 365.
7. B. Wu, Y.-Z. Wang, X.-L. Wang, K.-K. Yang, Y.-D. Jin and H. Zhao, *Polym. Degrad. Stab.*, 2002, **76**, 401.
8. R. Sablong, R. Duchateau, C. E. Koning, D. Pospiech, A. Korwitz, H. Komber, S. Starke, L. Haubler, D. Jehnichen and M. Auf der Landwehr, *Polym. Degrad. Stab.*, 2011, **96**, 334.
9. H.-B. Chen, X. Dong, D. A. Schiraldi, L. Chen, D.-Y. Wang and Y.-Z. Wang, *J. Anal. Appl. Pyrol.*, 2013, **99**, 40.
10. C. Hamciuc, T. Vlad-Bubulac, O. Petreus and G. Lisa, *Eur. Polym. J.*, 2007, **43**, 980.
11. T. Vlad-Bubulac, C. Hamciuc, O. Petreus and M. Bruma, *Polym. Adv. Technol.*, 2006, **17**, 647.
12. Y.-L. Liu, G.-H. Hsiue, C.-W. Lan, J.-K. Kuo, R.-J. Jeng and Y.-S. Chiu, *J. Appl. Polym. Sci.*, 1996, **63**, 875.
13. T. C. Chang, Y. C. Chen, S. Y. Ho and Y. S. Chiu, *Polymer*, 1996, **37**, 2963.
14. H. Vahabi, L. Ferry, C. Longuet, R. Sonnier, C. Negrell-Guirao, G. David and J.-M. Lopez-Cuesta, *Eur. Polym. J.*, 2012, **48**, 604.
15. H. Vahabi, C. Longuet, L. Fery, G. David, J.-J. Robin and J.-M. Lopez-Cuesta, *Polym. Int.*, 2012, **61**, 129.
16. D. Price, K. Pyrah, T. R. Hull, G. J. Milnes, J. R. Ebdon, B. J. Hunt and P. Joseph, *Polym. Degrad. Stab.*, 2002, **77**, 227.
17. D. Price, L. K. Cunliffe, K. J. Bullett, T. R. Hull, G. J. Milnes, J. R. Ebdon, B. J. Hunt and P. Joseph, *Polym. Adv. Technol.*, 2008, **19**, 710.
18. A. Dumitrascu and B. Howell, *Polym. Degrad. Stab.*, 2011, **96**, 342.
19. A. Dumitrascu and B. Howell, *Polym. Degrad. Stab.*, 2012, **97**, 2611.

20. Q. Tai, L. Song, Y. Hu, R. K. K. Yuen, H. Feng and Y. Tao, *Mater. Chem. Phys.*, 2012, **134**, 163.
21. Y.-L. Liu, G.-H. Hsiue, Y.-S. Chiu, R.-J. Jeng and C. Ma, *J. Appl. Polym. Sci.*, 1996, **59**, 1619.
22. O. Petreus, E. Avram and D. Serbezeanu, *Polym. Eng. Sci.*, 2010, **50**, 48.
23. M.-F. Lin, W.-C. Tsen, Y.-C. Shu and F.-S. Chuang, *J. Appl. Polym. Sci.*, 2001, **79**, 881.
24. Y.-L. Liu, G.-H. Hsiue, C.-W. Lan and Y.-S. Chiu, *J. Polym. Sci., Part A: Polym. Chem.*, 1997, **35**, 1769.
25. D.-J. Liaw and S.-P. Lin, *J. Appl. Polym. Sci.*, 1996, **61**, 315.
26. L. Tibiletti, L. Ferry, C. Longuet, A. Mas, J.-J. Robin and J.-M. Lopez-Cuesta, *Polym. Degrad. Stab.*, 2012, **97**, 2602.
27. S. Duquesne, J. Lefebvre, G. Seeley, G. Camino, R. Delobel and M. Le Bras, *Polym. Degrad. Stab.*, 2004, **85**, 883.
28. O. Fischer, D. Pospiech, A. Korwitz, K. Sahre, L. Haubler, P. Friedel, D. Fischer, C. Harnisch, Y. Bykov and M. Doring, *Polym. Degrad. Stab.*, 2011, **96**, 2198.
29. R. E. Lyon and R. N. Walters, *J. Anal. Appl. Pyrol.*, 2004, **71**, 27.
30. R. Sonnier, H. Vahabi, L. Ferry and J.-M. Lopez-Cuesta, *Fire and Polymers VI: New Advances in Flame Retardant Chemistry and Science*, ed. A. Morgan, C. Wilkie and G. Nelson, *ACS Symp. Ser.*, vol. 1118, American Chemical Society, Washington, 2012, p. 361.
31. C. Huggett, *Fire Mater.*, 1980, **4**, 61.
32. V. Babushok and W. Tsang, *Combust. Flame.*, 2000, **123**, 488.
33. R. Sonnier, B. Otazaghine, L. Ferry and J.-M. Lopez-Cuesta, *Combust. Flame.*, 2013, **160**, 2182.
34. R. E. Lyon, R. N. Walters and S. Crowley, presented at the *24th Annual Conference on Recent Advances in Flame Retardancy of Polymeric Materials*, May 2013, Stamford, CT, USA.
35. P. M. Hergenrother, C. M. Thompson, J. G. Smith Jr, J. W. Connell, J. A. Hinkley, R. E. Lyon and R. Moulton, *Polymer.*, 2005, **46**, 5012.
36. B. Schartel, U. Braun, A. I. Balabanovich, J. Artner, M. Ciesielski, M. Doring, R. M. Perez, J. K. W. Sandler and V. Altstadt, *Eur. Polym. J.*, 2008, **44**, 704.
37. R. N. Walters and R. E. Lyon, *J. Appl. Polym. Sci.*, 2003, **87**, 548.
38. R. E. Lyon, M. T. Takemori, N. Safronava, S. I. Stoliarov and R. N. Walters, *Polymer.*, 2009, **50**, 2608.
39. R. Sonnier, C. Negrell-Guirao, H. Vahabi, B. Otazaghine, G. David and J.-M. Lopez-Cuesta, *Polymer.*, 2012, **53**, 1258.
40. J. R. Ebdon, D. Price, B. J. Hunt, P. Joseph, F. Gao, G. J. Milnes and L. K. Cunliffe, *Polym. Degrad. Stab.*, 2000, **69**, 267.
41. D.-J. Liaw and W.-C. Shen, *Polymer.*, 1993, **34**, 1336.
42. http://www.frxpolymers.com/index.html.
43. A. Schafer, S. Seibold, O. Walter and M. Doring, *Polym. Degrad. Stab.*, 2008, **93**, 557.

44. M. Cochez, M. Ferriol, J. V. Weber, P. Chaudron, N. Oget and J. L. Mieloszynski, *Polym. Degrad. Stab.*, 2000, **70**, 455.
45. J. Canadell, B. J. Hunt, A. G. Cook, A. Mantecon and V. Cadiz, *J. Polym. Sci., Part A: Polym. Chem.*, 2006, **44**, 6728.
46. J.-Y. Shieh and C.-S. Wang, *J. Polym. Sci., Part A: Polym. Chem.*, 2002, **40**, 369.
47. A. Schafer, S. Seibold, W. Lohstroh, O. Walter and M. Doring, *J. Appl. Polym. Sci.*, 2007, **105**, 685.
48. L. Yan, Y. B. Zheng, X. Liang and Q. Ma, *J. Appl. Polym. Sci.*, 2010, **115**, 1032.
49. D. Price, K. Pyrah, T. R. Hull, G. J. Milnes, J. R. Ebdon, B. J. Hunt, P. Joseph and C. S. Konkel, *Polym. Degrad. Stab.*, 2001, **74**, 441.
50. Y. Morita, M. Hagiwara and K. Araki, *J. Appl. Polym. Sci.*, 1980, **25**, 2711.
51. R. Sonnier, B. Otazaghine, H. Vahabi, J.-M. Lopez-Cuesta, C. Negrell-Guirao, G. David, F. Iftene and B. Canniccioni, presented at the 7th MoDeST Conference, September 2012, Prague, Czech Republic.
52. Y.-L. Chang, Y.-Z. Wang, D.-M. Ban, B. Yang and G.-M. Zhao, *Macromol. Mater. Eng.*, 2004, **289**, 703.

CHAPTER 13

Proton Conducting Phosphonated Polymers and Membranes for Fuel Cells

BAHAR BINGÖL AND PATRIC JANNASCH*

Polymer & Materials Chemistry, Department of Chemistry,
Lund University, P.O. Box 124, Lund, SE-221 00, Sweden
*Email: patric.jannasch@chem.lu.se

13.1 Introduction

The demand for energy in our modern society necessitates significant progress when it comes to energy storage and conversion technologies. Moreover, these technologies need to be environmentally benign not to worsen the serious pollution issues in urban areas and the global climate changes. Critical to the success of many new power sources is the development of new advanced materials that improve the performance, increase the lifetime, and widen the operational window and the applicability of the systems. Solid-state polymer electrolytes play an important role as key materials in advanced fuel cells, electrolyzers, and batteries, and have received significant research investments over the last decade. Thus, new generations of advanced ion-containing polymers (ionomers) with controlled transport properties are sought for thin polymer electrolyte membranes (PEMs) for use in, for example, polymer electrolyte membrane fuel cells (PEMFCs).[1,2]

The function of the PEMFC membrane is to physically separate the two electrodes, and to facilitate the transport of protons from the anode to the

RSC Polymer Chemistry Series No. 11
Phosphorus-Based Polymers: From Synthesis to Applications
Edited by Sophie Monge and Ghislain David
© The Royal Society of Chemistry 2014
Published by the Royal Society of Chemistry, www.rsc.org

cathode. At the anode, hydrogen gas is electrocatalytically split into protons and electrons. The electrons are led through an outer electrical circuit, while the protons are conducted through the membrane. Finally, the electrons, protons, and oxygen are catalytically combined to form water at the cathode. PEMFCs operating on hydrogen fuel potentially give no emissions, and the use of petroleum-based fuels can thus be avoided. Electrolyzers have the same principle as the PEMFC, but are operated in the reverse mode, using surplus electrical energy to generate pure hydrogen gas.

PEMFC membranes are based on hydrated ionomers in the protonated form. These materials typically phase separate into a percolating network of hydrophilic nanopores embedded in a hydrophobic polymer-rich phase domain.[2,3] The nanopores contain water and the acidic moieties, and conductivity occurs *via* protolysis and transport of dissociated protons by the dynamics of the water which acts as a proton solvent. The hydrophobic phase domain provides mechanical strength by stabilizing the morphology of the membrane. Ionomer membranes have to fulfill a number of important demands, including high proton conductivity, low oxygen/hydrogen permeability, and mechanical, chemical, and thermal stability.[2] In order to reduce the complexity and the cost, and to significantly increase the efficiency of the systems, it is highly desirable to operate the PEMFCs at or above 100 °C at water vapor partial pressures below 0.5 bar or with no humidification at all.[2] This is especially important for automotive applications.[4] While the current state-of-the-art perfluorosulfonic acid (PFSA) membranes have quite good properties which enable rather efficient PEMFC operation, the major obstacle to large-scale commercial use is their softness, limited gas barrier properties, and rather low proton conductivity at high temperatures and low humidities.[2] The softness is a consequence of the low glass transition temperature of PFSA membranes and leads to mechanical creep, and eventual failure at temperatures exceeding 80 °C. The limited gas barrier properties typically lead to crossover of H_2 to the cathode, where aggressive radicals are formed which degrade the membrane and catalyst layer. The proton conductivity of PFSA membranes suffer greatly at temperatures above 80 °C because of loss of the water necessary for conduction. This requires the development of new materials beyond the PFSAs with high mechanical stability and high proton conductivity from subzero degrees up to 120 °C.

To devise polymeric systems that can conduct protons with little or no water, in addition to possessing good chemical and electrochemical stability above 100 °C, is very challenging and has been the focus of intensive research over the last decade. One of the most successful strategies has been to replace the water with less volatile compounds. For example, strong oxo acids, *e.g.* phosphoric acid[5,6] or sulfuric acid,[6] have been complexed with basic polymers such as polybenzimidazole (PBI). PEMs based on PBIs doped with phosphoric acid have been used successfully at temperatures well above 100 °C under very dry conditions.[7] However, there is still a risk that the acid will be partly leached out, especially at low temperature and high humidity.

Similar to phosphoric acid, phosphonic acids are intrinsically proton conductive.[8] However, phosphonic acids can be covalently attached to polymers in order to prevent leaching. Comparative studies of sulfonic acid, imidazole, and phosphonic acid functionalized model compounds have indicated that phosphonic acids possess attractive features for applications in high-temperature PEMFCs.[9] This has spawned extensive research in recent years to prepare and study phosphonated polymers as proton conductors under low humidity conditions.[10,11]

In the present chapter we discuss different approaches to phosphonated polymers and their properties as proton conducting PEMs. The focus is thus mainly on materials most relevant for PEMFC. First, we discuss the general properties of phosphonic acid in the context of PEMs. Then we review various synthetic strategies to phosphonated polymers. The synthetic strategies include chemical modification of polymers by attaching phosphonic acid *via* alkyl, aryl, or perfluoroalkyl spacer groups, as well as direct polymerization using phosphonated monomers. Finally, we point out some possible future directions for the development of new generation materials.

13.2 Molecular Design of Phosphonated Polymers for Proton Conducting Membranes

It is important to consider the specific properties of the phosphonic acid group, especially in relation to the more commonly employed sulfonic acid group ($pK_a < -1$). The two ionizable moieties of alkyl- and arylphosphonic acids typically have pK_a values between 2–3 and 7–8 for the first and second protons, respectively, and have a pronounced amphoteric character.[12,13] These compounds are usually associated in dynamic hydrogen-bonded networks with a high degree of proton self-dissociation, leading to a high concentration of charge carriers.[8,9] Phosphonic acids can facilitate structure diffusion of protons through fast hydrogen-bond breaking and forming processes in the hydrogen bonded network through a so-called Grotthus mechanism.[9] Thus, provided that the hydrogen bonded network is sufficiently large and has fast dynamics, phosphonated PEMs can reach high intrinsic proton conductivity at temperatures above 100 °C under dry conditions (Table 13.1).[9] However, under these conditions, phosphonic acids tend to form anhydrides *via* condensation and formation of water [$R-PO(OH)_2 + (HO)_2OP-R \rightarrow R-PO(OH)-O-(HO)OP-R + H_2O$]. This causes loss of acid concentration, but also a slower dynamics of the hydrogen bonded network because of the anhydride crosslinks. The overall result is a serious decrease in the proton conductivity. Still, the anhydride formation is reversible and the acid form can be regenerated by treatment in hot water. The proton conductivity increases considerably with the water concentration in the membrane, but the water uptake per acid unit ($\lambda = [H_2O]/[-PO_3H_2]$) is typically much lower than for the stronger sulfonic acids.[10] The potential of phosphonic acids as proton conductors has been assessed by analyzing the

Table 13.1 Proton conductivity of selected phosphonated polymers and
PEMs.

Polymer[a]	Phosphonic acid content (mmol P g^{-1})	Conductivity/Conditions[b]	Ref.
PVPA	9.25	3×10^{-3} mS cm^{-1} at 20 °C and 0.1 mS cm^{-1} at 100 °C under nominally dry conditions	20,21
PVPA	9.25	10^{-3} mS cm^{-1} at 10% RH and 10 mS cm^{-1} at 60% RH, 21 °C	20
PVPA	9.25	1 and 50 mS cm^{-1} at 100 and 161 °C, respectively, at a water partial pressure of 10^5 Pa	20
Poly(vinylbenzyl-phosphonic acid)	5.04	4.5×10^{-1} mS cm^{-1} at 160 °C under anhydrous conditions; 6 mS cm^{-1} at 110 °C under 50% RH	24
PS-*ran*-PVPA	0.7	7.16×10^{-13} mS cm^{-1} at ambient temperature under dry conditions; 10^{-6} mS cm^{-1} at ambient temperature under immersed conditions; 10^{-2} mS cm^{-1} at ambient temperature after doping with phosphoric acid	40
Poly(triazole-*co*-VPA)	8.03	1 mS cm^{-1} at 120 °C under anhydrous conditions	39
Poly(vinyl-imidazole-*co*-VPA)	5.3–3.7	10^{-3} and 10^{-9} mS cm^{-1} between room temperature and 140 °C under dry conditions	38
Phosphonated poly(4-phenoxybenzoyl-1,4-phenylene)	1.3	0.1 mS cm^{-1} at 80 °C and 90% RH	47
Phosphonated poly(phenyl sulfone)	1.98	1.8 mS cm^{-1} at 110 °C and 100% RH	44
Phosphonated poly(phenyl sulfone)	2.07	6 mS cm^{-1} at 80 °C and 95% RH	44
Poly(1,3-phenylene-5-phosphonic acid)	6.3	30 mS cm^{-1} at 220 °C and 3 mS cm^{-1} at 110 °C under dry conditions	53
Poly(1,3-phenylene-5-phosphonic acid)	6.3	30 mS cm^{-1} at 110 °C under 1 atm vapor pressure	53
Phosphonated poly(1,3,4-oxadiazoles)	0.5	4×10^{-4} mS cm^{-1} at 20 °C to 5×10^{-3} mS cm^{-1} at 90 °C at 100% RH	54
Phosphonated poly(arylene ether)	4.13	92 mS cm^{-1} at ambient temperature, fully hydrated	56
Phosphonated poly(arylene ether)	4.13	150 mS cm^{-1} at ambient temperature, fully hydrated	56
PAES-*g*-PVPA	1.72	4 mS cm^{-1} at 120 °C, fully immersed conditions	58
PAES-*g*-PVPA	3.16	25 mS cm^{-1}, fully immersed conditions	58
Benzoyl(difluoro-methylenephos-phonic acid) functional polysulfone	1.47	5 mS cm^{-1} at 100 °C, unspecified conditions	60

Table 13.1 (*Continued*)

Polymer[a]	Phosphonic acid content (mmol P g^{-1})	Conductivity/Conditions[b]	Ref.
Poly(trifluorovinyl ether) with phosphonated oligo(ethylene oxide) side chains	2.94	68 mS cm^{-1} at room temperature, 100% RH	64
Fluoro polymer with pendant phosphonic acid groups	2.7–8.4	0.02–20 mS cm^{-1} at room temperature and 95% RH	66
Fluoro polymer with pendant phosphonic acid groups	2.7–8.4	2.51–0.1 mS cm^{-1} between 90 and 120 °C at RH from 30 to 100%	66
Phosphonated poly(pentafluoro-styrene)	3.89	10–100 mS cm^{-1} at 10^5 Pa partial water pressure between 97 and 180 °C	63

[a] PVPA = poly(vinylphosphonic acid); PS = polystyrene; PAES = poly(arylene ether sulfone).
[b] RH = relative humidity.

temperature dependence of conductivity of model compounds such as heptylphosphonic and 1,3-propylenediphosphonic acid.[9] Contrary to the rapidly declining conductivity of Nafion® at temperatures above 80 °C, anhydrous heptylphosphonic acid and 1,3-propylenediphosphonic acid showed a steady rise in conductivity to reach 10 and 50 mS cm^{-1}, respectively, at 200 °C. A comparative study of hydrated heptylsulfonic acid and anhydrous heptylphosphonic acid further indicated the superior properties of the phosphonic acid, which included a higher conductivity in the anhydrous state and a greater thermal and electrochemical stability at elevated temperatures.[9] The acidity of phosphonic acids is significantly increased by inductive effects. For example, trifluoromethylphosphonic acid has $pK_{a1} = 1.2$ and $pK_{a2} = 3.9$.[13] By increasing the acidity, the water uptake increases, the propensity for anhydride formation is reduced, and the properties associated with the amphotericity are gradually lost.

Preparation of phosphonated membranes with high mechanical stability and high conductivity requires careful macromolecular design. High intrinsic conductivity requires a high local concentration of phosphonic acids to enable the formation of percolating hydrogen-bonded networks. This is usually not possible by combining ionic and non-ionic monomers to form random and alternating copolymers where the acid groups become too dispersed in the membrane. In contrast, block and graft copolymers typically self-assemble to form nanophase separated structures because of the connectivity and immiscibility of the dissimilar segments.[3,14] Thus, the self-assembly of block and graft copolymers containing phosphonated segments usually leads to nanostructured membranes with high local concentration of phosphonic acid. Moreover, the density and spatial distribution of phosphonic acids can be tuned by controlling the volume fraction, chemistry, and degree of polymerization of the blocks and grafts. This is a significant advantage since the morphology plays a key role for the water and proton transport properties of the membranes. The phosphonated segments need to be combined with rigid and high molecular weight non-ionic blocks to provide membranes with good water swelling resistance and high mechanical stability.

The interest in phosphonated polymers for PEMFC applications has resulted in the development of new synthetic pathways to these materials, either by direct polymerization of phosphonated monomers or by post-phosphonation of prepolymers. Until recently, most phosphonated polymers were prepared *via* the latter strategy, which usually requires multiple synthetic steps. This strategy is mainly used to phosphonate polymers with aromatic backbones, such as poly(arylene ether)s, poly(arylene ether sulfone)s (PAESs), and poly(arylene ether ketone)s (PAEKs). These high-performance polymers typically have excellent mechanical, chemical, and thermal stability. Polymerization of phosphonated monomers such as phosphonated vinyl monomers provides a more direct route to phosphonated polymers. However, the typical polymerization behavior of phosphonated monomers both in their ester and acid forms often leads to side reactions, and thus to polymers with limited molecular weight and broad molecular weight distribution. More recently, significant research efforts have been devoted to detailed investigations of the polymerization behavior of phosphonated monomers with the goal to overcome the challenges associated with the phosphonate group. In the following sections we review various approaches to phosphonated polymers and the properties of the materials in the context of fuel cells.

13.3   Phosphonated Non-Aromatic Hydrocarbon Polymers

Poly(vinylphosphonic acid) (PVPA) contains an extremely high concentration of phosphonic acid directly attached to a flexible polymer main chain (9.25 mmol P/dry membrane, Scheme 13.1a).[15] High molecular weight PVPA has been prepared either by radical polymerization of vinylphosphonic acid (VPA) in aqueous medium[16] or *via* anionic polymerization[17] of dialkyl vinylphosphonates in tetrahydrofuran, followed by hydrolysis of the ester groups. Radical and anionic polymerization have been reported to give average molecular weights of 62 and 814 kg mol^{-1}, respectively.[16,17] Microstructure analysis of PVPA obtained by radical polymerization suggested that the polymerization of VPA proceeds *via* cyclopolymerization of the VPA anhydride as intermediate.[16,18] As a result, the atactic polymer contains a substantial fraction of head-to-head and tail-to-tail configurations, in addition to the dominating head-to-tail arrangement. The polymerization mechanism of VPA largely explains the temperature and solvent dependence

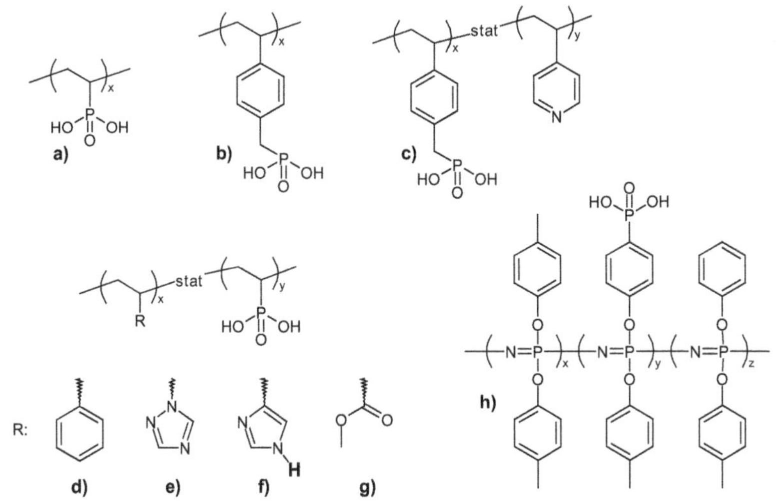

Scheme 13.1 Selected phosphonated non-aromatic hydrocarbon polymers: (a) PVPA,[16] (b) poly(vinylbenzylphosphonic acid),[32] (c) poly(vinylbenzyl-phosphonic acid-*co*-4-vinylpyridine),[41] (d) poly(VPA-*co*-styrene),[40] (e) poly(VPA-*co*-1-vinyl-1,2,4-triazole),[39] (f) poly(VPA-*co*-4-vinylimidazole),[38] (g) poly(VPA-*co*-methyl acrylate),[90] and (h) poly(aryloxyphosphazene).[34]

of the rate of polymerization, as well as the copolymerization behavior of VPA.[19] Although anionic polymerization has the potential to yield PVPA with controlled molecular weight and narrow molecular weight distribution, polymers obtained from anionic polymerization may be rather polydisperse (PDI > 2) because of side reactions.[17]

The proton conductivity of PVPA has been investigated to obtain a fundamental understanding of the intrinsic proton conductivity of phosphonated polymers.[17,20,21] PVPA prepared by radical and anionic mechanisms reached a similar conductivity, and the proton conductivity of PVPA under nominally dry conditions has been reported to be 3.0×10^{-3} and 1.0×10^{-1} mS cm^{-1} at 20 and 100 °C, respectively.[21] The conductivity is significantly affected by the water content. Under nominally dry conditions, when the material contained 30% of condensation products and had a water content corresponding to $\lambda = 1.5$, the conductivity reached 5×10^{-1} mS cm^{-1} at 130 °C,[22] while the conductivity reached 50 mS cm^{-1} at 130 °C under 10^5 Pa water vapor pressure.[20] The proton conductivity results from mobile phosphonic acid protons, and the ^{1}H motion of P–OH by hydrogen bonding, which allows proton transfer *via* a vehicle type of mechanism. The comparable conductivity of the polymer in its potassium form to that of its acid form also supports a vehicle type of proton transfer mechanism. Any contribution from structural diffusion would be much more affected by the introduction of foreign cations into the hydrogen bond network. Annealing PVPA at high temperatures leads to a dramatic decrease in conductivity because of anhydride formation. Condensation occurs at random locations in PVPA since there is no phase segregation observed between the normal acid and the acid anhydride.[21] Up to 45% relative humidity (RH), corresponding to 17 wt% water uptake, the sample remained as a powder, but above this humidity a transparent film was formed. With increasing humidity, the film became more sticky and eventually turned into a gel at 100% RH. The phosphonic anhydrides were hydrolyzed to the free acids when in contact with humid air or water. These studies provide useful information on the general physicochemical properties of PVPA and other related polymers.[20-22] Still, neat linear PVPA has poor film-forming properties and the high acid concentration makes it water soluble. This necessitates, for example, copolymerization to obtain useful membranes.

Styrene monomers with phosphonic acid placed on side chains with different spacer length have been polymerized by a radical mechanism to obtain membranes with controlled water uptake at high ion exchange capacity (IEC) values.[23-25] The proton conductivity of polymers containing ethyl and hexyl spacers between the polystyrene (PS) main chain and the phosphonic acid reached similar conductivity between -10 and 140 °C.[25] Although the polymer with the hexyl spacer had a lower concentration of phosphonated sites, it seemingly formed a more favorable morphology for charge transport. The conductivity of these polymers was measured to be approx. 3×10^{-5} mS cm^{-1} under nominally dry conditions at 25 °C,[25] which was much higher than that of poly(vinylbenzylphosphonic acid) at 1×10^{-7} to 1×10^{-8} mS cm^{-1}.[24,26] The much higher proton conductivity of the polymers with ethyl and hexyl spacers was attributed to the presence of highly aggregated acids which formed proton conducting hydrogen-bond networks. Small-angle X-ray scattering experiments indicated the existence of nanophase separated domains. Recently, Higashihara and co-workers have similarly prepared polyacrylates functionalized with phosphonic acid *via* long alkyl side chains.[27] In addition, hyperbranched phosphonated polyacrylates with a proton conductivity of 1.5×10^{-5} S cm^{-1} at 150 °C under dry conditions have been proposed for high-temperature fuel cells.[28] However, the presence of the ester links in these materials raises questions concerning the stability. A similar trend with the spacer length has previously been reported for imidazoles tethered to polymers *via* alkyl spacers.[29,30] The increase in conductivity with the alkyl spacer length was explained by a higher flexibility which enabled less restrictions for aggregation and hydrogen bonding of the imidazole group.[31]

Diethyl vinylbenzylphosphonate has been homopolymerised *via* atom transfer radical polymerization (ATRP) to obtain poly(diethyl vinylbenzylphosphonate), which was hydrolyzed to obtain the acid derivative (Scheme 13.1b).[32] The proton conductivity of poly(vinylbenzylphosphonic acid) was measured both at dry conditions between 20 and 160 °C and in an atmosphere with a defined RH. Under dry conditions, the polymer exhibited a proton conductivity with a linear dependence on temperature, and reached

4.5×10^{-1} mS cm^{-1} at 160 °C. At 50% RH, the conductivity increased by two orders of magnitude, reaching a maximum of 6 mS cm^{-1} at 110 °C.

Polyphosphazenes are flexible inorganic phosphorus-containing polymers which have been phosphonated and studied as electrolyte membranes.[33–35] For example, brominated poly(aryloxyphosphazene) has been phosphonated *via* lithiation and reaction with diphenyl chlorophosphonate (Scheme 13.1h). Membranes cast from *N,N*-dimethylformamide (DMF) had proton conductivities between 10 and 100 mS cm^{-1} after equilibration in water.[34]

Fukuzaki *et al.* have prepared highly phosphonated poly(*N*-phenylacrylamide).[36] The proton conductivity of a crosslinked membrane was 88 mS cm^{-1} at 95% RH and 80 °C, and it reached 1.9 and 4.7×10^{-2} mS cm^{-1} at 50 and 30% RH, respectively. Like with the polyacrylates, the hydrolytic stability of this material is questionable.

In addition to homopolymers, various random and block copolymers based on phosphonated monomers have been synthesized *via* radical and anionic polymerization.[17,37] The details of the synthesis and properties of membranes based on block and graft copolymers are described below in Section 13.6. Statistical copolymers of VPA with styrene and other vinyl monomers functionalized with heterocycles such as imidazole[38] and triazole[39] have been investigated by Bozkurt and co-workers (Scheme 13.1e,f). For example, when carried out at similar feed composition, free radical copolymerization of VPA with 1-vinyl-1,2,4-triazole[39] and 4-vinylimidazole[38] in DMF yielded statistical copolymers with similar VPA content. Investigating the proton conductivity of a series of copolymers with varying composition indicated an increase of conductivity with increasing VPA content, and a Grotthus-like mechanism was reported to be responsible for the proton transport. Poly(VPA-*co*-1-vinyl-1,2,4-triazole) containing 70 mol% VPA had a conductivity of 10^{-3} S cm^{-1} at 120 °C under dry conditions, which was about an order of magnitude higher than that of PVPA under the same conditions. The observed increase in conductivity was attributed to an increased mobility of the copolymer backbone compared to PVPA, which suggested that the triazole-containing copolymers provided defect sites to facilitate the proton conduction.[39] In addition, poly(VPA-*co*-1-vinyl-1,2,4-triazole) had improved film-forming properties compared to PVPA. These polymers showed much higher conductivity compared to poly(VPA-*co*-4-vinylimidazole), which reached conductivities between 10^{-12} and 10^{-6} S cm^{-1} in the temperature range −2 to 100 °C.[38]

Poly(VPA-*co*-styrene) has been obtained by free radical polymerization of styrene and dimethyl vinylphosphonate in the bulk, followed by hydrolysis.[40] The resulting copolymers, shown in Scheme 13.1d, had a low VPA content (5–14 mol%). Despite the difference between the reactivity ratios of the comonomers, poly(styrene-*co*-dimethyl vinylphosphonate) presented only a single glass transition temperature, indicating a near random connectivity between the styrene and dimethyl vinylphosphonate units in the copolymers.[40] The copolymers were stable under air and inert atmosphere up to 300 °C. The conductivity under dry and immersed conditions was 7×10^{-13} and 1×10^{-6} mS cm^{-1}, respectively, which could be increased to 10^{-2} mS cm^{-1} after doping with phosphoric acid. Thus, the conductivity of the copolymers was too low to be considered for fuel cell applications.

Statistical copolymers from diisopropyl *p*-vinylbenzylphosphonate and 4-vinylpyridine were prepared *via* ATRP.[41] The resulting copolymers were hydrolyzed to obtain acid derivatives (Scheme 13.1c). The proton conductivity of a copolymer containing 90 mol% 4-vinylpyridine was investigated at 25 °C under varying RH. The proton conductivity rose by several orders of magnitude with increasing water content. At 60% RH, the proton conductivity reached 10 mS cm^{-1} at 25 °C.

Markova *et al.* have prepared phosphonated polyalkenes where the acid groups were precisely sequenced along an unsaturated polyethylene backbone.[42] The authors reported proton conductivities of ~ 10 mS cm^{-1} under anhydrous as well as humidified conditions.

13.4 Phosphonated Aromatic Backbone Polymers

Several synthetic strategies have been developed to prepare phosphonated aromatic polymers, where the acid groups are attached either directly[7,43–45] or *via* spacers to an aromatic backbone.[44,46–48] These polymers can be prepared either by post-phosphonation of prepolymers *via*, *e.g.* transition metal catalyzed Michaelis-Arbuzov reactions and lithiation chemistry, or by direct polymerization of phosphonated monomers *via* polycondensation. Both synthetic strategies then require hydrolysis of the esters to obtain the free acid. The former strategy requires the formation of C–P bonds in the polymer structure, and the latter necessitates the synthesis and purification of suitable monomers.

13.4.1 Aromatic Polymers Directly Phosphonated on the Main Chain

Polymeric materials with phosphonic acid groups directly linked to an aromatic ring are of great interest for PEMFC applications due to the generally high chemical and thermal stability of aromatic polymers. One important route involves the reaction of polymeric aryl bromides and iodides with trialkyl phosphites *via* metal-catalyzed Michaelis-Arbuzov reactions.[47,49,50] A higher degree of phosphonation at low reaction temperatures was achieved when palladium was used instead of nickel as the catalyst. Aryl bromide sites on brominated poly(phenyl sulfone)s have been reacted with a H-phosphonate diester in the presence of Pd(PPh$_3$)$_4$ and triethylamine to obtain the corresponding phosphonated derivative.[44,50] Complete phosphonation was only possible at high catalyst loadings (30–50 mol%). Still, the degree of phosphonation achieved in these reactions were quite low (1.3–2.07 mmol P g^{-1}) due to the difficulties in forming the C–P bond.

For one of the resulting polymers the proton conductivity was measured to be 100 mS cm^{-1} under 90% RH at 80 °C.[47]

Lithiation represents a non-catalytic strategy to form C–P bonds as an alternative to catalyzed Michaelis–Arbuzov reactions.[51] The formation of C–P bonds through aryllithium reagents has been known since the 1950s.[52] Various phosphonic acid functional aromatic hydrocarbon polymers have been prepared using this chemistry. PAESs with quite a low degree of phosphonation (0.76 mmol P g^{-1}) in *ortho*-sulfone positions were obtained by lithiation using *n*-butyllithium followed by addition of diaryl and dialkyl chlorophosphonates.[45] The low phosphonation level was attributed to the steric hindrance due to the presence of bulky sulfone groups next to the reaction sites, in combination with a low reactivity of the *ortho*-to-sulfone aryllithium sites. A higher degree of phosphonation (0.95 mmol P g^{-1}) was achieved when *ortho*-ether lithiated sites were reacted with the chlorophosphonates. PAESs phosphonated in *ortho*-sulfone positions formed films with a higher mechanical stability compared to the *ortho*-ether analogs. The membranes obtained from these polymers were not further characterized for fuel cell use because of the low water uptake of the acid derivatives (∼2 wt%).

Aromatic polymers with phosphonated sites directly linked to the main chain have also been prepared by direct polymerization (Scheme 13.2a–d).[53–56] For example, phosphonated poly(1,3,4-oxadiazole)s have been prepared by direct polycondensation of 1,4-dicarboxy-2-phenylphosphonic acid and their dihydrazides in an ionic liquid using triphenyl phosphite as a reaction activator.[54] The membranes obtained from these polymers were thermally and chemically stable and possessed proton conductivities in the range between 4×10^{-4} and 5×10^{-3} mS cm^{-1} at 100% RH, which is three orders of magnitude lower than the state-of-the-art PFSA membrane Nafion®.

Phosphonic acid functional poly(arylene ether)s have been prepared from phosphonated aromatic diols and aromatic difluorides *via* nucleophilic aromatic substitution at high temperatures, followed by hydrolysis of the ester groups.[56] These polymers formed membranes with a conductivity of 150 mS cm^{-1} at 100 °C under fully hydrated and 2.2 mS cm^{-1} at 120 °C under dry conditions. The proton conductivity reached ∼100 mS cm^{-1} upon doping with phosphoric acid. Kallitsis and co-workers have reported on homopolymers and copolymers with moderate molecular weights based on aromatic polyethers bearing diphosphonate and diphosphonic acid groups.[55] Blends with pyridine-modified polyether were prepared in order to fabricate membranes. Using the blended membranes, it was possible to reach higher doping levels with phosphoric acid than with the neat pyridine-modified polyether membrane. Rager and co-workers prepared poly(*m*-phenylenephosphonic acid) by nickel-catalyzed reductive coupling.[53] The conductivity of the polymer, depicted in Scheme 13.2c, reached 6 mS cm^{-1} at 115 °C under full humidification and decreased to 2 mS cm^{-1} at 161 °C. Reproducible conductivity measurements were possible without

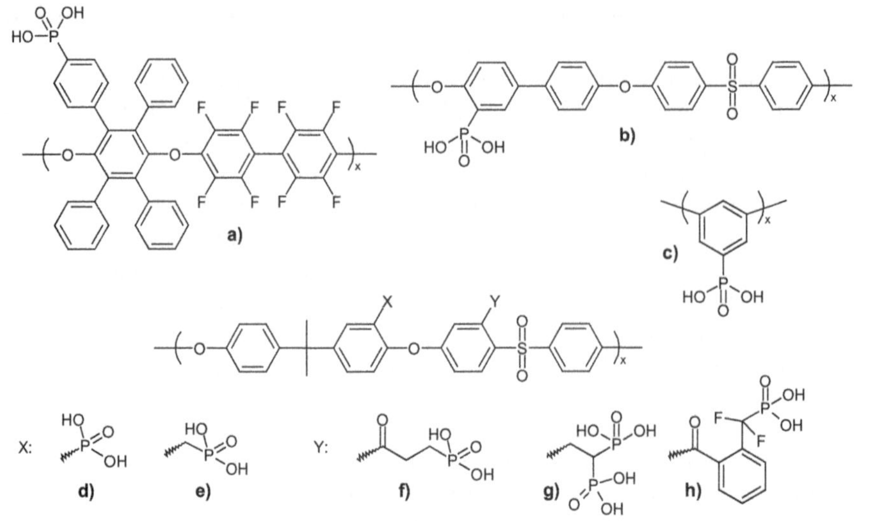

Scheme 13.2 Selected phosphonated aromatic backbone polymers: (a) phosphonated poly(arylene ether),[43] (b) phosphonated poly(phenyl sulfone),[44] (c) poly(m-phenylene-5-phosphonic acid),[53] (d) PAES with a
phosphonic acid group directly attached to the aromatic backbone,[45] (e) methylenephosphonic acid functionalized PAES, (f)
PAES with a phosphonated alkyl chain,[58] (g) PAES bearing a
bis(phosphonic acid) on short alkyl chains,[58] and (h) PAES with a
pendant phenyl-CF$_2$PO$_3$H$_2$ group.[60]

humidification, which was attributed to irreversible condensation reactions
under these conditions. These results combined with its high thermo-oxidative stability indicated the promise of poly(m-phenylenephosphonic acid)
as a proton conductor under sufficiently high RH.

13.4.2 Aromatic Polymers with Phosphonated Side Chains

The phase separation of the acid groups from the polymer backbone may be
enhanced in the membrane if the phosphonic acid groups are placed on
short side chains.[57] Aromatic polymers with the phosphonic acid group attached to the main chain via various side chains (Scheme 13.2f–h) can be
obtained by phosphonation of prepolymers via catalyzed Michaelis–Arbuzov
reactions and by lithiation chemistry. The synthesis does not require the
formation of C–P bonds, which typically enables higher degrees of phosphonation compared to aromatic polymers with phosphonic acids directly
attached to the polymer backbone. For example, reacting ortho-sulfone
lithiated PAESs with diethyl 3-phosphonopropionate or tetraisopropyl viny-
lidenediphosphonate produced polymers with a degree of phosphonation of
0.8 and 1.4 mmol P g^{-1}, respectively (Scheme 13.2f,g).[58] Hydrolysis of these
materials gave PAESs with short alkyl chains terminated with mono- and

bisphosphonic acids, respectively. Membranes obtained from polymers with monophosphonated side chains only took up moderate amounts of water (<10 wt%) and reached a maximum conductivity of 4 mS cm^{-1} at 120° C under fully immersed conditions. In contrast, membranes obtained from bisphosphonic acid functionalized PAESs displayed much higher water uptake, ~28 wt% at 93% hydrolysis, and conductivity, 25 mS cm^{-1}, than monophosphonic acid functionalized polymers under identical conditions, most probably because of higher acidity and higher acid concentrations. Post-phosphonation of chloromethyl-functionalized PAESs in *ortho*-ether positions *via* the Michaelis–Arbuzov reaction yielded polymers having a methylene group in-between the aromatic chain and the phosphonic acid group, as shown in Scheme 13.2e.[59] Membranes formed by these polymers took up a high amount of water, 52 wt%, and reached a conductivity of 1.2 mS cm^{-1} under fully immersed conditions. Scheme 13.2h shows PAESs functionalized with pendant phenyl-CF$_2$PO$_3$H$_2$ groups, which were synthesized by functionalization of lithiated PAESs with iodobenzoyl groups, then reacting with [(diethoxyphosphinyl)difluoromethyl]zinc bromide.[60] The α,α-difluorophosphonic acid induced a high acidity and the polymers consequently reached a rather high conductivity, 5 mS cm^{-1} at 100 °C.

13.5 Phosphonated Fluoropolymers

By using hydrophobic fluorinated polymer backbones, the phase separation can be made more distinct in phosphonated membranes, which in turn can enhance the proton conductivity. DesMarteau and co-workers have studied the proton transport characteristics of model perfluoroacid compounds functionalized with phosphonic, phosphinic, sulfonic, and carboxylic acids.[61] The results indicated that the proton transfer in phosphonic and phosphinic acids occurs *via* structural diffusion rather than by a vehicle mechanism. The findings suggested that fluoroalkylphosphonic and - phosphinic acids are good candidates for further development as anhydrous, high-temperature proton conductors.

Phosphonated polymers with fluorinated backbones have been obtained by reacting fluorinated poly(arylene ether)s with a H-phosphonate diester in the presence of Pd(PPh$_3$)$_4$ (Scheme 13.3d).[43,62] Using fluorinated aromatic polymers as the backbone can be expected to improve the dimensional stability of the membranes compared to aromatic polymers with hydrocarbon backbones because of the increased hydrophobicity induced by the presence of the fluorocarbon. The phosphonated polymers with fluorinated aromatic backbones reached a conductivity of 6 mS cm^{-1} at 95% RH at 80 °C and 2.6 mS cm^{-1} in water at room temperature,[62] which was below that of Nafion® 117.

Recently, Atanasov and Kerres reported on the phosphonation of poly(2,3,4,5,6-pentafluorostyrene) *via* the Michaelis–Arbuzov route using tris(trimethylsilyl) phosphite as the phosphonating agent, and the phosphonic

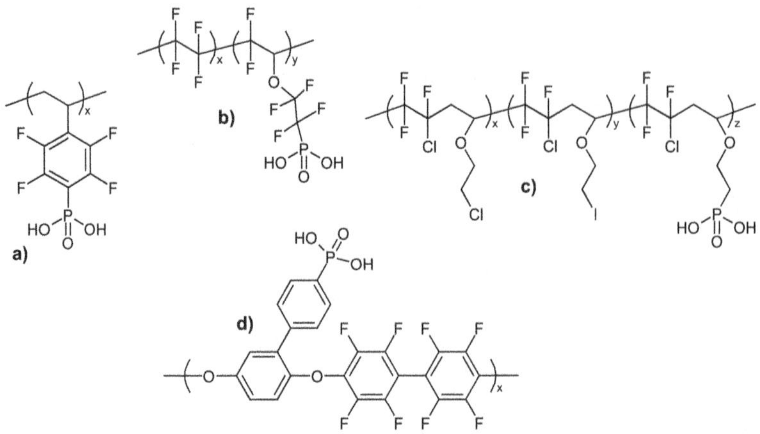

Scheme 13.3 Phosphonated fluorinated polymers: (a) poly(2,3,5,6-tetrafluoro-4-vinylphenylphosphonic acid),[63] (b) phosphonated perfluorocarbon polymer,[91] (c) phosphonated fluorinated copolymer,[66] (d) phosphonated fluorinated poly(aryl ether).[62]

acid derivatives were obtained after hydrolysis.[63] A high degree of phosphonation was achieved and the water uptake of the resulting poly(2,3,5,6-tetrafluoro-4-vinylphenylphosphonic acid), shown in Scheme 13.3a, exceeded that of Nafion® 117. However, the λ values of the phosphonated polymer and Nafion® were quite similar. The anhydride formation was reported to be suppressed, and the higher level of water uptake gave a high proton conductivity. As seen in Figure 13.1, the proton conductivity of poly(2,3,5,6-tetrafluoro-4-vinylphenylphosphonic acid) was found to be a factor of 4–10 higher than phosphonated hydrocarbon polymers such as PVPA and poly(*m*-phenylenephosphonic acid). This was even higher than that of Nafion® 117, up to two times higher at 155 °C. Consequently, the conductivity of poly(2,3,5,6-tetrafluoro-4-vinylphenylphosphonic acid) is among the highest proton conductivities reported for phosphonated polymers. The high conductivity is presumably due to the high acidity induced by the four fluorine atoms in the aromatic ring, which depresses the anhydride formation and leads to a high level of water uptake and acid dissociation. Still, the polymer is water soluble and therefore needs to be immobilized in the membrane before use in fuel cells.[63]

He *et al.* have prepared phosphonated aryl trifluorovinyl ethers containing flexible oligo(ethylene oxide) units and subsequently homopolymerized these monomers by thermal cyclopolymerization in bulk at 180 °C.[64] The resulting polymer was treated with bromotrimethylsilane to obtain the phosphonic acid derivative. The phosphonated trifluorovinyl ether monomers[65] were also copolymerized with tetrafluoroethylene *via* free radical polymerization in an autoclave at 70 °C using 1,1,2-trichloro-1,2,2-trifluoroethane as solvent. Moreover, radical terpolymerizations of the same trifluorovinyl ether monomers were performed with mixtures of

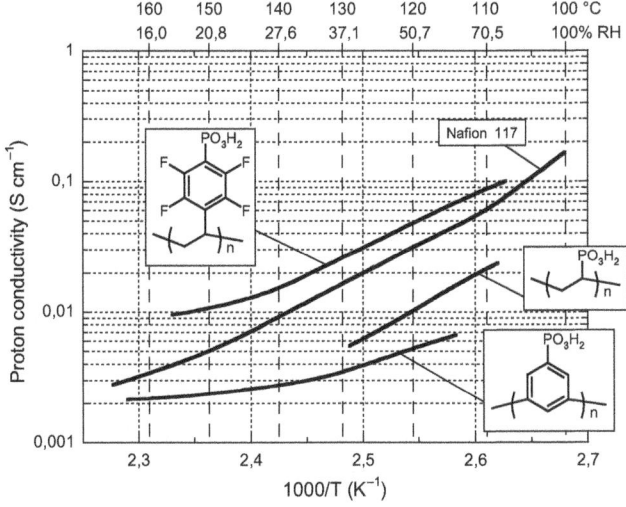

Figure 13.1 Arrhenius conductivity diagram showing the proton conductivities of phosphonated polypentafluorostyrene, PVPA, poly(*m*-phenylenephosphonic acid), and Nafion® 117 measured as a function of temperature at a water partial pressure of 1 atm (105 Pa). (Adapted from Atanasov and Kerres[63] with permission from the American Chemical Society.)

tetrafluoroethylene and perfluoro(propyl vinyl ether) at 70 °C. Membranes fabricated from these polymers displayed proton conductivities in the range of 36–68 mS cm^{-1} at room temperature under fully hydrated conditions.

Roualdes and co-workers have prepared phosphonated polymers by synthesizing fluorinated polymers *via* radical polymerization of fluorinated alkenes such as chlorotrifluoroethylene and vinyl ethers such as 2-chloroethyl vinyl ether, followed by phosphonation *via* the Arbuzov reaction.[66] The resulting polymers depicted in Scheme 13.3c had a phosphonic acid content ranging from 18 to 47 mol%. The level of conductivity was found to be in the range 0.02–20 mS cm^{-1}, which was comparable with Nafion® 115 at about 50 mS cm^{-1} at 25 °C and 95% RH.

13.6 Phosphonated Block and Graft Copolymers

Various phosphonated block and graft copolymers have been prepared in order to obtain membranes with high local concentrations of phosphonic acid. The block copolymers have mainly been based on phosphonated and non-phosphonated vinyl monomers, whereas the graft copolymers have been based on polymers with aromatic backbones such as PAESs, poly(phenylene oxide) (PPO), and PBI, from which phosphonated vinyl monomers have been grafted.

Scheme 13.4(a) shows poly(styrene-*b*-VPA) which has been synthesized *via* sequential anionic polymerization of styrene with diethyl[37] or diisopropyl

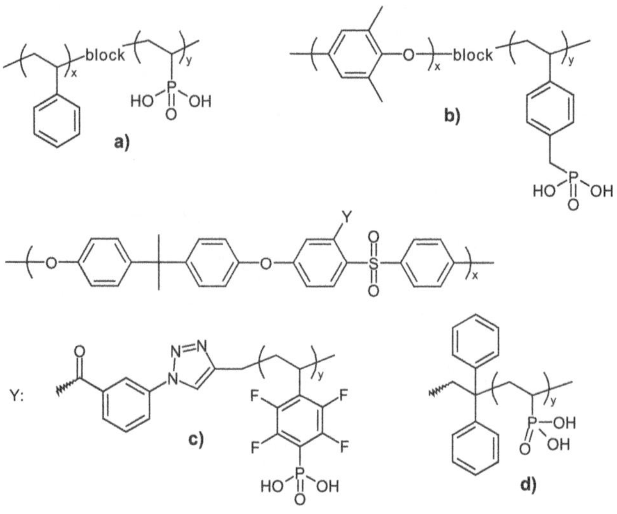

Scheme 13.4 Phosphonated block and graft copolymers: (a) poly(VPA-*b*-styr-ene),[17,37] (b) poly(phenylene oxide-*b*-vinylbenzylphosphonic acid),[69] (c) PAES-*g*-PVPA,[58] and (d) PAES-*g*-poly(2,3,5,6-tetrafluoro-4-vinylphe-nylphosphonic acid).[73]

vinylphosphonates,[17] followed by hydrolysis. The block copolymerization of styrene with diethyl vinylphosphonate led to high molecular weights (12–57 kg mol^{-1}). When diisopropyl vinylphosphonate was used as the comonomer, the growing chain precipitated at an early stage of the poly-merization due to its limited solubility in the reaction solvent. This limited the conversion and thus the molecular weight of the polymers to below 10 kg mol^{-1}. The VPA content of poly(styrene-*b*-VPA) from diisopropyl vinylphosphonate ranged from 32 to 82 wt% with block lengths between 1.8 and 3.8 kg mol^{-1},[17] while block copolymers from diethyl vinylphosphonate contained 30 and 70 wt% phosphonic acid with block lengths between 5 and 15 kg mol^{-1}.[37] Block copolymers obtained from diethyl or diisopropyl vinylphosphonate showed two distinct glass transition temperatures to in-dicate the microphase separation of the blocks. The glass transition tem-perature of PS and PVPA blocks were observed to shift to higher temperatures upon annealing the material, where the shift was much more pronounced for the PVPA block.[37] This shift was attributed to the change in segmental mobility of both blocks as a result of anhydride formation.

Block copolymers obtained from diethyl vinylphosphonate self-assembled into phase separated morphologies with continuous proton-conducting PVPA phase domains. For example, a block copolymer containing 51 wt% VPA formed a lamellar-like morphology, with domain sizes between 15–25 nm.[37] Although the phases were distinctly separated, the block co-polymers displayed rather irregular morphologies, which may be more ad-vantageous for the proton conductivity compared to periodic morphologies.[67]

The details of the self-assembly of the block copolymer from diisopropyl vinylphosphonate were not described. The block copolymers obtained both from diethyl or diisopropyl vinylphosphonate formed mechanically stable and transparent films. Among the block copolymers obtained from diethyl vinylphosphonate, the one with the highest PVPA content (69 wt%, 68 mol%) took up 137 wt% water (corresponding to $\lambda = 4$–16) at 98% RH and 25 °C, and among those obtained from diisopropyl vinylphosphonate the maximum water uptake was 50 wt% at 95% RH and 25 °C, as observed with a block copolymer containing 82 mol% PVPA. Block copolymers obtained from diethyl and diisopropyl vinylphosphonate with similar PVPA content showed similar conductivity properties. The conductivity of the block copolymer from diisopropyl vinylphosphonate was 40 mS cm^{-1} at 100 °C, while the block copolymer obtained from the diethyl derivative containing 71 wt% PVPA reached 30 mS cm^{-1}. At a high PS content ($\sim$65 %) the conductivity decreased dramatically. This was attributed to the loss of connectivity of the conducting PVPA phase domain.[17] In addition to relatively high proton conductivities, both block copolymers displayed good mechanical and thermal stability. For example, a block copolymer containing 65 wt% PS almost did not change its dimensions and kept its mechanical integrity, even though taking up 40 wt% water at 95% RH. Chain degradation of block copolymers obtained from diisopropyl vinylphosphonate was observed at 240 °C, while the degradation of polymer obtained from diethyl vinylphosphonate was observed at higher temperatures, 300–370 °C. Phosphonic acid anhydride formation was noted between 120–150 °C for both block copolymers.

Jannasch and co-workers have grafted PVPA from PAESs[58] (Scheme 13.4d) and PPO[48] by lithiation, followed by anionic polymerization of diethyl vinylphosphonate. Hydrolysis of the ester groups gave the corresponding acid derivatives. The PAES and PPO graft copolymers had similar PVPA contents, 32–57 and 26–64 wt%, respectively. PAES-*g*-PVPA displayed two glass transition temperatures, between 184 and 191 °C for the PAES phase and 250 °C for the PVPA phase, as expected from graft copolymers with immiscible segments. However, only a single glass transition temperature was observed at 166–186 °C for the PPO-*g*-PVPA copolymers. This glass transition was attributed to the PVPA phase since it was observed to increase with increasing temperatures during thermal cycling, as a result of anhydride formation. Both graft copolymers were thermally stable up to 400 °C, with only loss of water due to anhydride formation. The copolymers formed transparent, flexible, and mechanically strong membranes with good thermal and thermo-oxidative stability, and a phase-separated morphology. The membrane obtained from PPO-*g*-PVPA with a VPA content of 64 wt% (IEC = 6 mol P g^{-1}) achieved a conductivity of 80 mS cm^{-1} at 120 °C, which is very close to the value of 93 mS cm^{-1} reached by the membrane from PAES-*g*-PVPA with a VPA content of 59 wt% under hydrated conditions. These conductivities were comparable to that of Nafion® 115 under the same conditions, 188 mS cm^{-1}. Recently, Jannasch *et al.* have selectively grafted

PVPA from the PAES blocks of aromatic multiblocks.[68] In comparison to the previous studies of PAES and PPO grafted with PVPA, the grafted multiblock copolymers reached similar proton conductivity at a given IEC, but at a significantly lower water content.

Müllen *et al.* have used PBI functionalized with allyl or *p*-vinylbenzyl chloride as a macroinitiator for controlled radical polymerization to produce PBI-*g*-PVPA.[69] Membranes containing 12 vinylphosphonic acid units per benzimidazole unit had a maximum conductivity of 28 mS cm^{-1} at 160 °C under dry conditions. The membranes showed a high thermal and mechanical stability, and good film forming properties. Gubler *et al.* have studied PBI containing an interpenetrating network of PVPA designed for application in direct methanol fuel cells.[70] With the blend membrane, the fuel cell performance could be sustained to higher methanol feed concentration at around half the methanol crossover rate of Nafion® 117. However, the ohmic resistance of the membrane and the cathode/membrane interface showed higher losses than Nafion® at increasing current density.

Proton exchange membranes with pendent phosphonic acid groups have been synthesized by Schmidt-Naake *et al.* by using pre-irradiation grafting from vinylbenzyl chloride from fluorinated ethylene propylene (FEP) and ethylene tetrafluoroethylene (ETFE) films with subsequent Arbuzov phosphonation.[71] The membranes were doped with phosphoric acid to produce acid-base composite materials that exhibited high proton conductivities up to 100 mS cm^{-1} at 120 °C and no humidification.[72] Recently, Dimitrov *et al.* grafted PAES with phosphonated polypentafluorostyrene (Scheme 13.4c) by using controlled radical polymerization and click chemistry.[73] The copolymer membranes showed high thermal stability and proton conductivity up to 83 mS cm^{-1} at 120 °C under immersed conditions.

13.7 Phosphonated Inorganic–Organic Membranes

Highly phosphonated polymers can be efficiently immobilized by forming inorganic-organic composites. These materials are often prepared using sol–gel procedures and thus the structure and properties are usually critically dependent on the precise preparation method. Several research groups have investigated various hybrid systems based on silica and silane chemistry.[74–82] For example, phosphonic acid-grafted hybrid inorganic-organic PEMs have been synthesized using a sol–gel process, and the proton conductivities of the membranes increased with $-PO_3H_2$ group content and RH, reaching 62 mS cm^{-1} at 100 °C and 100% RH.[81] This was comparable to Nafion® under similar conditions. The proton conductivity of the membranes in the anhydrous state was enhanced by substitution of the $-CH_2PO_3H_2$ groups with $-CF_2PO_3H_2$.[81]

Recently, Herring *et al.* prepared clear and flexible hybrid organic-inorganic PEMS by copolymerization of vinylphosphonic acid and zirconium vinylphosphonate *via* a radical mechanism.[83] Dispersion of the zirconium phosphonate in the PVPA matrix increased the proton dissociation of the polymer, resulting in a polymer with much higher proton conductivity at low RH than the parent PVPA. The resulting membrane demonstrated a very high proton conductivity under dry conditions, up to 50 mS cm^{-1} at 80 °C. This suggested that the Grotthus mechanism dominated under dry conditions through structural diffusion *via* continuous hydrogen-bonded networks. Upon hydration, the ionic conductivity greatly increased due to combined Grotthus and vehicle transport mechanisms, yielding a proton conductivity higher than 100 mS cm^{-1} under 95% RH at 80 °C.

Organic–inorganic PEMs were prepared from (3-glycidoxypropyl)-trimethoxysilane and 1-hydroxyethane-1,1-diphosphonic acid *via* a sol–gel process. The hybrid membranes were stable up to 250 °C.[84] The hybrid membranes with a molar ratio of (3-glycidoxypropyl)trimethoxysilane: 1-hydroxyethane-1,1-diphosphonic acid of 2:1 reached a proton conductivity of 1 mS cm^{-1} under anhydrous conditions at 130 °C.[84] In addition, self-standing and transparent hybrid membranes were obtained by co-polymerization of VPA, dimethylethoxyvinylsilane, and (3-glycidoxypropyl)-trimethoxysilane, followed by sol–gel processing.[85] The proton conductivity of the resulting membranes ranged from 3.0×10^{-3} to 15 mS cm^{-1} with an increase in RH from 32 to 100%.

13.8   Possible Future Directions

The number of reports on fuel cell membranes based on phosphonated polymers is still very limited in comparison to that of sulfonated polymer membranes. However, phosphonated membranes may potentially show important advantages, especially for PEMFC operation under low-humidity conditions at high temperatures. Here, phosphonated polymers generally show a higher hydrolytic and thermal stability due to the strength of the C–P bond, as well as higher proton conductivity, provided that the formation of anhydride can be suppressed. In this respect, there still exist unexplored opportunities to tune the acidity of the phosphonic acid, between the rather low acidity of the alkylphosphonic acid and the high acidity of the per-fluoroalkylphosphonic acid, in order to optimize the performance in terms of, for example, intrinsic proton conductivity. Moreover, the relationships between phosphonated polymer structures and the resulting morphology and proton conductivity remain largely unknown, especially in relation to sulfonated systems. In order to achieve high intrinsic proton conductivity, the aggregation of phosphonic acids is important for proton transfer. In this context, recent studies of low molecular weight aromatic multipho-sphonated compounds[86] can provide some guidance on the molecular design of macromolecular systems.

Not surprisingly, the focus in this area has been on the development of phosphonated PEMFC membranes. However, phosphonated membranes can also be of interest for use as electrolytes in other electrochemical devices such as flow batteries.[87] In addition, there are now a few reports on the preparation and study of polymers functionalized with phosphonium cations for use as hydroxide ion conducting electrolyte membranes for emerging alkaline membrane fuel cells.[88,89] Ultimately, in order to critically assess the possibility of using phosphonated membranes in fuel cells, it is of vital importance to investigate their performance under different operational conditions and evaluate their durability under realistic situations.

References

1. Y. Wang, K. S. Chen, J. Mishler, S. C. Cho and X. C. Adroher, *Appl. Energy*, 2011, **88**, 981.
2. J. Zhang, Z. Xie, Y. Tang, C. Song, T. Navessin, Z. Shi, D. Song, H. Wang, D. P. Wilkinson, Z.-S. Liu and S. Holdcroft, *J. Power Sources*, 2006, **160**, 872.
3. C. H. Park, C. H. Lee, M. D. Guiver and Y. M. Lee, *Progr. Polym. Sci.*, 2011, **36**, 1443.
4. C. Wieser, *Fuel Cells*, 2004, **4**, 245.
5. J.-T. Wang, R. F. Savinell, J. Wainright, M. Litt and H. Yu, *Electrochim. Acta*, 1996, **41**, 193.
6. X. Glipa, B. Bonnet, B. Mula, D. J. Jones and J. Roziere, *J. Mater. Chem.*, 1999, **9**, 3045.
7. Q. Li, J. O. Jensen, R. F. Savinell and N. J. Bjerrum, *Progr. Polym. Sci.*, 2009, **34**, 449.
8. M. Schuster, K. D. Kreuer, H. Steininger and J. Maier, *Solid State Ionics*, 2008, **179**, 523.
9. M. Schuster, T. Rager, A. Noda, K. D. Kreuer and J. Maier, *Fuel Cells*, 2005, **5**, 355.
10. B. Lafitte and P. Jannasch, On the prospects for phosphonated polymers as proton-exchange fuel cell membranes, in *Advances in Fuel Cells*, ed. T. S. Zhao, K. D. Kreuer, and T. Van Nguyen, Elsevier, Oxford, 2007, vol. 1, p. 119.
11. A. Rusanov, P. Kostoglodov, M. M. Abadie, V. Voytekunas and D. Likhachev, *Adv. Polym. Sci.*, 2008, **216**, 125.
12. P. C. Crofts and G. M. Kosolapoff, *J. Am. Chem. Soc.*, 1953, **75**, 3379.
13. L. D. Freedman and G. O. Doak, *Chem. Rev.*, 1957, **57**, 479.
14. T. J. Peckham and S. Holdcroft, *Adv. Funct. Mater.*, 2010, **22**, 4667.
15. L. Macarie and G. Ilia, *Progr. Polym. Sci.*, 2010, **35**, 1078.
16. B. Bingöl, W. H. Meyer, M. Wagner and G. Wegner, *Macromol. Rapid. Commun.*, 2006, **27**, 1719.
17. T. Wagner, A. Manhart, N. Deniz, A. Kaltbeitzel, M. Wagner, G. Brunklaus and W. H. Meyer, *Macromol. Chem. Phys.*, 2009, **210**, 1903.
18. M. Millaruelo, V. Steinert, H. Komber, R. Klopsch and B. Voit, *Macromol. Chem. Phys.*, 2008, **209**, 366.
19. B. Bingöl, C. Strandberg, A. Szabo and G. Wegner, *Macromolecules*, 2008, **41**, 2785.
20. A. Kaltbeitzel, S. Schauff, H. Steininger, B. Bingöl, G. Brunklaus, W. H. Meyer and H. W. Spiess, *Solid State Ionics*, 2007, **178**, 469.

21. Y. J. Lee, B. Bingöl, T. Murakhtina, D. Sebastiani, W. H. Meyer, G. Wegner and H. W. Spiess, *J. Phys. Chem. B*, 2007, **111**, 9711.
22. H. Steininger, M. Schuster, K. D. Kreuer, A. Kaltbeitzel, B. Bingöl, S. Schauff, G. Brunklaus, W. H. Meyer, J. Maier and H. W. Spiess, *Phys. Chem. Chem. Phys.*, 2007, **9**, 1764.
23. Y. Tamura, L. Sheng, S. Nakazawa, T. Higashihara and M. Ueda, *J. Polym. Sci., Part A: Polym. Chem.*, 2012, **50**, 4334.
24. F. Jiang, A. Kaltbeitzel, B. Fassbender, G. Brunklaus, H. Pu, W. H. Meyer, H. W. Spiess and G. Wegner, *Macromol. Chem. Phys.*, 2008, **209**, 2494.
25. S.-I. Lee, K.-H. Yoon, M. Song, H. Peng, K. A. Page, C. L. Soles and D. Y. Yoon, *Chem. Mater.*, 2012, **24**, 115.
26. F. Jiang, A. Kaltbeitzel, B. Fassbender, G. Brunklaus, H. Pu and W. H. Meyer, *Macromolecules*, 2008, **41**, 3081.
27. T. Higashihara, N. Fukuzaki, Y. Tamura, Y. Rho, K. Nakabayashi, S. Nakazawa, S. Murata, M. Ree and M. Ueda, *J. Mater. Chem. A*, 2013, **1**, 1457.
28. T. Itoh, K. Hirai, M. Tamura, T. Uno, M. Kubo and Y. Aihara, *J. Solid State Electrochem.*, 2012, **14**, 2179.
29. M. F. H. Schuster, W. H. Meyer, M. Schuster and K. D. Kreuer, *Chem. Mater.*, 2004, **16**, 329.
30. J. C. Persson and P. Jannasch, *Chem. Mater.*, 2003, **15**, 3044.
31. K. D. Kreuer, *ChemPhysChem*, 2002, **3**, 771.
32. D. Markova, A. Kumar, M. Klapper and K. Müllen, *Polymer*, 2009, **50**, 3411.
33. F. Hacivelioglu, E. Okutan, S. U. Celik, S. Yesilot, A. Bozkurt and A. Kilic, *Polymer*, 2012, **53**, 3659.
34. H. R. Allcock, M. A. Hofmann, C. M. Ambler, S. N. Lvov, X. Y. Y. Zhou, E. Chalkova and J. Weston, *J. Membr. Sci.*, 2002, **201**, 47.
35. H. R. Allcock, M. A. Hofmann, C. M. Ambler and R. V. Morford, *Macromolecules*, 2002, **35**, 3484.
36. N. Fukuzaki, K. Nakabayashi, S. Nakazawa, S. Murata, T. Higashihara and M. Ueda, *J. Polym. Sci., Part A: Polym. Chem.*, 2011, **49**, 93.
37. R. Perrin, M. Elomaa and P. Jannasch, *Macromolecules*, 2009, **42**, 5146.
38. A. Bozkurt, W. H. Meyer, J. Gutmann and G. Wegner, *Solid State Ionics*, 2003, **164**, 169.
39. S. Ü. Çelik, Ü. Akbey, R. Graf, A. Bozkurt and H. W. Spiess, *Phys. Chem. Chem. Phys.*, 2008, **10**, 6058.
40. L. E. P. Santos, L. S. Hanamoto, R. P. Pereira, A. M. Rocco and M. I. Felisberti, *J. Appl. Polym. Sci.*, 2011, **119**, 460.
41. F. Jiang, A. Kaltbeitzel, W. H. Meyer, H. Pu and G. Wegner, *Macromolecules*, 2008, **41**, 3081.
42. D. Markova, K. L. Opper, M. Wagner, M. Klapper, K. B. Wagener and K. Mullen, *Polym. Chem.*, 2013, **4**, 1351.
43. K. Miyatake and A. S. Hay, *J. Polym. Sci., Part A: Polym. Chem.*, 2001, **39**, 3770.

44. E. Parcero, R. Herrera and S. P. Nunes, *J. Membr. Sci.*, 2006, **285**, 206.
45. B. Lafitte and P. Jannasch, *J. Polym. Sci., Part A: Polym. Chem.*, 2005, **43**, 273.
46. T. Bock, H. Möhwald and R. Mülhaupt, *Macromol. Chem. Phys.*, 2007, **208**, 1324.
47. S. Yanagimachi, K. Kaneko, Y. Yakeoka and M. Rikukawa, *Synth. Met.*, 2003, **135**, 69.
48. M. Ingratta, M. Elomaa and P. Jannasch, *Polym. Chem.*, 2010, **1**, 739.
49. T. Hirao, T. Masunaga, N. Yamada, Y. Ohshiro and T. Agawa, *Bull. Chem. Soc. Jpn.*, 1982, **55**, 909.
50. K. Jakoby, K. V. Peinemann and S. P. Nunes, *Macromol. Chem. Phys.*, 2003, **204**, 61.
51. C. S. Demmer, N. Krogsgaard-Larsen and L. Bunch, *Chem. Rev.*, 2011, **111**, 7981.
52. A. Burger and N. D. Dawson, *J. Org. Chem.*, 1951, **16**, 1250.
53. T. Rager, M. Schuster, H. Steininger and K. D. Kreuer, *Adv. Mater.*, 2007, **19**, 3317.
54. A. S. Shaplov, E. I. Lozinskaya, I. L. Odinets, K. A. Lyssenko, S. A. Kurtova, G. I. Timofeeva, C. Iojoiu, J.-Y. Sanchez, M. J. M. Abadie, V. Yu. Voytekunas and Y. S. Vygodskii, *React. Funct. Polym.*, 2008, **68**, 208.
55. K. D. Papadimitriou, A. K. Andreopoulu and J. K. Kallitsis, *J. Polym. Sci., Part A: Polym. Chem.*, 2010, **48**, 2817.
56. E. Abouzari-Lotf, H. Ghassemi, A. Shockravi and T. Zawodzinski, *Polymer*, 2011, **52**, 4709.
57. E. P. Jutemar and P. Jannasch, *J. Membr. Sci.*, 2010, **351**, 87.
58. J. Parvole and P. Jannasch, *Macromolecules*, 2008, **41**, 3893.
59. N. Y. Abu-Thabit, S. A. Ali and S. M. J. Zaidi, *J. Membr. Sci.*, 2010, **360**, 26.
60. B. Lafitte and P. Jannasch, *J. Polym. Sci., Part A: Polym. Chem.*, 2007, **45**, 269.
61. M. B. Herath, S. E. Creager, A. Kitaygorodskiy and D. D. DesMarteau, *ChemPhysChem*, 2010, **11**, 2871.
62. B. Liu, G. P. Robertson, M. D. Guiver, Z. Shi, T. Navessin and S. Holdcroft, *Macromol. Rapid. Commun.*, 2006, **27**, 1411.
63. V. Atanasov and J. Kerres, *Macromolecules*, 2011, **44**, 6416.
64. Y. Zhu, H. Chen and C. He, *J. Polym. Res.*, 2011, **18**, 1409.
65. M. Yamabe, K. Akiyama, Y. Akatsuka and M. Kato, *Eur. Polym. J.*, 2000, **36**, 1035.
66. R. Tayouo, G. David, B. Ameduri, J. Roziere and S. Roualdes, *Macromolecules*, 2010, **43**, 5269.
67. E. M. W. Tsang, Z. Zhang, Z. Shi, T. Soboleva and S. Holdcroft, *J. Am. Chem. Soc.*, 2007, **129**, 15106.
68. A. Sannigrahi, S. Takamuku and P. Jannasch, *Polym. Chem.*, 2013, **4**, 4207.
69. P. R. Sukumar, W. Wu, D. Markova, O. Unsal, M. Klapper and K. Müllen, *Macromol. Chem. Phys.*, 2007, **208**, 2258.

70. L. Gubler, D. Kramer, J. Belack, O. Unsal, T. J. Schmidt and G. G. Scherer, *J. Electrochem. Soc.*, 2007, **154**, B981.
71. G. Schmidt-Naake, M. Bohme and A. Cabrera, *Chem. Eng. Technol.*, 2005, **28**, 720.
72. C. Schmidt and G. Schmidt-Naake, *Macromol. Mater. Eng.*, 2007, **292**, 1164.
73. I. Dimitrov, S. Takamuku, K. Jankova, P. Jannasch and S. Hvilsted, *Macromol. Rapid Commun.*, 2012, **33**, 1368.
74. E. Labalme, G. David, P. Buvat, J. Bigarre and T. Boucheteau, *J. Polym. Sci., Part A: Polym. Chem.*, 2012, **50**, 1308.
75. L. D. Lou and H. T. Pu, *Int. J. Hydrogen Energy*, 2011, **36**, 3123.
76. J. Umeda, M. Suzuki, M. Kato, M. Moriya, W. Sakamoto and T. Yogo, *J. Power Sources*, 2010, **195**, 5882.
77. H. T. Pu, H. Y. Pan, Y. J. Qin, D. C. Wan and J. J. Yuan, *Mater. Lett.*, 2010, **64**, 1510.
78. J. Umeda, M. Moriya, W. Sakamoto and T. Yogo, *Electrochim. Acta*, 2009, **55**, 298.
79. M. Kato, W. Sakamoto and T. Yogo, *J. Membr. Sci.*, 2007, **303**, 43.
80. M. Kato, S. Katayama, W. Sakamoto and T. Yogo, *Electrochim. Acta*, 2007, **52**, 5924.
81. S. W. Li, Z. Zhou, H. Abernathy, M. L. Liu, W. Li, J. Ukai, K. Hase and M. Nakanishi, *J. Mater. Chem.*, 2006, **16**, 858.
82. V. V. Binsu, R. K. Nagarale and V. K. Shahi, *J. Mater. Chem.*, 2005, **15**, 4823.
83. G. J. Schlichting, J. L. Horan, J. D. Jessop, S. E. Nelson, S. Seifert, Y. Yang and A. M. Herring, *Macromolecules*, 2012, **45**, 3874.
84. C. Shen, Z. Guo, C. Chen and S. Gao, *J. Appl. Polym. Sci.*, 2012, **126**, 954.
85. H. Onizuka, M. Kato, T. Shimura, W. Sakamoto and T. Yogo, *J. Sol-Gel Sci. Technol.*, 2008, **46**, 107.
86. L. Jimenez-Garcia, A. Kaltbeitzel, V. Enkelmann, J. S. Gutmann, M. Klapper and K. Mullen, *Adv. Funct. Mater.*, 2011, **21**, 2216.
87. A. Z. Weber, M. M. Mench, J. P. Meyers, P. N. Ross, J. T. Gostick and Q. H. Liu, *J. Appl. Electrochem.*, 2011, **41**, 1137.
88. K. J. T. Noonan, K. M. Hugar, H. A. Kostalik, E. B. Lobkovsky, H. D. Abruna and G. W. Coates, *J. Am. Chem. Soc.*, 2012, **134**, 18161.
89. S. Gu, R. Cai, T. Luo, K. Jensen, C. Contreras and Y. S. Yan, *ChemSusChem*, 2010, **3**, 555.
90. B. Bingöl, G. Hart-Smith, C. Barner-Kowollik and G. Wegner, *Macromol. Rapid Commun.*, 2008, **41**, 1634.
91. M. Yamabe, K. Akiyama, Y. Akatsuka and M. Kato, *Eur. Polym. J.*, 2000, **36**, 1035.

Subject Index

References to tables and charts are in **bold** type